PARALLEL ROBOTICS: RECENT ADVANCES IN RESEARCH AND APPLICATION

JINSONG WANG
AND
XIN-JUN LIU
EDITORS

Nova Science Publishers, Inc.
New York

For permission to use material from this book please contact us:
Telephone 631-231-7269; Fax 631-231-8175
Web Site: http://www.novapublishers.com

LIBRARY OF CONGRESS CATALOGING-IN-PUBLICATION DATA

Parallel robotics : recent advances in research & application / Jinsong Wang and Xin-Jun Liu, Editors.
p. cm.
ISBN 978-1-60456-859-2 (hardcover)
1. Robotics. I. Wang, Jinsong. II. Liu, Xin-Jun.
TJ211.P36 2008
629.8'92--dc22 2008023750

Published by Nova Science Publishers, Inc. ✦ New York

PARALLEL ROBOTICS: RECENT ADVANCES IN RESEARCH AND APPLICATION

CONTENTS

PREFACE

Parallel robots are the systems with closed-loop chains. They have the advantage of high speed, high stiffness, accuracy, compactness, time-saving of machining, high load/weight ratio, and low moving inertia. They have potential applications in the fields of industrial robots, wire robots, motion simulators, medical equipments, haptic devices and parallel kinematic machines. This new book presents the latest research in the field of parallel robotics concerning synthesis, optimum design, and applications.

Chapter 1 - This chapter introduces a geometric approach to the analysis and synthesis of parallel manipulator mechanisms with lower mobility based on the screw theory. In section 1, some basic concepts such as kinematic pair screws, restraining screws, and output degrees of freedom of parallel manipulators are given firstly. In section 2, a procedure of analyzing kinematic and constraint characteristics of spatial parallel mechanisms with lower mobility is developed with the concepts mentioned above. In section 3, the motion and constraint characteristics of three kinds of parallelogram linkages (4R, 4U and 4S) are discussed. These parallelogram linkages are often used as complex joints of spatial parallel mechanisms. In such a way, a Delta-I robot, $R(4U)_{RPP}$, is thought to be equivalent to a general 3-RRPP mechanism, and a Delta-II robot , $R(4S)_{RRPP}$, equivalent to a general 3-RRRPP mechanism, which makes the analysis of motion and constraint characteristics of spatial parallel mechanisms with hybrid limbs simple. In section 4, a systematic procedure for the synthesis of manipulators with lower mobility based on the screw theory is presented. The new approach employs the concepts of linearly independence of screws and uses the relationship between a screw system and its reciprocal systems, which converts a mechanism design into a screw algebra operation. It results in the production of new manipulator mechanisms. In a sense, the approach is an algebraic interpretation of direct geometrical synthesis based on experience and observation. It can aid designers to find more required mechanisms than the direct geometrical method. A typical study is contributed to the synthesis of parallel manipulator mechanisms with four degrees of freedom. Following the four steps, i.e. determining restraint wrenches, identifying reference screws reciprocal to the restraint wrenches, constructing serial limbs by linearly combining the reference screws and allocating the serial limbs, new parallel mechanisms such as 2-CRP/2-PSS, 3-URU/SPS and 4-UPU mechanisms are hence developed.

Chapter 2 - A parallel mechanism (PM) with less than six degrees of freedoms (DOFs) is called a lower-mobility PM. A well-optimized lower-mobility parallel mechanism (PM) may replace a 6-DOF PM in many tasks which do not require all the six DOFs with reduced cost in design process, manufacturing process, control system, actuation system, Delta robot (Clavel 1988) is one typical example of successful application of lower-mobility PM. Hence, lower-mobility PMs have attracted much attention in recent years.

Mobility analysis and type synthesis are two basic and closely-related topics in research on lower-mobility PMs. A PM comprises a fixed base, a moving platform, and several limbs connecting the base and the moving platform. The limb can be a serial concatenation of kinematic pairs and composite joints like parallelograms. The mobility analysis aims to specify the number of independent variable parameters needed to determine the position and orientation of the moving platform. Often, one also expects to know the motion pattern of the moving platform, namely, rotational or translational DOF, or a combination of both. For example, the moving platform can rotate about the x and y axis, and translate along the x and y axis. The type synthesis aims to invent new architectures based on pre-specified DOFs and motion patterns, which is a reverse process of mobility analysis.

According to the motion pattern, that is, rotational (R) or translational (T), the lower-mobility PMs can be classified. The 5-DOF PMs can be sorted into two subcategories: 3R2T and 2R3T. The 4-DOF PMs can be sorted into three subcategories: 3R1T, 1R3T, and 2R2T. The 3-DOF PMs can be sorted into four subcategories: 1R2T, 2R1T, 3T, and 3R. The 2-DOF PMs can be sorted into three subcategories: 2R, 2T, and 1R1T.

The research on type synthesis of lower-mobility PMs can be dated back to 1983, when Hunt (1983) proposed some architectures, including some planar PMs, 3-DOF PMs, 6-DOF PMs and some asymmetrical 4- and 5-DOF PMs. Clavel (1988) invented the famous successful 3-DOF translational Delta Robot. Hervé and Sparacino (1991) proposed the family of 3-DOF translational PMs with 4-DOF limbs. Tsai (1996) proposed the 3-UPU translational PM, which attracted much attention in following years because of its special motion properties. Pierrot and Company (1999) invented the 4-DOF H4 parallel robot, using parallelograms to eliminate rotations. Zhao and Huang (2000) proposed the 4-URU PM with three translational DOFs and one rotational DOF along the norm of the moving platform. Zlatov and Gosselin (2001) proposed a 4-DOF PM with three rotational DOFs and one translational DOF. Huang and Li (2001) invented symmetrical 5-DOF PMs which had been recognized as being vacant for a long time.

Other important progress in inventing new types of lower-mobility PMs include the linear translational PMs (Kong and Gosselin, 2002), 3-URC wrist (Di Gregorio, 2001), singularity-free fully-isotropic translational PMs (Carricato and Parenti-Castelli 2002), fully isotropic 4-DOF parallel mechanisms for Schoenflies motion (Carricato 2005), 3-DOF fully PMs with high rotational capability (Liu, Wang and Pritschow 2005), 3-DOF translational PM with doubly planar limbs (Lee and Hervé 2006) and two-mode 3-DOF PMs (Refaat et.al 2007).

One traditional method for type synthesis of PMs is based on the general Grübler-Kutzbach mobility criterion. Tsai (1999) applied this method to the type synthesis of a kind of 3-DOF PMs. However, this method failed to describe the geometrical arrangement of kinematic pairs, hence invalid results were often obtained (Merlet 2002).

Another method, group-based synthesis method, is based on the pioneering works on the algebraic properties of displacement subgroup theory (Hervé 1978,). The group-based synthesis method is applicable to all kinds of lower-mobility PMs (Hervé and Sparacino1991, Hervé 1999, Li, Huang, and Hervé 2004, Huynh and Hervé 2005a, Karouia and Hervé 2002, Karouia and Hervé 2005, Lee and Hervé 2006). Other researchers (Angeles 1982, 2002, Rico *et al* 2006, Kong and Gosselin 2007, Meng, Liu and Li 2007) also made important contributions related with the group-based synthesis method. Because the displacement subgroup deals with finite motion, PMs synthesized by the group-based method are inherently with full-cycle mobility. Hence, the group-based synthesis method is at the level of finite motion.

The constraint-synthesis method is based on screw theory, particularly, the reciprocal screw theory. Because a screw is inherently instantaneous, the constraint-synthesis method is at the level of instantaneous motion. Hence, one indispensable step of the constraint-synthesis method is to identify if the synthesized mechanism is instantaneous or not, which sometimes is difficult for some complicated architectures. In most cases, finite motion conditions can be established by checking the geometrical conditions of the PM. The constraint-synthesis method has been applied to all kinds of lower-mobility PMs (Huang, Kong and Fang 1997, Frisoli *et al.* 2000, Huang and Li 2002, 2003, Kong and Gosselin 2004a, 2004b, Fang and Tsai 2002, 2004). Numerous novel PMs have been invented using this method, including the symmetrical 5-DOF PMs.

In addition, Jian and Yang (2004) proposed a type synthesis method based single-unit open chains and synthesized some novel mechanisms. Gao, Zhang and Li (2005) proposed the synthesis method based on the Gf coordinates. Gogu (2004, 2007) applied a synthesis method based on linear transform theory. Carricato and Parenti-Castelli (2002) synthesized a family of Singularity-free fully isotropic translational PMs. One common feature of these methods is the use of velocity variables. Hence, these methods are called a motion-synthesis approach for convenience, though not very accurately.

However, there are still some gaps in the knowledge of type synthesis of 4-DOF 2R2T PMs with identical limbs. Little has been addressed on the systematical type synthesis of 4-DOF 2R2T PMs (Li and Huang 2003). This paper aims to apply the constraint-synthesis method to the type synthesis of 4-DOF 2R2T PMs systematically. In section 1.2, the authors review the fundamentals of constraint-synthesis method briefly. In section 1.3 they develop the mechanism and limb constraint-wrench system and identify the structural geometrical conditions under which the combined effect of limb constraint-wrench systems is equal to the mechanism constraint-wrench system. In section 1.4, limb kinematic chains are generated by linear combination of twists reciprocal to the limb constraint-wrench system. On the basis of structural geometrical conditions, numerous 2R2T 4-DOF parallel mechanisms are constructed with limb kinematic chains obtained above. In section 1.5, 2R2T 4-DOF parallel mechanisms with full-cycle mobility are identified based on the integrity of structural geometrical conditions.

Chapter 3 - A hybrid machine is able to be regarded as a special type of parallel kinematic mechanical system which utilizes characteristically different actuators simultaneously to drive a multi-degree-of-freedom (DOF) mechanism for a programmable output motion with a large power. The uncontrollable actuator(s) (e.g., a constant speed motor) is/are expected to provide the main part of the output motion with a large power, while the controllable actuator(s) (e.g., a small servomotor) is/are responsible for the adjustablility/flexibility of the output motion. In this chapter, the design is outlined for this type of machine by taking a new controllable mechanical press as example.

Chapter 4 - Parallel manipulators are always welcome in conditions where large payload capability, high stiffness and fine precision are needed. They have been studied intensively for more than a couple of decades and still attract attention from scholars of academe and industry [1-5]. To work in more complicated situations, many parallel manipulators with specified numbers and types of DOF (degree of freedom) have been proposed, and many researchers focused on the type synthesis of parallel manipulators. Several approaches for the type synthesis of parallel manipulators were also developed, such as the method based on the screw theory [6], the method based on the displacement group theory [7] and the special motion coordinate technique [8, 9].

However, the workspace of the parallel manipulators is relatively smaller than the serial manipulators. This results in lower translational or rotational capability and larger equipment fixing space. Many researchers have focused on new methods to avoid these disadvantages. Liu, Wang and Pritschow proposed a new family of spatial 3-DoF fully-parallel manipulators with high rotational capability [10]. Merlet's research showed that the joints' layout could have large influence on the workspace volume for robots [11]. Many serial-parallel manipulators were developed to achieve larger workspace [12, 13]. Also, Kim proposed an Eclipse-II parallel manipulator that used a circular leader rail to achieve the continuous 360-degree spinning [14]. Except the Eclipse-II, parallel manipulators with the mobile platform rotating continuously are seldom.

In this chapter, the authors will discuss three types of parallel manipulators with special topology structures and special limbs to make their mobile platforms rotate continuously. The first types of parallel manipulators possess redundant actuators and multi-actuators in each limb (at least in one limb). The second types of parallel manipulators possess a special limb, a RUPU (R-revolute joint, U-universal joint, and P-prismatic joint) limb or a RP^+ (P^+-two P joints configured in vertical directions) limb. The third types of parallel manipulators are those where their limbs are embedded together. Mechanisms of these parallel manipulators are described respectively, and some topology mechanisms of these types of parallel manipulators are given. Moreover, their kinematic equations are developed to analyze their input-output characteristics.

Chapter 5 - The dynamic analysis is a very important aspect on design, analysis and control of the parallel manipulators and parallel kinematic machine tools. The dynamics of any parallel manipulators can also be divided into the inverse problem and the forward (or direct) problem as its kinematics does. The inverse dynamic analysis gives the driving forces and driving moments of all active joints according to the desired trajectory of the movable platform, which provides a prerequisite for the design of control strategy. The forward dynamic analysis calculates the real dynamic response of the movable platform and all legs according to the forces and moments applied to the system. The goal of forward dynamic modeling is to simulate the velocity fluctuation, vibration and other dynamic behavior and to investigate their effect on positioning or machining quality when the parallel manipulator performs a certain task such as machining or picking up and placing.

However, contrast to the vast literature on the synthesis and kinematics of the parallel manipulators, studies on its dynamics, especially the forward dynamics, are relatively few. The forward dynamic modeling and simulation of the parallel manipulators are more complex than other mechanisms because of two main factors: 1) any parallel manipulator has closed loop topology, the modeling of such a topology is harder than that of a tree or chain topology when the mechanism is considered as a multi-body system (MBS); 2) calculation of the kinematics of the mechanism is needed at each step in dynamic analysis, however, not all the parallel manipulators have a closed-form solution for forward kinematics; moreover, the assembly problem caused by numerical fault is inevitable when solutions of different legs have errors resulting from a certain numerical integration algorithm. Considering these characteristics of the parallel manipulators, this chapter puts the emphasis on the forward dynamics of parallel manipulators and improving the efficiency of analysis.

A general modeling and simulation method for the forward dynamics of parallel manipulators is proposed in this chapter, which is based on the contracted Newton-Euler formulation with generalized coordinates. The chapter consists of six sections. section one is an introduction part; in section two, the state of the art will be presented on research progress on dynamics of parallel manipulators; the aforementioned general method for the forward dynamics of parallel manipulators will be proposed and depicted in detail in section three; in section four, a complete analysis of the forward and inverse dynamics of a 3-DOF (degree of freedom) parallel manipulator will be presented; and in section five, numerical solution will be given as a case study to show the feasibility and application of the method; section six will give the conclusion of the whole chapter.

Chapter 6 - This paper treats the problems of kinematic analysis and optimization of the 5R symmetrical parallel mechanism taking into account the force transmissibility. In the design process, the theoretical workspace cannot be used practically due to the inside singularity. The usable workspace without singularity inside cannot still be useful in an optimization problem since a mechanism at and near the boundary of the workspace will be out of control. The local conditioning index (LCI), which is widely used in the field of serial robots and has been applied in many optimizations, has the problem of physical significance and difficulty of measuring the distance to a singularity of a parallel mechanism. For such reasons, the concept of transmission angle is introduced in this paper to the optimization of the planar 5R parallel mechanisms. Based on this concept, three new indices are proposed. After presenting the performance charts, the optimization process of the mechanisms is finally presented by considering the workspace and force transmissibility.

Chapter 7 - As a new promising mechanism type, flexure-based compliant parallel kinematic machines (FCPKMs) gather the advantages of both compliant mechanism and parallel architecture. It is surely a primary choice in designing an ultra high-accuracy micromanipulator or micropositioner/nanopositioner. Accordingly design theory and technology on the FCPKMs have become one of the most important issues to be discussed in the past few years. This paper will demonstrate a comprehensive design process for a general FCPKM, which covers topology and size synthesis of the FCPKMs, design of compliant joints employed in the FCPKMs, kinetostatic analysis for the FCPKMs, and design methodology for the general FCPKMs. At last, in terms of the proposed theoretical analysis and design method, some macro-scale compliant in-parallel stages are designed and used in various applications as the high-precision micromanipulators.

Chapter 8 - Parallel kinematic machine (PKM) is recognized as an important supplement of traditional serial mechanisms, and will play a key role in reconfigurable manufacturing system (RMS) since they can be quickly and efficiently reconfigured.

A reconfigurable PKM consists of an inventory of interchangeable modules. This chapter presents a kind of structural design of reconfigurable PKMs. The reconfigurable PKM is established in either telescopic drive form or railway linear drive form with the same modules but different assembly modes. In the architecture of the PKM, an angle-variable module is designed to change the setting-angle in X and Z orientations for working more efficiently. Due to the modularity and reconfigurability, a PKM constructed in this mode can have difference structures, degree of freedoms (DOFs), and performance. In the workspace analysis of the PKMs with two drive modes, inverse kinematics solution is given, and the effect of setting-angle of spherical joint on PKM has been investigated and the spherical joint setting-angle selection method has been proposed. A comparison result is supplied to show the difference of workspace between these two kinds of PKMs.

Reconfiguration planning is recognized as an important factor for reducing the cost in manufacturing reconfigurable products, and the associated main task is to generate a set of optimal or approximate optimal reconfiguration sequences by using some effect algorithms. After structure design of PKMs, a method of modularizing and integrating is proposed to design modules. Modules' interface information is described by the combination of interface status number and undirected graph. Based on this description, a Petri Net is applied to the modeling reconfiguration process in reconfiguration task planning. A reconfiguration cost matrix(RCM) is used to figure out dynamic cost in reconfiguration. Then a one-direction search algorithm and a two-direction search algorithm, which combine graph-subgraph recognition and heuristic algorithms, are presented to search optimal reconfiguration path aiming at minimizing total cost.

In experiment section, reconfiguration and cutting experiments of several reconfigurable PKMs are presented to verify the theory and method provided in this chapter. The results of two searching algorithms are compared with other ones. The same structure searching algorithm is realized by Visual C++.

In: Parallel Robotics: Recent Advances in Research... ISBN: 978-1-60456-859-2
Editors: Jinsong Wang and Xin-Jun Liu, pp. 1-32

Chapter 1

THE GEOMETRIC ANALYSIS AND SYNTHESIS OF PARALLEL MANIPULATOR MECHANISMS WITH LOWER MOBILITY WITH SCREW THEORY

Tieshi Zhao*

Robotics Research Center, Yanshan University, Qinhuangdao, 066004, Hebei Province, China

This chapter introduces a geometric approach to the analysis and synthesis of parallel manipulator mechanisms with lower mobility based on the screw theory. In section 1, some basic concepts such as kinematic pair screws, restraining screws, and output degrees of freedom of parallel manipulators are given firstly. In section 2, a procedure of analyzing kinematic and constraint characteristics of spatial parallel mechanisms with lower mobility is developed with the concepts mentioned above. In section 3, the motion and constraint characteristics of three kinds of parallelogram linkages (4R, 4U and 4S) are discussed. These parallelogram linkages are often used as complex joints of spatial parallel mechanisms. In such a way, a Delta-I robot, $R(4U)_{RPP}$, is thought to be equivalent to a general 3-RRPP mechanism, and a Delta-II robot , $R(4S)_{RRPP}$, equivalent to a general 3-RRRPP mechanism, which makes the analysis of motion and constraint characteristics of spatial parallel mechanisms with hybrid limbs simple. In section 4, a systematic procedure for the synthesis of manipulators with lower mobility based on the screw theory is presented. The new approach employs the concepts of linearly independence of screws and uses the relationship between a screw system and its reciprocal systems, which converts a mechanism design into a screw algebra operation. It results in the production of new manipulator mechanisms. In a sense, the approach is an algebraic interpretation of direct geometrical synthesis based on experience and observation. It can aid designers to find more required mechanisms than the direct geometrical method. A typical study is contributed to the synthesis of parallel manipulator mechanisms with four degrees of freedom. Following the four steps, i.e. determining restraint wrenches, identifying reference screws reciprocal to the restraint

* Email: tszhao@ysu.edu.cn

wrenches, constructing serial limbs by linearly combining the reference screws and allocating the serial limbs, new parallel mechanisms such as 2-CRP/2-PSS, 3-URU/SPS and 4-UPU mechanisms are hence developed.

1. Basic Concepts

1.1. Screws and Linear Independence

A screw is a geometric element representing a spatial straight line (screw axis) to which a scalar quantity, the pitch of the screw, is added. A unit screw can be expressed as $\boldsymbol{S}=(\boldsymbol{s};\ \boldsymbol{s}_0)^{\mathrm{T}}=(\mathbf{s};\boldsymbol{r}\times\mathbf{s}+h\mathbf{s})^{\mathrm{T}}$, where, $\boldsymbol{s}$ is a unit vector indicating the direction of the screw axis, $\boldsymbol{r}$ is a vector from the reference origin to the screw axis, h is the pitch of the screw , The cross product $r \times s$ is the moment of the screw axis relative to the origin of the reference frame. Since both the screw axis and the moment are directed at right angle to one another, their dot product is identically zero. In this chapter, a unit screw indicates a screw whose axis has a unit length, and is called a screw for short. When carrying a velocity amplitude, a screw describes a spatial motion of a rigid body, and is called a twist. It can be expressed as $\omega\boldsymbol{S}=(\omega;\ \boldsymbol{v})^{\mathrm{T}}=(\omega;\ \boldsymbol{r}\times\boldsymbol{\omega}+h_{\omega}\boldsymbol{\omega})^{\mathrm{T}}$, where, $\boldsymbol{\omega}=\omega\boldsymbol{s}$ and $h_{\omega}=\dfrac{\boldsymbol{\omega}\cdot\boldsymbol{v}}{\boldsymbol{\omega}\cdot\boldsymbol{\omega}}$. Similarly, a wrench is a screw with a force intensity, expressed as $f\boldsymbol{S}^{r}=(f;\ \boldsymbol{m})^{\mathrm{T}}=(\boldsymbol{f};\ \boldsymbol{r}\times\boldsymbol{f}+h_{f}\ \boldsymbol{f})^{\mathrm{T}}$, where, $\boldsymbol{f}=f\ \boldsymbol{s}$ and $h_{f}=\dfrac{\boldsymbol{f}\cdot\boldsymbol{m}}{\boldsymbol{f}\cdot\boldsymbol{f}}$. When the pitch of a screw is zero or infinite, the screw degenerates to a line vector $\boldsymbol{S}=(\boldsymbol{s};\boldsymbol{r}\times\boldsymbol{s})$ or a free vector $\boldsymbol{S}=(\mathbf{0};\boldsymbol{s})$. A line vector or a free vector carrying a velocity amplitude ω represents a pure rotation $(\omega;\boldsymbol{r}\times\boldsymbol{\omega})$ or translation $(\mathbf{0};\boldsymbol{v})$ of a rigid body, respectively. A line vector or a free vector carrying a force intensity f gives a pure force $(\boldsymbol{f};\boldsymbol{r}\times\boldsymbol{f})$ or couple $(\mathbf{0};\boldsymbol{c})$ applied to a rigid body.

If n screws $(n\le 6)$ are given as $\boldsymbol{S}_1,\boldsymbol{S}_2,\cdots\boldsymbol{S}_n$, and n real numbers, $a_1,a_2,\cdots a_n$, which, without all of them being zero simultaneously, can not be found to satisfy the equation, $a_1\boldsymbol{S}_1+a_2\boldsymbol{S}_2+\cdots+a_n\boldsymbol{S}_n=0$, then the n screws are said to be linearly independent. Otherwise, they are said to be linearly dependent. The linear independence can be checked by finding the order of a group of screws, i.e. calculating the rank of the matrix formed by the screws in the group. If the order equals the number of screws, these screws are linearly independent. Otherwise, they are linear dependent. Generally, the number of linearly independent screws in a screw group is six. However, in special geometric arrangements, the order of a screw group is fewer than six, that is, the number of linearly independent screws is fewer than six. For examples , when being common in a plane, the linear independence number is 3 for line vectors and 5 for screws; when being concurrent on a plane, the linear independence number is 2 for line vectors and 3 for screws; when being concurrent in a space, the linear independence number is 3 for line vectors and 3 for screws.

1.2. KP-Screw and Equivalent KP-Screw Groups

A spatial mechanism can be represented with different methods. The screw theory, as one of geometric tools, has been widely used in robotics and mechanics. A screw is a geometrical element representing a spatial straight line to which a scalar quantity, the pitch of the screw, is added. When a screw moves into the domain of mechanics, it can carry a twist or a wrench. A twist, $\boldsymbol{T} = \omega \boldsymbol{S} = (\boldsymbol{\omega}; \boldsymbol{v}) = (\boldsymbol{\omega}; \boldsymbol{r} \times \boldsymbol{\omega} + h_{\omega}\boldsymbol{\omega})$, about an instantaneous screw, $\boldsymbol{S} = (\boldsymbol{s}; \boldsymbol{s}_0) = (\boldsymbol{s}; \boldsymbol{r} \times \boldsymbol{s} + h_{\omega}\boldsymbol{s})$, describes a general spatial motion of a rigid body. When the twist describes the relative movement between two rigid bodies connected with a joint, the instantaneous screw specifies the type of the kinematic pair such as a revolute joint, a prismatic joint or a helical joint. So screw $\boldsymbol{S}$ is called a kinematic-pair screw (KP-screw) in this chapter. A serial kinematic chain or manipulator with n degrees of freedom can be represented by a KP-screw group $\{\boldsymbol{S}_1,\ldots\boldsymbol{S}_j,\ldots\boldsymbol{S}_n\}$. Any motion of an end-effector can be described by linear combination of n KP-screws. Suppose n KP-screws are linearly independent and $n<6$, they comprise a basis of a n-dimensional screw subspace. With linear transformation theory, the same motion of an end-effector can also be obtained by linearly combining other KP-screw groups as the bases of the screw subspace. These KP-screw groups come from linear transformations of the inertial basis. It indicates that serial limbs represented by all KP-screw groups as bases of the same screw subspace are equivalent to initial serial limb. Hence, these KP-screw groups are called equivalent KP-screw group in this chapter.

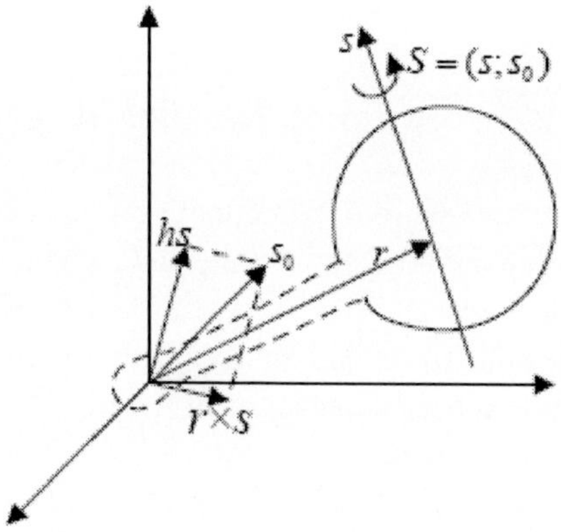

Figure 1. A screw and its components

1.3. Restraining-Screw and Equivalent Restraining-Screw Groups

A constraint between two links of a mechanism can be represented by a restraining wrench $\boldsymbol{W} = f\,\boldsymbol{S}^r$. Where, f is the density of a wrench, screw $\boldsymbol{S}^r = (\boldsymbol{s}^r; \boldsymbol{s}_0^r)$ describes the position, direction and pitch of a constraint. So screw $\boldsymbol{S}^r$ is termed as a restraining-screw in this chapter. Restraining-screws incurred by a serial kinematical limb can be obtained by calculating the reciprocal screws of corresponding KP-screws. A screw $\boldsymbol{S}^r$ is the reciprocal screw of a group of screws $\{\boldsymbol{S}_1,\ldots,\boldsymbol{S}_j,\ldots\boldsymbol{S}_n\}$ (n<6), when their reciprocal products or virtual coefficients are zero, i.e. $\boldsymbol{S}_j \circ \boldsymbol{S}^r = \mathbf{s}_j \cdot \mathbf{s}_0^r + \mathbf{s}^r \cdot \mathbf{s}_{0j} = 0 \quad (j = 1,2,\ldots,n)$. The number of screws reciprocal to a linearly independent n-screw group is 6–n. When a rigid body is subject to a force line vector $\boldsymbol{S}^r$ (pure force), the velocity (or component of velocity) of the body along $\boldsymbol{S}^r$ must be zero. In this case, the rigid body can only translate in the directions orthogonal to $\boldsymbol{S}^r$, and rotate about the axes sharing the same plane with $\boldsymbol{S}^r$. It implies that when a body is subject to a force line vector, its rotations will be restricted as well. When a rigid body is subject to two force line vectors $\boldsymbol{S}_1^r$ and $\boldsymbol{S}_2^r$, it can translate in a direction orthogonal to $\boldsymbol{S}_1^r$ and $\boldsymbol{S}_2^r$, and rotate about the axes which are intersecting or parallel with $\boldsymbol{S}_1^r$ and $\boldsymbol{S}_2^r$. When a rigid body is subject to three force line vectors, generally, all its translational motions in three dimensions are restrained, and its possible motions are three rotations. However, any of these available rotational axes must intersect or be parallel with the three line vectors.

When the end-effector of a manipulator is exerted with r linearly independent restraining-screws, the r restraining screws form a basis of a r-dimensional restraining-screw subspace. From linear transformation principle, any constraint in the restraining-screw subspace can be obtained by the combination of screws as bases of the restraining-screw subspace. So all bases of initial restraining-screw subspace are called equivalent restraining-screw groups in this chapter.

1.4. Output Degrees-of-Freedom of Parallel Manipulators

The concepts of mobility and degree-of-freedom of a mechanism have been discussed in a great deal of literature. The mobility of a kinematic chain or a mechanism is used to describe the number of independent variables that must be specified in order to locate all members of the mechanism relative to one another. The mobility, sometimes, is called the number of degrees of freedom of a mechanism or of a kinematic chain. The relative freedom between two links in a mechanism is termed their connectivity. For a manipulator mechanism, the connectivity between the end-effector and the base is essential. In most cases, it is the same as the mobility of a manipulator mechanism. However, in some cases, they are different. For examples, the mobility of a serial manipulator with seven kinematic-pairs is seven, but its end-effector has connectivity equal to six, because the end-effector has three translations and three rotations.

Connectivity between the output link and the base of a mechanism describes the degrees of freedom that a mechanism contributes to a given task, and is termed the output-freedom of a mechanism for short. It equals to the number of degrees of freedom minus the number of constraints on the output link. Generally, the output link has the largest number of degrees of freedom in all links of a mechanism. The output-freedom of a serial manipulator indicates the degrees of freedom of its end-effector with respect to its base. For a parallel manipulator, it gives the degrees of freedom of the moving platform.

In spatial cases, a rigid body has six degrees of freedom relative to a reference base, and $(n-1)$ rigid bodies have $6(n-1)$ degrees of freedom. When n rigid bodies are linked through g joints, each having f_i degrees of freedom, the mobility of the mechanism from Kutzbach-Grübler criterion is

$$M = 6(n-g-1) + \sum_{i=1}^{g} f_i \tag{1}$$

When joint axes of a mechanism form a screw system of order d, Waldron proposed a mobility coefficient that varies with the order of the screw system as

$$M = d(n-g-1) + \sum_{i=1}^{g} f_i \tag{2}$$

The order d equals $(6-\lambda)$, where λ indicates the number of common restraint wrenches acting on the mechanism. When using the above equations, if there exist overconstraints in the mechanism, the result needs to add the number of overconstraints. Generally, if there exist idle degrees of freedom in the mechanism, the number of idle degrees of freedom should be subtracted from the result. In this chapter, linear dependence of screws is employed to identify overconstraints and idle degrees of freedom.

2. Geometric Analysis of Overconstrained Parallel Manipulators with Lower Mobility

2.1. Equivalent Screw Group Method of Overconstraint Analysis

With the concepts given above, a procedure of analyzing kinematic and constraint characteristics of spatial parallel mechanisms with less than six degrees of freedom is developed as follows.

Step 1. Write KP-screw groups of all serial limbs with vectors specifying axis directions and positions of joints;

Step 2. Obtain an equivalent KP-screw group of every serial limb, and note idle degrees of freedom[(31)];

Step 3. Obtain restraining-screws acting on the moving platform contributed by every serial limb, and find an equivalent restraining-screw group;

Step 4. Obtain an equivalent restraining-screw group of restraining-screws contributed by all serial limbs;

Step 5. Calculate the degree of freedom of a parallel mechanism.

2.2. Geometric Analysis of 3-DOF Parallel Mechanisms

2.2.1 Geometric Analysis of 3-RRC Mechanisms

A 3-RRC platform mechanism[4] is shown in figure 2. It has three limbs; each of them consists of two revolute joints and one cylindric joint. The axes or centerline of three kinematic pairs in a limb are parallel to each other, and all of them are parallel to coordinate plane *XOY*. The centerlines of the three cylindric pairs locate at the same plane, i.e. the moving platform plane, and are not parallel to the same line. Reference frame *O-XYZ* is the one fixed at the base platform. When analyzing the motion of this mechanism, a cylindric pair can be seen as a revolute joint and a prismatic pair. So that each limb is equivalent to one having four single-freedom kinematic pairs, and the mechanism is equivalent to a 3-RRRP parallel platform mechanism.

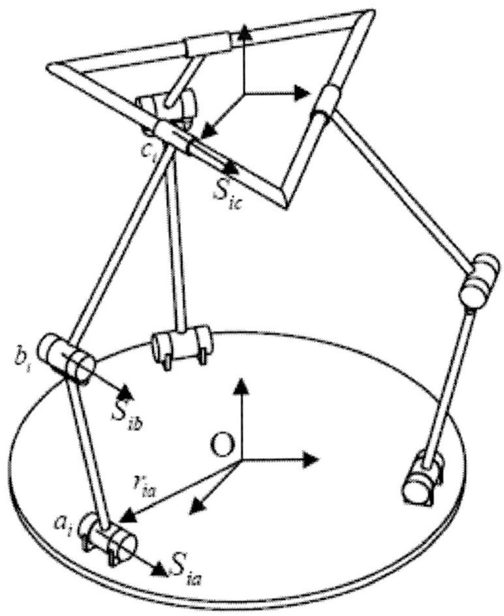

Figure 2. 3-RRC parallel platform mechanism.

A KP-screw group representing such a RRRP serial limb can be given as

$$\boldsymbol{S}_{i1}=(\boldsymbol{s}_{ia};\ \boldsymbol{r}_{ia}\times\boldsymbol{s}_{ia}),\quad \boldsymbol{S}_{i2}=(\boldsymbol{s}_{ib};\ \boldsymbol{r}_{ib}\times\boldsymbol{s}_{ib});\boldsymbol{S}_{i3}=(\boldsymbol{s}_{ic};\ \boldsymbol{r}_{ic}\times\boldsymbol{s}_{ic}),\quad \boldsymbol{S}_{i4}=(\boldsymbol{0};\ \boldsymbol{s}_{ic}), \tag{3}$$

where, $\boldsymbol{s}_{ia}$, $\boldsymbol{s}_{ib}$ and $\boldsymbol{s}_{ic}$ are the direction vectors of axes or centerline, and they are parallel to each other; Vector $\boldsymbol{r}_{ia}$, $\boldsymbol{r}_{ib}=\boldsymbol{r}_{ia}+l_{iab}\boldsymbol{s}_{iab}$ and $\boldsymbol{r}_{ic}=\boldsymbol{r}_{ib}+l_{ibc}\boldsymbol{s}_{ibc}$ are position vectors of axes or centerline of three joints; $\boldsymbol{s}_{iab}$, $\boldsymbol{s}_{ibc}$ and l_{iab}, l_{ibc} are the direction vectors and lengths of two links, respectively, subscript i indicates the ith limb, i=1,2,3. With the above geometric relations, a equivalent KP-screw group can be obtained form a linear transformation,

$$\boldsymbol{S}_{ie1}=(\boldsymbol{s}_{ia};\ \boldsymbol{r}_{ia}\times\boldsymbol{s}_{ia}),\quad \boldsymbol{S}_{ie2}=(\boldsymbol{0};\ \boldsymbol{s}_{iab}\times\boldsymbol{s}_{ib});\ \boldsymbol{S}_{ie3}=(\boldsymbol{0};\ \boldsymbol{s}_{ibc}\times\boldsymbol{s}_{ic}),\quad \boldsymbol{S}_{ie4}=(\boldsymbol{0};\ \boldsymbol{s}_{ic}), \tag{4}$$

where, $\boldsymbol{S}_{ie1} = \boldsymbol{S}_{i1}$, $\boldsymbol{S}_{ie2} = \boldsymbol{S}_{i2} - \boldsymbol{S}_{i1}$, $\boldsymbol{S}_{ie3} = \boldsymbol{S}_{i3} - \boldsymbol{S}_{i2}$ and $\boldsymbol{S}_{ie4} = \boldsymbol{S}_{i4}$. This gives a serial limb with one revolute joint and three prismatic joints, indicating such a RRC serial limb is equivalent to a RPPP serial limb. Since $\boldsymbol{s}_{ia}$ is parallel to plane *XOY*, the restraining-screws acting the moving platform can be obtained by solving the screws reciprocal to KP-screws in equations (2) and (3) as follows,

$$\boldsymbol{S}_{i1}^{r} = (\boldsymbol{0};\ \boldsymbol{z}) \text{ and } \boldsymbol{S}_{i2}^{r} = (0;\ \boldsymbol{z} \times \boldsymbol{s}_{ia}) \tag{5}$$

There are two restraint couples about axe $\mathbf{z}$ and $\mathbf{z} \times \mathrm{s}_{ia}$. Six such restraining-screws can be obtained from the three limbs to constrain the moving platform. Since three restraining screws, $\boldsymbol{S}_{i2}^{r}$, *i*=1,2,3, are linearly dependent, and restrain the rotations about axes parallel to plane *XOY*, they can be represented with

$$\boldsymbol{S}_{1}^{r} = (\boldsymbol{0};\ \boldsymbol{x}), \text{and } \boldsymbol{S}_{2}^{r} = (\boldsymbol{0};\ \boldsymbol{y}) , \tag{6}$$

indicating there exist one overconstraint. Similarly, $\boldsymbol{S}_{i1}^{r}$, *i*=1,2,3, can be denoted as

$$\boldsymbol{S}_{3}^{r} = (\boldsymbol{0};\ \boldsymbol{z}) , \tag{7}$$

indicating there is one common constraint on the moving platform, so the order of this machanism is 5. Both equations (6) and (7) demonstrate that the moving platform loses all rotational degrees of freedom, remains only three translational degrees of freedom. Because of one overconstraint, the moving platform of a 3-RRC mechanism has three degrees of freedom with respect to the base platform, i.e. $M = 5(8-9-1)+1\times 6+2\times 3+1=3$.

2.2.2 Kinematic Characteristics of 3-CPR Mechanisms

A 3-CPR platform mechanism is given in figure 3. The mechanism comprises three serial limbs each with one cylindric joint, one prismatic joint and one revolute joint. The centerline of cylindric joint is parallel to the axis of revolute joint in the same limb, i.e. $\boldsymbol{s}_{ia} // \boldsymbol{s}_{ib}$. The centerlines, $\boldsymbol{s}_{1a}$, $\boldsymbol{s}_{2a}$ and $\boldsymbol{s}_{3a}$, of three cylindric joints are parallel to the same plane, *XOY*, and are arranged as $\boldsymbol{s}_{a1} // \boldsymbol{s}_{a3}$, and $\boldsymbol{s}_{a1} \perp \boldsymbol{s}_{a2}$.

A KP-screw group representing such a CPR serial limb, also a RPPR serial limb, is given as

$$\boldsymbol{S}_{i1} = (\boldsymbol{s}_{ia};\ \boldsymbol{r}_{ia} \times \boldsymbol{s}_{ia}),\ \boldsymbol{S}_{i2} = (\boldsymbol{0};\ \boldsymbol{s}_{ia});\ \boldsymbol{S}_{i3} = (\boldsymbol{0};\ \boldsymbol{s}_{iab}),\ \boldsymbol{S}_{i4} = (\boldsymbol{s}_{ib};\ \boldsymbol{r}_{ib} \times \boldsymbol{s}_{ib}), \tag{8}$$

where, $\boldsymbol{s}_{ia}$, $\boldsymbol{s}_{iab}$ and $\boldsymbol{s}_{ib}$ are the direction vectors of axes or centerlines; $\boldsymbol{r}_{ia}$ and $\boldsymbol{r}_{ib} = \boldsymbol{r}_{ia} + l_{iab}\boldsymbol{s}_{iab}$ are the position vectors of points a_i and b_i which are the intersection between

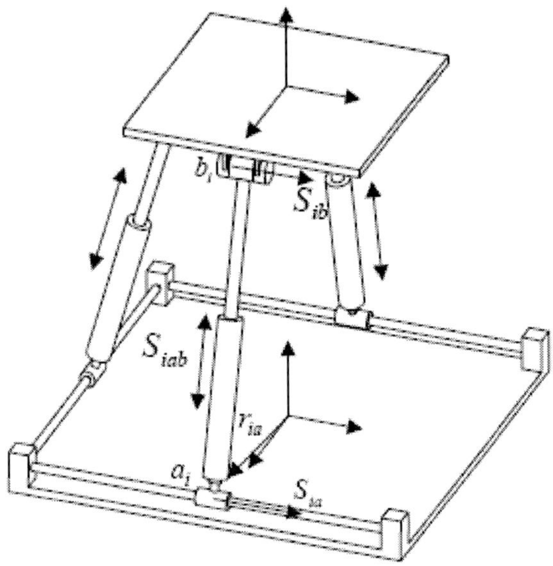

Figure 3. 3-CPR parallel platform mechanism.

lines $\boldsymbol{s}_{ia}$, $\boldsymbol{s}_{iab}$ and $\boldsymbol{s}_{ib}$; l_{iab} is the displacement from a_i to b_i. The KP-screw group is equivalent to

$$\boldsymbol{S}_{ie1} = (\boldsymbol{s}_{ia};\ \boldsymbol{r}_{ia} \times \boldsymbol{s}_{ia}),\ \boldsymbol{S}_{ie2} = (\mathbf{0};\ \boldsymbol{s}_{ia});\ \boldsymbol{S}_{ie3} = (\mathbf{0};\ \boldsymbol{s}_{iab}),\ \boldsymbol{S}_{ie4} = (\mathbf{0};\ \boldsymbol{s}_{iab} \times \boldsymbol{s}_{ia}), \tag{9}$$

where, $\boldsymbol{S}_{ie1} = \boldsymbol{S}_{i1}$, $\boldsymbol{S}_{ie2} = \boldsymbol{S}_{i2}$, $\boldsymbol{S}_{ie3} = \boldsymbol{S}_{i3}$ and $\boldsymbol{S}_{ie4} = (\boldsymbol{S}_{i4} - \boldsymbol{S}_{i1})/l_{iab}$. This also presents a serial limb with one revolute joint and three prismatic joints, indicating such a CPR serial limb is also equivalent to a RPPP serial limb. This contributes two restraining screws on the moving platform. Since $\boldsymbol{s}_{ia}$ is parallel to plane XOY, the two restraining-screws can be represented as

$$\boldsymbol{S}_{i1}^r = (\mathbf{0};\ \boldsymbol{z}) \text{ and } \boldsymbol{S}_{i2}^r = (0;\ \boldsymbol{z} \times \boldsymbol{s}_{i1}) \tag{10}$$

Three CPR serial limbs contribute six restraining screws on the moving platform. Since three restraining screws, $\boldsymbol{S}_{i2}^r$, i=1,2,3, are linearly dependent, they can be represented with

$$\boldsymbol{S}_1^r = (\mathbf{0};\ \boldsymbol{x}) \text{ and } \boldsymbol{S}_2^r = (0;\ \boldsymbol{y}), \tag{11}$$

indicating one of the three restraining screws is redundant, which represents one overconstraint. Further, $\boldsymbol{S}_{i1}^r$, i=1,2,3, are denoted as

$$\boldsymbol{S}_3^r = (\mathbf{0};\ \boldsymbol{z}), \tag{12}$$

indicating there is one common constraint on the moving platform.so the order of this machanism is 5. The above two equations give an equivalent representation of restraining-screws acting on the moving platform. This demonstrates that the moving platform only translates and does not have rotation. Because of one overconstraint, the moving platform of a 3-CPR mechanism has three degrees of freedom, i.e. $M = 5(8-9-1)+1\times 6+2\times 3+1=3$.

2.3 Geometric Analysis of 4-Dof Parallel Mechanisms

A 4-RRUR platform mechanism is shown in figure 4. It has four RRUR limbs. Each RRUR limb consists of one universal joint and three revolute joint. Because a universal joint can be seen as two revolute joints, a RRUR limb can be also denoted as $R_1R_2R_3R_4R_5$. Revolute joint R_1 connects with a base platform, and R_5 with the moving platform. The axes of revolute joints R_1, R_2 and R_3 are parallel to each other, and parallel to the base platform. The axes of the later two revolute joints intersect at a common point.

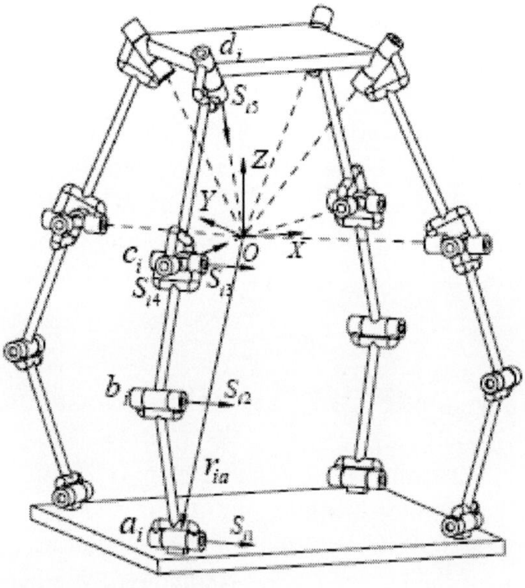

Figure 4. 4-RRUR parallel platform mechanism.

A KP-screw group representing such a RRUR serial limb can be given as

$$\boldsymbol{S}_{i1} = (\boldsymbol{s}_{i1};\ \boldsymbol{r}_{ia}\times\boldsymbol{s}_{i1}), \boldsymbol{S}_{i2} = (\boldsymbol{s}_{i2};\ \boldsymbol{r}_{ib}\times\boldsymbol{s}_{i2}), \boldsymbol{S}_{i3} = (\boldsymbol{s}_{i3};\ \boldsymbol{r}_{ic}\times\boldsymbol{s}_{i3}), \boldsymbol{S}_{i4} = (\boldsymbol{s}_{i4}\ ;0), \boldsymbol{S}_{i5} = (\boldsymbol{s}_{i5}\ ;0) \quad (13)$$

where, $\boldsymbol{s}_{ij}$, (i=1,2,3,4, j=1,2…5) are unit vectors indicating revolute joint axes, vector $\boldsymbol{r}_{ia}$, $\boldsymbol{r}_{ib} = \boldsymbol{r}_{ia} + l_{iab}\boldsymbol{s}_{iab}$ and $\boldsymbol{r}_{ic} = \boldsymbol{r}_{ib} + l_{ibc}\boldsymbol{s}_{ibc}$ are position vectors of axes; $\boldsymbol{s}_{iab}$, $\boldsymbol{s}_{ibc}$ and l_{iab}, l_{ibc} are the direction vectors and lengths of two links, respectively, subscript i indicates the ith limb, j the jth joint. With the above geometric relations, $s_{i1} = s_{i2} = s_{i3}$, an equivalent KP-screw group can be obtained,

$$\boldsymbol{S}_{ie1}=(\boldsymbol{s}_{i1};\ \boldsymbol{r}_{ia}\times\boldsymbol{s}_{i1}),\boldsymbol{S}_{ie2}=(0;\ \boldsymbol{s}_{iab}\times\boldsymbol{s}_{i1}),\boldsymbol{S}_{ie3}=(0;\ \boldsymbol{s}_{ibc}\times\boldsymbol{s}_{i1}),\boldsymbol{S}_{ie4}=(\boldsymbol{s}_{i4}\ ;0),\boldsymbol{S}_{ie5}=(\boldsymbol{s}_{i5}\ ;0) \quad (14)$$

where, $\boldsymbol{S}_{ie1}=\boldsymbol{S}_{i1}$, $\boldsymbol{S}_{ie2}=(\boldsymbol{S}_{i2}-\boldsymbol{S}_{i1})/l_{iab}$, $\boldsymbol{S}_{ie3}=(\boldsymbol{S}_{i3}-\boldsymbol{S}_{i2})/l_{ibc}$, $\boldsymbol{S}_{ie4}=\boldsymbol{S}_{i4}$ and $\boldsymbol{S}_{ie5}=\boldsymbol{S}_{i5}$. The KP-screw group represents a serial limb with three revolute joints and two prismatic joints, which indicates such a RRUR serial limb is equivalent to a RRRPP serial limb. The restraining-screw acting the moving platform can be obtained by solving the screw reciprocal to KP-screws in equations (13) and (14),

$$\boldsymbol{S}_i^r=(\boldsymbol{s}_{i1};0) \quad (15)$$

It represents a restraint force passes the origin point and lies along axis s_{i1} ,. Four such restraining-screws can be obtained from the four limbs to constrain the moving platform. Since s_{i1} , is parallel to the base platform, $\boldsymbol{S}_i^r$, i=1,2,3,4 are linearly dependent, and constrain the translations about axes parallel to plane XOY, they can be represented with

$$\boldsymbol{S}_1^r=(x;\ \boldsymbol{0}),\text{ and } \boldsymbol{S}_2^r=(y;\ \boldsymbol{0}), \quad (16)$$

which indicates there exists two overconstraint. The above equation reveals that the moving platform loses two translational degrees of freedom, remains three rotational degrees of freedom and one translational degree of freedom along axis z. Because two overconstraints exist, the moving platform of a 4-RRUR mechanism has four degrees of freedom with respect to the base platform, i.e. $M=6(18-20-1)+1\times12+4\times2+2=4$.

2.4 Geometric Analysis of 5-Dof Parallel Mechanisms

A 3- RRCR platform mechanism is given in figure 5. It has three same RRCR limbs each with one cylindrical joint and three revolute joint. It can also be denoted as a 3-$R_1R_2P_3R_4R_5$ mechanism. Revolute joint R_1 connects with a base platform, and R_5 with the moving platform. The axes, $\boldsymbol{s}_{i1}$, of R_1 and $\boldsymbol{s}_{i2}$, of R_2 are parallel to each other and perpendicular to the base platform. The axis, $\boldsymbol{s}_{i3}$, of P_3 is parallel to the base platform ,at the same time, axes $\boldsymbol{s}_{i3}$ is perpendicular to axes, $\boldsymbol{s}_{i1}$. The axes of revolute joints R_4 and R_5 are intersect at a common point.

A KP-screw group representing such a RRCR serial limb is

$$\boldsymbol{S}_{i1}=(\boldsymbol{s}_{i1};\ \boldsymbol{r}_{ia}\times\boldsymbol{s}_{i1}),\boldsymbol{S}_{i2}=(\boldsymbol{s}_{i2};\ \boldsymbol{r}_{ia}\times\boldsymbol{s}_{i2}),\boldsymbol{S}_{i3}=(\boldsymbol{0};\ \boldsymbol{s}_{i3}),\boldsymbol{S}_{i4}=(\boldsymbol{s}_{i4};\ 0),\boldsymbol{S}_{i5}=(\boldsymbol{s}_{i5};\ 0) \quad (17)$$

where, $\boldsymbol{s}_{ij}$, (i=1,2,3, j=1,2…5) are unit vectors indicating revolute joint's axes or prismatic joint's centerline, vector $\boldsymbol{r}_{ia}$ and $\boldsymbol{r}_{ib}=\boldsymbol{r}_{ia}+l_{iab}\boldsymbol{s}_{iab}$ are position vectors of axes; $\boldsymbol{s}_{iab}$ and l_{iab} are

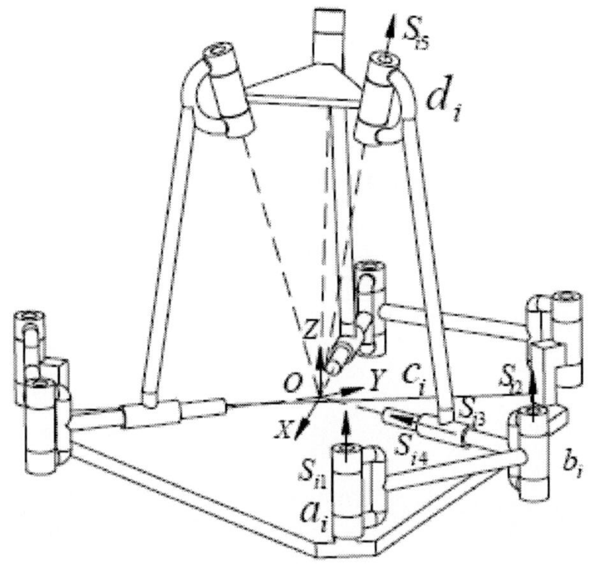

Figure 5. 3- RRCR parallel platform mechanism.

the direction vector and length between point a_i to point b_i. From geometric relations $\boldsymbol{s}_{i1} = \boldsymbol{s}_{i2}$ and $\boldsymbol{s}_{i3} = \boldsymbol{s}_{i4}$, an equivalent KP-screw group can be obtained,

$$\boldsymbol{S}_{ie1} = (\boldsymbol{s}_{i1};\ \boldsymbol{r}_{ia} \times \boldsymbol{s}_{i1}), \boldsymbol{S}_{ie2} = (0;\ \boldsymbol{s}_{iab} \times \boldsymbol{s}_{i1}), \boldsymbol{S}_{ie3} = (0;\ \boldsymbol{s}_{i3}), \boldsymbol{S}_{ie4} = (\boldsymbol{s}_{i4};\ 0), \boldsymbol{S}_{ie5} = (\boldsymbol{s}_{i5};\ 0) \quad (18)$$

where, $\boldsymbol{S}_{ie1} = \boldsymbol{S}_{i1}$, $\boldsymbol{S}_{ie2} = (\boldsymbol{S}_{i2} - \boldsymbol{S}_{i1})/l_{iab}$, $\boldsymbol{S}_{ie3} = \boldsymbol{S}_{i3}$, $\boldsymbol{S}_{ie4} = \boldsymbol{S}_{i4}$ and $\boldsymbol{S}_{ie5} = \boldsymbol{S}_{i5}$. The KP-screw group represents a serial limb with two prismatic joints and three revolute joints. The restraining-screw acting the moving platform can be obtained by solving the screw reciprocal to KP-screws in equations (17) and (18),

$$\boldsymbol{S}_i^r = (\boldsymbol{s}_{i1};\ 0) \quad (19)$$

It represents a restraint force along axis $\boldsymbol{s}_{i1}$ through the origin piont. Three such restraining-screws can be obtained from the three limbs. Since s_{i1} is perpendicular to the base platform, the three restraining-screws, $\boldsymbol{S}_i^r$, i=1,2,3, are the same, they can be represented with

$$\boldsymbol{S}_1^r = (z;\ 0), \quad (20)$$

which indicates there exist one common constraint, the order of this machanism is 5. The above equation reveals that the moving platform loses one translational degree of freedom, remains two translational degrees of freedom and three rotational degrees of freedom. So the moving platform of a 3-RRCR mechanism has five degrees of freedom with respect to the base platform, i.e. $M = 5(14-15-1)+1\times 15 = 5$.

3. Equivalent Representation of Parallelogram Linkages

With the above equivalent screw group method, this section discusses motion and constraint characteristics of three kinds of parallelogram linkages. These parallelogram linkages are often used as complex joints of spatial parallel mechanisms.

3.1 4R Parallelogram Linkage Units

A 4R parallelogram unit, shown in figure 6, is a planar mechanism with one degree of freedom. Let link 1 and link 4 be its base and output links. Two serial kinematic limbs, R_aR_c and R_bR_d, restrain link 4. Their KP-screws are

$$\begin{aligned} \boldsymbol{S}_{11} &= (\boldsymbol{s}_a;\ \boldsymbol{r}_a \times \boldsymbol{s}_a),\quad \boldsymbol{S}_{12} = (\boldsymbol{s}_c;\ \boldsymbol{r}_c \times \boldsymbol{s}_c); \\ \boldsymbol{S}_{21} &= (\boldsymbol{s}_b;\ \boldsymbol{r}_b \times \boldsymbol{s}_b),\quad \boldsymbol{S}_{22} = (\boldsymbol{s}_d;\ \boldsymbol{r}_d \times \boldsymbol{s}_d), \end{aligned} \tag{21}$$

where, $\boldsymbol{s}_a // \boldsymbol{s}_b // \boldsymbol{s}_c // \boldsymbol{s}_d$ are the direction vectors of axes; $\boldsymbol{r}_a$, $\boldsymbol{r}_b = \boldsymbol{r}_a + l_{ab}\boldsymbol{s}_{ab}$, $\boldsymbol{r}_c = \boldsymbol{r}_a + l_{ac}\boldsymbol{s}_{ac}$ and $\boldsymbol{r}_d = \boldsymbol{r}_a + l_{ab}\boldsymbol{s}_{ab} + l_{bd}\boldsymbol{s}_{bd}$ are position vectors of points *a*, *b*, *c*, *d*; $\boldsymbol{s}_{ac}$, $\boldsymbol{s}_{ab}$, $\boldsymbol{s}_{bd}$ and l_{ac}, l_{ab}, l_{bd} are the direction vectors and lengths of lines *ac*, *ab* and *bd*, respectively.

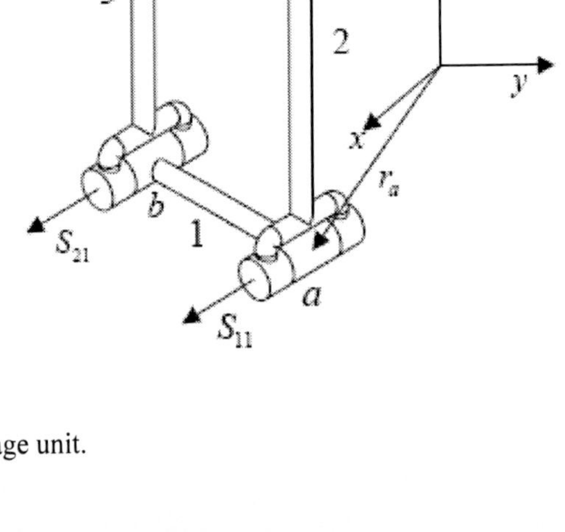

Figure 6. 4-R parallel linkage unit.

Two KP-screw groups equivalent to that in the above equation,

$$\begin{aligned} &\boldsymbol{S}_{e11}=(\boldsymbol{s}_a;\ \boldsymbol{r}_a\times\boldsymbol{s}_a),\quad \boldsymbol{S}_{e12}=(\boldsymbol{0};\ \boldsymbol{s}_{ac}\times\boldsymbol{s}_a);\\ &\boldsymbol{S}_{e21}=(\boldsymbol{s}_a;\ (\boldsymbol{r}_a+l_{ab}\boldsymbol{s}_{ab})\times\boldsymbol{s}_a),\quad \boldsymbol{S}_{e22}=(\boldsymbol{0};\ \boldsymbol{s}_{ac}\times\boldsymbol{s}_a), \end{aligned} \tag{22}$$

are obtained by linear transformation $\boldsymbol{S}_{e11}=\boldsymbol{S}_{11}$, $\boldsymbol{S}_{e12}=\boldsymbol{S}_{12}-\boldsymbol{S}_{11}$, $\boldsymbol{S}_{e21}=\boldsymbol{S}_{21}$ and $\boldsymbol{S}_{e22}=\boldsymbol{S}_{22}-\boldsymbol{S}_{21}$. The restraining-screws reciprocal to KP-screw group $\boldsymbol{S}_{11}$, $\boldsymbol{S}_{12}$ and $\boldsymbol{S}_{21}$, $\boldsymbol{S}_{22}$ are

$$\boldsymbol{S}_{11}^r=(\boldsymbol{0};\boldsymbol{s}_{ac}\times\boldsymbol{s}_a),\boldsymbol{S}_{12}^r=(\boldsymbol{0};\boldsymbol{s}_{ab}\times\boldsymbol{s}_a),\boldsymbol{S}_{13}^r=(\boldsymbol{s}_a;\boldsymbol{0}),\boldsymbol{S}_{14}^r=(\boldsymbol{s}_{ac};\boldsymbol{0}), \tag{23}$$

and

$$\boldsymbol{S}_{21}^r=(\boldsymbol{0};\boldsymbol{s}_{ac}\times\boldsymbol{s}_a),\boldsymbol{S}_{22}^r=(\boldsymbol{0};\boldsymbol{s}_{ab}\times\boldsymbol{s}_a),\boldsymbol{S}_{23}^r=(\boldsymbol{s}_a;\boldsymbol{0}),\boldsymbol{S}_{14}^r=(\boldsymbol{s}_{ac};\boldsymbol{0}), \tag{24}$$

There are five restraining-screws being linearly independent in the above eight restraining-screws. So a restraining-screws acting on link 4, which is equivalent to those in equations (22) and (23), is given as

$$\boldsymbol{S}_{e1}^r=(\boldsymbol{0};\boldsymbol{s}_{ac}\times\boldsymbol{s}_a),\boldsymbol{S}_{e2}^r=(\boldsymbol{0};\boldsymbol{s}_{ab}\times\boldsymbol{s}_a),\boldsymbol{S}_{e3}^r=(\boldsymbol{0};\boldsymbol{s}_{ab}\times\boldsymbol{s}_{ac}),\boldsymbol{S}_{e4}^r=(\boldsymbol{s}_a;\boldsymbol{0}),\boldsymbol{S}_{e5}^r=(\boldsymbol{s}_{ac};\boldsymbol{0}) \tag{25}$$

Under constraints of the five restraining-screws, the realizable motion of link 4 can be represented with a twist reciprocal to the five restraining-screws, i.e.

$$\omega\boldsymbol{S}=\omega(0;\boldsymbol{s}_{ac}\times\boldsymbol{s}_a) \tag{26}$$

It shows that link 4 can only perform translations along the direction $\boldsymbol{s}_{ac}\times\boldsymbol{s}_a$, which means a 4R parallelogram unit is a complex prismatic joint $(4\mathrm{R})_\mathrm{P}$ with equivalent KP-screw, $\boldsymbol{S}=(0;\boldsymbol{s}_{ac}\times\boldsymbol{s}_a)$.

3.2 4U Parallelogram Linkage Units

A 4U parallelogram linkage, in figure 7a, comprises four links connected by four universal joints with following structural characteristics. The axes of two revolute joints in both link 1 (base member) and link 4 are collinear and denoted as $\boldsymbol{s}_{ab}$ and $\boldsymbol{s}_{cd}$, respectively. The axes of two revolute joints in link 2 are parallel to each other and denoted as $\boldsymbol{s}_a$ and $\boldsymbol{s}_c$ respectively, and link 2 which connects the central points of two universal joints, *a* and *c,* is orthogonal to two revolute joint axes. This is similar in link 3. The links connecting the central points of the four universal joints, *a* ,*b*, *c* and *d*, form a parallelogram. Let link 1 and link 4 be the base and the output link of the linkage, respectively.

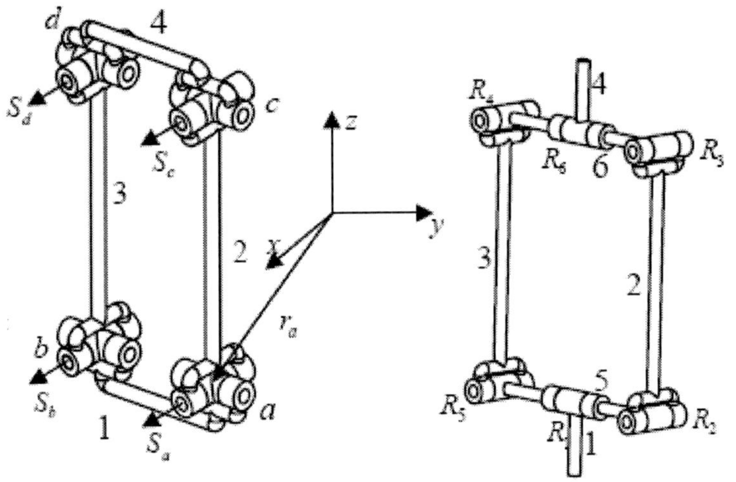

Figure 7. 4-U parallelogram Linkage unit and equivalent R(P)$_{4R}$R hybrid limb.

Since the 4U linkage unit is assembled into a parallelogram, the rotary axes of four universal joints are $s_a // s_b // s_c // s_d$, $s_{ab} // s_{cd}$ and links $s_{ac} // s_{bd}$, and the lengths of links are $l_{ab} = l_{cd}$ and $l_{ac} = l_{bd}$. The two KP-screw groups are given as

$$\begin{aligned}&\boldsymbol{S}_{11} = (\boldsymbol{s}_{ab};\ \boldsymbol{r}_a \times \boldsymbol{s}_{ab}), \qquad \boldsymbol{S}_{12} = (\boldsymbol{s}_a;\ \boldsymbol{r}_a \times \boldsymbol{s}_a),\\ &\boldsymbol{S}_{13} = (\boldsymbol{s}_c;\ (\boldsymbol{r}_a + l_{ac}\boldsymbol{s}_{ac}) \times \boldsymbol{s}_c), \boldsymbol{S}_{14} = (\boldsymbol{s}_{cd};\ (\boldsymbol{r}_a + l_{ac}\boldsymbol{s}_{ac}) \times \boldsymbol{s}_{cd}),\end{aligned} \tag{27}$$

and

$$\begin{aligned}&\boldsymbol{S}_{21} = (\boldsymbol{s}_{ab};\ \boldsymbol{r}_a \times \boldsymbol{s}_{ab}), \qquad \boldsymbol{S}_{22} = (\boldsymbol{s}_b;\ (\boldsymbol{r}_a + l_{ab}\boldsymbol{s}_{ab}) \times \boldsymbol{s}_b),\\ &\boldsymbol{S}_{23} = (\boldsymbol{s}_d;\ (\boldsymbol{r}_a + l_{ab}\boldsymbol{s}_{ab} + l_{bd}\boldsymbol{s}_{bd}) \times \boldsymbol{s}_d), \boldsymbol{S}_{24} = (\boldsymbol{s}_{cd};\ (\boldsymbol{r}_a + l_{ab}\boldsymbol{s}_{ab} + l_{bd}\boldsymbol{s}_{bd}) \times \boldsymbol{s}_{cd}).\end{aligned} \tag{28}$$

With the above geometric relations, two equivalent KP-screw groups yield,

$$\boldsymbol{S}_{e11} = (\boldsymbol{s}_{ab};\ \boldsymbol{r}_a \times \boldsymbol{s}_{ab}), \boldsymbol{S}_{e12} = (\boldsymbol{s}_a;\ \boldsymbol{r}_a \times \boldsymbol{s}_a), \boldsymbol{S}_{e13} = (\boldsymbol{0};\ \boldsymbol{s}_{ac} \times \boldsymbol{s}_c), \boldsymbol{S}_{e14} = (\boldsymbol{0};\ \boldsymbol{s}_{ac} \times \boldsymbol{s}_{cd}), \tag{29}$$

and

$$\boldsymbol{S}_{e21} = (\boldsymbol{s}_{ab};\ \boldsymbol{r}_a \times \boldsymbol{s}_{ab}), \boldsymbol{S}_{22} = (\boldsymbol{s}_a;\ (\boldsymbol{r}_a + l_{ab}\boldsymbol{s}_{ab}) \times \boldsymbol{s}_a), \boldsymbol{S}_{e23} = (\boldsymbol{0};\ \boldsymbol{s}_{ac} \times \boldsymbol{s}_a), \boldsymbol{S}_{e24} = (\boldsymbol{0};\ \boldsymbol{s}_{ac} \times \boldsymbol{s}_{ab}) \tag{30}$$

The restraining-screws acting on link 4 contributed by the two limbs can be obtained by evaluating the reciprocal screws of equations (29) and (30), i.e.

$$\boldsymbol{S}_{11}^r = (\boldsymbol{s}_{ac};\ \boldsymbol{r}_a \times \boldsymbol{s}_{ac}), \qquad \boldsymbol{S}_{12}^r = (0;\ \boldsymbol{s}_{ab} \times \boldsymbol{s}_a) \tag{31}$$

$$\boldsymbol{S}_{21}^r = (\boldsymbol{s}_{ac};\ (\boldsymbol{r}_a + l_{ab}\boldsymbol{s}_{ab}) \times \boldsymbol{s}_{ac}), \boldsymbol{S}_{22}^r = (0;\ \boldsymbol{s}_{ab} \times \boldsymbol{s}_a) \tag{32}$$

It can be found that the above four restraining-screws in above two equations are linearly dependent, indicating link 4 subject to an overconstraint. Since $\mathrm{s}_{ab} \times \mathrm{s}_{ac} = \mathrm{s}_a$, an equivalent restraining-screw group acting on link 4 is given as

$$\boldsymbol{S}_{e1}^{r} = (\boldsymbol{s}_{ac};\ 0),\ \boldsymbol{S}_{e2}^{r} = (0;\ \boldsymbol{s}_a),\ \boldsymbol{S}_{e3}^{r} = (0;\ \boldsymbol{s}_{ab} \times \boldsymbol{s}_a). \tag{33}$$

These form one restraint force and two restraint couples to constrain one translation along direction $\boldsymbol{s}_{ac}$ and two rotations about axes s_a and $\mathrm{s}_{ab} \times \mathrm{s}_a$ of link 4. As $\boldsymbol{s}_{ab}$ is always orthogonal to $\boldsymbol{s}_a$, an equivalent KP-screw group reciprocal to those in equation (33) can be obtained

$$\boldsymbol{S}_{e1} = (\boldsymbol{s}_{ab};\ 0),\quad \boldsymbol{S}_{e2} = (0; \boldsymbol{s}_{ac} \times \boldsymbol{s}_a),\ \boldsymbol{S}_{e3} = (0; \boldsymbol{s}_{ab} \times \boldsymbol{s}_a). \tag{34}$$

The above screws indicate that the 4U parallelogram unit can be seen as a complex joint, $(4U)_{RPP}$, with one rotary degree of freedom and two translational degrees of freedom. From the equivalent KP-screw group, further equivalent KP-screw group can be obtain by letting $\boldsymbol{S}'_{e1} = \boldsymbol{S}_{e1}$, $\boldsymbol{S}'_{e2} = \boldsymbol{S}_{e3}$, $\boldsymbol{S}'_{e3} = l_{ab}\boldsymbol{S}_1 + l_{e2}\boldsymbol{S}_{e2} + l_{e3}\boldsymbol{S}_{e3}$, that is

$$\boldsymbol{S}'_{e1} = (\boldsymbol{s}_{11};\ 0),\ \boldsymbol{S}'_{e2} = (0; \boldsymbol{s}_{ab} \times \boldsymbol{s}_{12}),\ \boldsymbol{S}'_{e3} = (\boldsymbol{s}_{11}; l_{ab}\boldsymbol{s}_{ab} \times \boldsymbol{s}_{12}), \tag{35}$$

where, let $\boldsymbol{r}_c \times \boldsymbol{S}_{ab} = l_{e2}\boldsymbol{S}_{e2} + l_{e3}\boldsymbol{S}_{e3}$ by choosing l_{e2} and l_{e3}. Equation (34) represents a hybrid serial limb with two rotational joints and one translational joint, where, the translational joint $\boldsymbol{S}'_{e2}$ is a match for complex joint $(4R)_P$ given in the above section. Hence the 4U parallelogram unit is equivalent to a hybrid serial limb, $R(4R)_PR$, as shown in figure 7b. From equation (33), it can be also obtained that link 4 is parallel to link 1, and link 2 parallel to link 3, indicating that the lines connecting the central points (*a* ,*b*, *c* and *d*) of the four universal joints keep being a parallelogram. Since there is one overconstraint, the output link of a 4U parallelogram unit has three degrees of freedom, i.e. $M_{(4U)} = 6(4-4-1)+8+1=3$.

3.3. 4S Parallel Linkage Unit

Figure 8 shows a 4S linkage. In special configuration, the central points (*a*, *b*, *c* and *d*) of four spherical joints form a parallelogram, as shown in figure 7a. Link 1 and link 4 are appointed as the base and output link, respectively.

For a single 4S linkage, it can't keep forming a parallelogram in a general configuration. As shown in figure 8b, link 4 is constrained by two force line vectors along centerlines of links 2 and 3, i.e.

$$\boldsymbol{S}_1^r = (\boldsymbol{s}_{ac}; \boldsymbol{r}_a \times \boldsymbol{s}_{ac}),\ \boldsymbol{S}_2^r = (\boldsymbol{s}_{bd}; (\boldsymbol{r}_a + l_{ab}\boldsymbol{s}_{ab}) \times \boldsymbol{s}_{bd}), \tag{36}$$

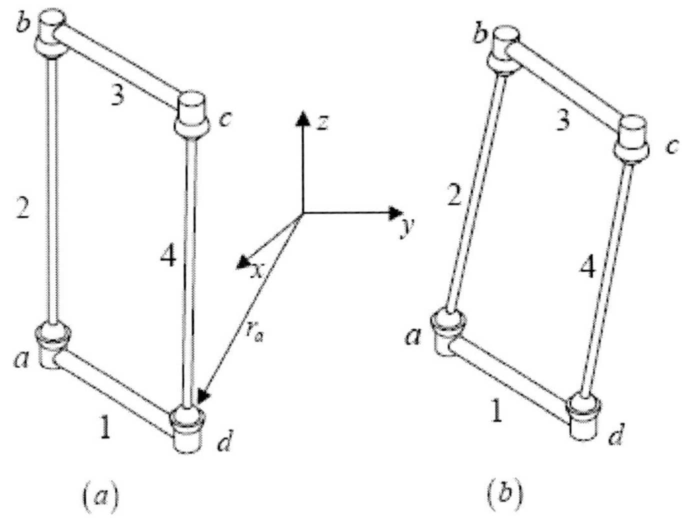

Figure 8. 4-S parallel linkage unit.

where, (s_{ab}, s_{ac} , s_{bd}) and (l_{ab}, l_{ac}, l_{bd}) are the direction vectors and lengths of lines *ab*, *dc* and *ad*, respectively. The two restraining-screws are linearly independent, restrain the translations of link 4 along directions s_{ac} and s_{bd}. For output link 4, the rotations of link 2 and link 3 about the centerlines of themselves are local motions, which are no effect on the output motion of link 4. So they are idle degrees of freedom. Hence, link 4 has four degrees of freedom with respect to link 1, i.e. $M_{(4\mathrm{S})} = 6(4-4-1)+4\times 3-2=4$.

An equivalent KP-screw group can be obtained by solving the screws reciprocal to the two restraining-screws, i.e.

$$\boldsymbol{S}_{e1}=(\boldsymbol{s}_{ab};\ \boldsymbol{r}_a\times\boldsymbol{s}_{ab}), \boldsymbol{S}_{e2}=(\boldsymbol{s}_{bd};\ (\boldsymbol{r}_a+l_{ab}\boldsymbol{s}_{ab})\times\boldsymbol{s}_{bd}), \boldsymbol{S}_{e3}=(0;\ \boldsymbol{s}_{ac}\times\boldsymbol{s}_{bd}),\ \boldsymbol{S}_{e4}=(\boldsymbol{s}_{cd};\boldsymbol{r}_d\times\boldsymbol{s}_{cd}). \quad (37)$$

where, $\boldsymbol{r}_d=\boldsymbol{r}_a+l_{ab}\boldsymbol{s}_{ab}+l_{bd}\boldsymbol{s}_{bd}=\boldsymbol{r}_a+l_{ac}\boldsymbol{s}_{ac}+l_{cd}\boldsymbol{s}_{cd}$. Equation (37) means that a 4S linkage unit in general configurations can be seen as a complex joint with three rotary degrees of freedom and one translational degree of freedom, $(4\mathrm{S})_{\mathrm{RRRP}}$.

When the lines connecting four center points, a b c and d, form a parallelogram, as shown in figure 8a, direction vectors are $s_{ac}//s_{bd}$ and $s_{ab}//s_{cd}$, respectively. The two restraining-screws acting on link 4 are

$$\boldsymbol{S}_1^r=(\boldsymbol{s}_{ac};\boldsymbol{r}_a\times\boldsymbol{s}_{ac}), \boldsymbol{S}_2^r=(\boldsymbol{s}_{bd};(\boldsymbol{r}_a+l_{ab}\boldsymbol{s}_{ab})\times\boldsymbol{s}_{bd}). \quad (38)$$

Their equivalent restraining-screws are given as

$$\boldsymbol{S}_{e1}^r=(\boldsymbol{s}_{ac};\boldsymbol{r}_a\times\boldsymbol{s}_{ac}), \boldsymbol{S}_{e2}^r=(\boldsymbol{0};\boldsymbol{s}_{ab}\times\boldsymbol{s}_{ac}). \quad (39)$$

They represent a restraint force along s_{ac} and a restraint couple about axis $s_{ab} \times s_{ac}$. So link 4 can't translate along s_{ac} and rotate about the axis parallel to $s_{ab} \times s_{ac}$. The KP-screw group reciprocal to those in equations (38) and (39) is

$$S_{e1} = (s_{ab};\ r_a \times s_{ab}), S_{e2} = (s_{ac};\ r_a \times s_{ac}), S_{e3} = (0;\ s_{ab} \times s_{ac}),\ S_{e4} = (0; (s_{ac} \times s_{ab}) \times s_{ac}). \quad (40)$$

Any available motion of link 4 can be presented by a linear combination of the four KP-screws. So a 4S linkage unit in parallelogram configurations can be seen as a complex joint with two rotary degrees of freedom and two translational degrees of freedom, $(4S)_{RRPP}$.

3.4. Overconstraint Analysis of Delta Robots

In this section, two parallel platform mechanisms, called Delta robots, 3-R(4U)RPP and 3-R(4S)RRPP are discussed, which shows the equivalent screw group method can discover the special geometric relations incurring overconstraints in a complicated spatial hybrid mechanisms.

The procedure presented in section 2 can be further developed for the geometric analysis of constraint characteristics of a parallel manipulator with hybrid limbs, that is,

Step 1a. Using equivalent screw group method, analyze the parallelogram linkage unit in a hybrid limb to obtain a restraining-screw group acting on output link and an equivalent KP-screw group of the parallelogram linkage unit, and note the numbers of overconstraints and idle degrees of freedom; The procedure can refer to the above ***step 1*** to ***5***;

Step 2a. Obtain KP-screw group representations of all hybrid limbs with vectors specifying axis directions and positions of joints, in which the equivalent KP-screw group representations of parallel linkage units are used;

Step 3a. Obtain a simplified equivalent KP-screw group of every hybrid limb, and note idle degrees of freedom;

Step 4a. Obtain a restraining-screws acting on the moving platform contributed by every hybrid limb, and simplify to a simple equivalent restraining-screw group;

Step 5a. Obtain an equivalent restraining-screw group of restraining-screws contributed by all hybrid limbs, and note to find overconstraints;

Step 6a. Calculate the degree of freedom of the hybrid mechanism.

3.4.1 Delta Robot-I: 3-R(4U)$_{RPP}$ Mechanism

A 3-R(4U)$_{RPP}$ mechanism[(2)], in figure 9, comprises a base, a moving platform and three same R(4U)$_{RPP}$ hybrid kinematical limbs. In each hybrid kinematic limb, the unit vectors are $s_{O_i} // s_{a_i b_i} // s_{c_i d_i}$, $s_{a_i} // s_{b_i} // s_{c_i} // s_{d_i}$, and the link lengths are $l_{a_i c_i} = l_{b_i d_i}$, $l_{a_i b_i} = l_{c_i d_i}$. Where, a_i, b_i, c_i and d_i, denotes the central points of universal joints. Suppose the axes, S_{O_i}, of the first revolute joints of three hybrid limbs are on the coordinate plane XOY. The three axes $\boldsymbol{S}_{O_1}$, $\boldsymbol{S}_{O_2}$ and $\boldsymbol{S}_{O_3}$ fixed at the base platform form a triangle $\Delta B_1B_2B_3$, the axes $\boldsymbol{S}_{c_1d_1}$, $\boldsymbol{S}_{c_2d_2}$ and $\boldsymbol{S}_{c_3d_3}$ fixed at the moving platform form a triangle $\Delta P_1P_2P_3$. The two are similar triangles, usually, both equilateral triangles.

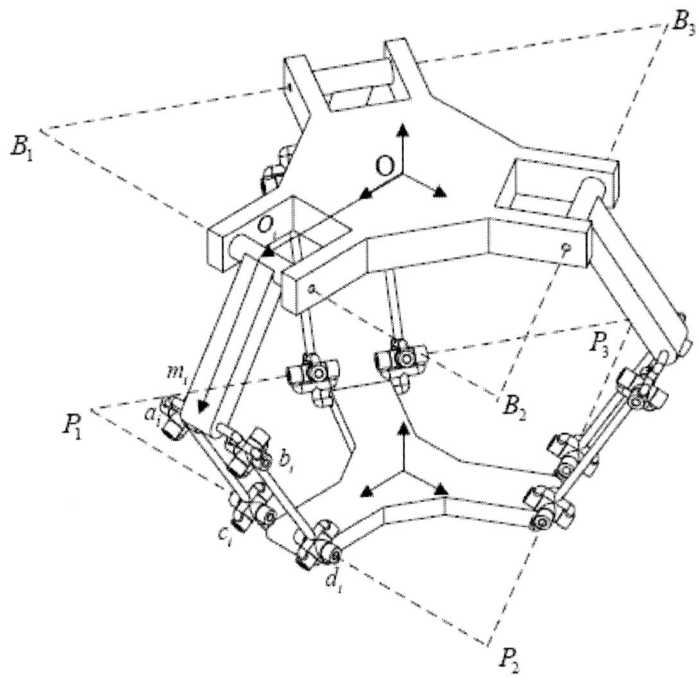

Figure 9. Delta Robot-I: 3-R(4U) parallel platform mechanism.

With the representation of equivalent KP-screw group in equation (34), each hybrid limb of the platform mechanism can be represented geometrically as

$$\boldsymbol{S}_{O_i}=(\boldsymbol{s}_{O_i};\boldsymbol{r}_{O_i}\times\boldsymbol{s}_{O_i}),\boldsymbol{S}_{ie1}=(\boldsymbol{s}_{a_ib_i};\ \boldsymbol{r}_{a_i}\times\boldsymbol{s}_{a_ib_i}),\ \boldsymbol{S}_{ie2}=(0;\boldsymbol{s}_{a_ic_i}\times\boldsymbol{s}_{a_i}),\ \boldsymbol{S}_{ie3}=(\boldsymbol{s}_{a_ib_i};\boldsymbol{r}_{c_i}\times\boldsymbol{s}_{a_ib_i}),\ (41)$$

where, $\boldsymbol{r}_{O_i}$, $\boldsymbol{r}_{a_i}$ and $\boldsymbol{r}_{c_i}$ are the position vectors of points O_i, a_i and c_i, respectively. With the geometric relations, $\boldsymbol{s}_{O_i}\,//\,\boldsymbol{s}_{a_ib_i}$, $\boldsymbol{r}_{a_i}=\boldsymbol{r}_{O_i}+l_{O_im_i}\boldsymbol{s}_{O_im_i}+l_{a_im_i}\boldsymbol{s}_{a_ib_i}$ and $\boldsymbol{r}_{c_i}=\boldsymbol{r}_{a_i}+l_{a_ic_i}\boldsymbol{s}_{a_ic_i}$, an equivalent KP-screw group of the above equation can be given

$$\boldsymbol{S}_{iE1}=(\boldsymbol{s}_{O_{iE1}};\boldsymbol{r}_{O_i}\times\boldsymbol{s}_{O_i}),\boldsymbol{S}_{iE2}=(\boldsymbol{0};\ \boldsymbol{s}_{O_im_i}\times\boldsymbol{s}_{O_i}),\ \boldsymbol{S}_{iE3}=(\boldsymbol{0};\boldsymbol{s}_{a_ic_i}\times\boldsymbol{s}_{a_i}),\ \boldsymbol{S}_{iE4}=(\boldsymbol{0};\boldsymbol{s}_{a_ic_i}\times\boldsymbol{s}_{O_i}).\ (42)$$

For the mechanism, the four KP-screws are linearly independent, indicating that hybrid limb is equivalent to a serial limb with one revolute joint and three prismatic joints, i.e. RPPP serial limb. The restraining-screws acting on the moving platform contributed by this limb can be obtained by solving the reciprocal screws of four KP-screws in the above equation. Since axis $\boldsymbol{S}_{O_i}$ is on plane *XOY*, two restraining-screws can be obtained

$$\boldsymbol{S}_{i1}^{r} = (\boldsymbol{0}; \boldsymbol{z} \times \boldsymbol{s}_{O_i}), \boldsymbol{S}_{i2}^{r} = (\boldsymbol{0}; \boldsymbol{r}_{O_i} \times \boldsymbol{s}_{O_i}) = (\boldsymbol{0}; \boldsymbol{z}), \quad (43)$$

where, z is a unit vector denoting coordinate axis Z. These are two couple line vectors about axes $\boldsymbol{z} \times \boldsymbol{s}_{O_i}$ and z. From three such limbs, six couple line vectors restrain the moving platform. Since $\boldsymbol{z} \times \boldsymbol{s}_{O_1}$, $\boldsymbol{z} \times \boldsymbol{s}_{O_2}$ and $\boldsymbol{z} \times \boldsymbol{s}_{O_3}$ are all three-dimensional vectors parallel to coordinate plane XOY, restraining-screws $\boldsymbol{S}_{11}^{r}$, $\boldsymbol{S}_{21}^{r}$ and $\boldsymbol{S}_{31}^{r}$ are linearly dependent, and equivalent to two linearly independent restraining-screws $\boldsymbol{S}_{x}^{r} = (\boldsymbol{0}; \boldsymbol{x})$ and $\boldsymbol{S}_{y}^{r} = (\boldsymbol{0}; \boldsymbol{y})$, indicating there is one overconstraint. Similarly, $\boldsymbol{S}_{12}^{r}$, $\boldsymbol{S}_{22}^{r}$ and $\boldsymbol{S}_{32}^{r}$ indicate two overconstraints. Hence the six restraining-screws acting on the moving platform are equivalent to a restraining-screw group

$$\boldsymbol{S}_{B1}^{r} = (\boldsymbol{0}; \boldsymbol{x}), \boldsymbol{S}_{B2}^{r} = (\boldsymbol{0}; \boldsymbol{y}), \boldsymbol{S}_{B3}^{r} = (\boldsymbol{0}; \boldsymbol{z}). \quad (44)$$

The three restraint couples restrain three pure rotations of the moving platform that remains only three pure translational degrees of freedom. Since there exist other three overconstraints in three 4U parallelogram units, as in section 5, the moving platform of a 3-R(4U)$_{RPP}$ has three degrees of freedom with respect to the base platform, i.e. $M_{a\,\text{Delta-I}} = 6(11 - 15 - 1) + 3 \times 1 + 12 \times 2 + 6 = 3$.

3.4.2. Delta Robot-II: 3-R(4S)$_{RRPP}$ Mechanism

A 3-R(4S)$_{RRPP}$ spatial hybrid parallel mechanisms, shown in figure 10, is an evolvement of the above 3-R(4U)$_{RPP}$ mechanism by changing its all universal joints into spherical joints.

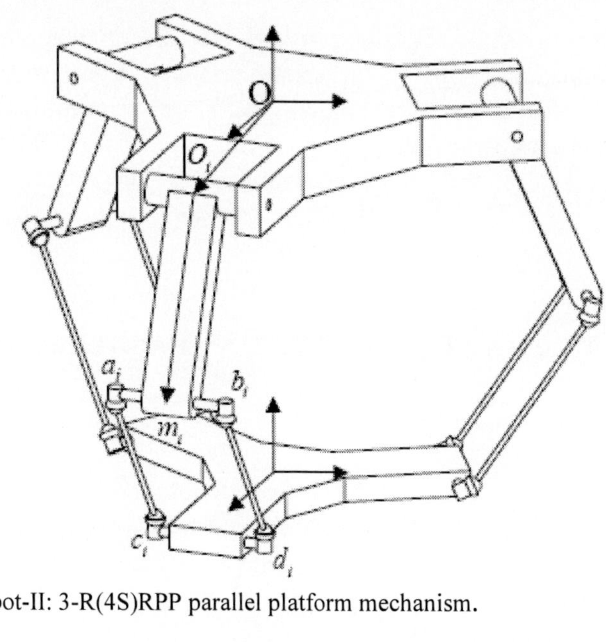

Figure 10. Delta Robot-II: 3-R(4S)RPP parallel platform mechanism.

Suppose all 4S parallel linkage units are parallelograms in initial assembly configuration, with the representation of equivalent KP-screw group in equation (41), a R(4S) hybrid kinematic limb can be represented as

$$\begin{array}{l} \boldsymbol{S}_{O_i} = \left(\boldsymbol{s}_{O_i}; \boldsymbol{r}_{O_i} \times \boldsymbol{s}_{O_i}\right), \boldsymbol{S}_{ie1} = \left(\boldsymbol{s}_{a_ib_i}; \boldsymbol{r}_{a_i} \times \boldsymbol{s}_{a_ib_i}\right), \boldsymbol{S}_{ie2} = \left(\boldsymbol{s}_{a_ic_i}; \boldsymbol{r}_{a_i} \times \boldsymbol{s}_{a_ic_i}\right), \\ \boldsymbol{S}_{ie3} = \left(\boldsymbol{0}; \boldsymbol{s}_{a_ib_i} \times \boldsymbol{s}_{a_ic_i}\right), \boldsymbol{S}_{ie4} = \left(\boldsymbol{0}; \left(\boldsymbol{s}_{a_ic_i} \times \boldsymbol{s}_{a_ib_i}\right)\boldsymbol{s}_{a_ic_i}\right) \end{array} . \tag{45}$$

Where, $\boldsymbol{r}_{O_i}$, and $\boldsymbol{r}_{a_i} = \boldsymbol{r}_{O_i} + l_{O_im_i}\boldsymbol{s}_{O_im_i} + l_{a_im_i}\boldsymbol{s}_{a_ib_i}$ are the position vectors of points O_i and a_i, respectively. With geometric relation $\boldsymbol{s}_{O_i} // \boldsymbol{s}_{a_ib_i}$, an equivalent KP-screw group yields

$$\begin{array}{l} \boldsymbol{S}_{iE1} = \left(\boldsymbol{s}_{O_i}; \boldsymbol{r}_{O_i} \times \boldsymbol{s}_{O_i}\right), \boldsymbol{S}_{iE2} = \left(\boldsymbol{0}; \boldsymbol{s}_{O_im_i} \times \boldsymbol{s}_{O_i}\right), \boldsymbol{S}_{iE3} = \left(\boldsymbol{s}_{a_ic_i}; \boldsymbol{r}_{a_i} \times \boldsymbol{s}_{a_ic_i}\right), \\ \boldsymbol{S}_{iE4} = \left(\boldsymbol{0}; \boldsymbol{s}_{a_ib_i} \times \boldsymbol{s}_{a_ic_i}\right), \boldsymbol{S}_{iE5} = \left(\boldsymbol{0}; \left(\boldsymbol{s}_{a_ic_i} \times \boldsymbol{s}_{a_ib_i}\right)\boldsymbol{s}_{a_ic_i}\right) \end{array} . \tag{46}$$

It represents a RPRPP serial limb. The restraining-screw acting on the mobile platform contributed by this limb is

$$\boldsymbol{S}_i^r = \left(\boldsymbol{0}; \boldsymbol{s}_{O_i} \times \boldsymbol{s}_{a_ic_i}\right). \tag{47}$$

Considering three such limbs, three couples restrain the mobile platform

$$\boldsymbol{S}_1^r = \left(\boldsymbol{0}; \boldsymbol{s}_{O_1} \times \boldsymbol{s}_{a_1c_1}\right), \boldsymbol{S}_2^r = \left(\boldsymbol{0}; \boldsymbol{s}_{O_2} \times \boldsymbol{s}_{a_2c_2}\right), \boldsymbol{S}_3^r = \left(\boldsymbol{0}; \boldsymbol{s}_{O_3} \times \boldsymbol{s}_{a_3c_3}\right). \tag{48}$$

In a general configuration, since vectors $\boldsymbol{s}_{O_1} \times \boldsymbol{s}_{a_1c_1}$, $\boldsymbol{s}_{O_2} \times \boldsymbol{s}_{a_2c_2}$ and $\boldsymbol{s}_{O_3} \times \boldsymbol{s}_{a_3c_3}$ are linearly independent, the three restraining-screws are linearly independent. They restrain completely the orientation of the mobile platform that remains only with three purely translational degrees of freedom. In section 6.3, it has been pointed that there exist two idle degrees of freedom in a 4S parallel linkage unit. Hence there exist six idle degrees of freedom in a 3-R(4S)$_{\text{RRPP}}$ mechanism, the moving platform has three degrees of freedom with respect to the base platform, i.e. $M_{a\,\text{Delta-II}} = 6(11-15-1)+3\times1+12\times3-6=3$.

4. Geometric Synthesis of Spatial Parallel Mechanisms

4.1 Synthesis Procedure

With the principle of equivalent KP-screw groups and equivalent restraining-screw groups, a procedure for the synthesis of parallel mechanisms with lower mobility can be developed. When this method is used to synthesize a parallel mechanism with constant constraints, the synthesis process will be simple and many mechanisms meeting the required tasks can be obtained.

Step 1. Describe restraining-screws acting on the moving platform

For specific tasks, some motions are necessary and others are unnecessary even not permitted. When a parallel manipulator is employed to perform such a task, the moving platform has to be subject to constraints. The constraints can be described with a group of restraining-screws. Let

$$\begin{aligned} &\boldsymbol{S}_X^r=(\boldsymbol{x};\boldsymbol{0}),\ \boldsymbol{S}_Y^r=(\boldsymbol{y};\boldsymbol{0}),\ \boldsymbol{S}_Z^r=(\boldsymbol{z};\boldsymbol{0}),\\ &\boldsymbol{S}_{RX}^r=(\boldsymbol{0};\boldsymbol{x}),\boldsymbol{S}_{RY}^r=(\boldsymbol{0};\boldsymbol{y}),\boldsymbol{S}_{RZ}^r=(\boldsymbol{0};\boldsymbol{z}), \end{aligned} \tag{49}$$

denote six reference restraining-screws, where, screw $\boldsymbol{S}_X^r$, $\boldsymbol{S}_Y^r$ and $\boldsymbol{S}_Z^r$ carry wrenches restraining three-dimensional translations, and screw $\boldsymbol{S}_{RX}^r$, $\boldsymbol{S}_{RY}^r$ and $\boldsymbol{S}_{RZ}^r$ carry pure couples restraining the three rotations. When the constraints are constants with respect to the configuration of the moving platform, they can be easily described by combining the six reference restraining-screws. For example, when the moving platform needs only three-dimensional rotations, the constraints are directly given as

$$\boldsymbol{S}_X^r=(\boldsymbol{x};\boldsymbol{0}),\ \boldsymbol{S}_Y^r=(\boldsymbol{y};\boldsymbol{0}),\ \boldsymbol{S}_Z^r=(\boldsymbol{z};\boldsymbol{0}). \tag{50}$$

When the constraints are variables with respect to the configuration of the moving platform, linear combinations of the six reference restraining-screws are needed. For example, when the moving platform is subject to a restraint wrench that is a function of the position and orientation parameters $(X_p,Y_p,Z_p;\alpha,\beta,\gamma)$ of the moving platform, the corresponding restraining-screw can be described by a linear combination,

$$\boldsymbol{S}^r=C_1\boldsymbol{S}_X^r+C_2\boldsymbol{S}_Y^r+C_3\boldsymbol{S}_Z^r+C_4\boldsymbol{S}_{RX}^r+C_5\boldsymbol{S}_{RY}^r+C_6\boldsymbol{S}_{RZ}^r, \tag{51}$$

where, C_i (i=1,2,…6) are the functions of $(X_p,Y_p,Z_p;\alpha,\beta,\gamma)$. The functions are given from the corresponding constraint geometry. When the moving platform is subject to more than one such constraint, a group of such restraining-screws can be employed.

Step 2. Identify basic KP-screw groups

To obtain serial kinematic limbs contributing the required constraints on the moving platform, the following linear equation group needs to be solved,

$$\boldsymbol{S}\circ\boldsymbol{S}_i^r=\boldsymbol{s}\cdot\boldsymbol{s}_{0i}^r+\boldsymbol{s}_0\cdot\boldsymbol{s}_i^r=0\quad(i=1,2,...,k), \tag{52}$$

where, k is the number of constraints, $\boldsymbol{S}^r$ denote restraining-screws, $\boldsymbol{S}$ denote KP-screws. The 6-k solution screws of the above equation group compose a KP-screw group. It represents a serial kinematic limb that is available for constructing the required parallel mechanism. Further, the solution screw-group of the following equation group is also available for the required parallel mechanism,

$$\boldsymbol{S} \square \boldsymbol{S}_i^r = \boldsymbol{s} \square \boldsymbol{s}_{0i}^r + \boldsymbol{s}_0 \square \boldsymbol{s}_i^r = 0 \quad (i = 1,2,...,e), \tag{53}$$

where, $e = 1,2,..(k-1)$. Following the above example, the solution screw-groups can be obtained. The serial limbs represented by the solution screw-groups contribute parts of required constraints to the moving platform. Such that many basic KP-screw groups as solutions of the above equation groups can be obtained. They represent available basic serial limb for the required parallel mechanism. When the moving platform is exerted by constraints with variable components, the basic KP-screw groups may be complex. The complexity will depend on the constraints. Sometime, complex joints may be needed.

Step 3. Construct equivalent serial limbs

Basic KP-screw groups describe the elementary serial limbs available for a required parallel mechanism. However, when the basic serial limbs are used to constitute the required mechanism, the performances of resulted mechanisms may be unsatisfactory. Hence, it is necessary and possible to obtain more available serial limbs from the principle of equivalent KP-screw groups. In the above example, for a parallel mechanism with three-dimensional translations, available serial limbs with three joints can only be PPP serial limbs denoted by $\{\boldsymbol{S}_X, \boldsymbol{S}_Y, \boldsymbol{S}_Z\}$, where, PPP represents a serial limb with three prismatic joints. No other type of serial limbs with three joints is equivalent. For a basic serial limb with four joints, it is not difficult to construct equivalent serial limbs. For example, from basic serial limb PPPR denoted by $\{\boldsymbol{S}_X, \boldsymbol{S}_Y, \boldsymbol{S}_Z, \boldsymbol{S}_{RX}\}$, equivalent serial limbs, such as PPRR and PRRR denoted by $\{\boldsymbol{S}_X, \boldsymbol{S}_Z, \boldsymbol{S}_{RX}, \boldsymbol{S}'_{\boldsymbol{RX}}\}$ and $\{\boldsymbol{S}_X, \boldsymbol{S}_{RX}, \boldsymbol{S}'_{\boldsymbol{RX}}, \boldsymbol{S}''_{\boldsymbol{RX}}\}$ respectively, can be constructed, where, $\boldsymbol{S}'_{RX} = \boldsymbol{S}_{RX} + q'\boldsymbol{S}_Y + r'\boldsymbol{S}_Z$, i.e. $\boldsymbol{S}'_{RX} = (\boldsymbol{x}; q'\boldsymbol{y} + r'\boldsymbol{z})$ and $\boldsymbol{S}''_{RX} = \boldsymbol{S}_{RX} + q''\boldsymbol{S}_Y + r''\boldsymbol{S}_Z$, i.e. $\boldsymbol{S}''_{RX} = (\boldsymbol{x}; q''\boldsymbol{y} + r''\boldsymbol{z})$, where the coefficients q', r', q'' and r'' are the functions of platform's position parameters. With analogous operations, many equivalent serial limbs can be obtained. For parallel mechanisms with variable constraints, constructions of equivalent serial limbs are generally complex unless in the cases with special geometric conditions.

Step 4. Allocate available serial limbs

There are three or more than three serial limbs in a parallel mechanism. Mostly, the number of limbs equals the number of degrees of freedom, so that each limb mounts one actuator. Besides available serial limbs obtained above, serial limbs with six degrees of freedom, such as SPS limbs, are also available for required parallel mechanisms. New parallel mechanisms hence can be obtained by allocating these available serial limbs, in which geometric symmetry and mechanics performances need consider in the meantime. It is most essential that restraining-screws contributed by all serial limbs must be equivalent to that described in step 1. When a moving platform exists overconstraints, the parallel mechanism generally has fewer joints and simpler architecture than that exists no overconstraints.

The above steps give a procedure in synthesis of parallel mechanisms with fewer than six degrees of freedom. Although it is still a complicated work to synthesize a parallel mechanism with variable constraints, the method is useful for designer to find a required mechanism. When this method is used to synthesize a parallel mechanism with constant constraints, the operations will be simple and straightforward.

4.2 Syntheses of Four-DOF Parallel Mechanisms

The above synthesis procedure is demonstrated by producing parallel mechanisms that perform three translations and one rotation about z-axis.

4.2.1 Describe Restraining-Screws Acting on the Moving Platform

For a parallel mechanism that performs three translations and one rotation about z-axis, two pure couples restrain two rotations of the moving platform. The two pure couples are denoted by wrenches on the following restraining-screws

$$\boldsymbol{S}_{RX}^{r} = (\boldsymbol{0}; \boldsymbol{x}) \text{ and } \boldsymbol{S}_{RY}^{r} = (\boldsymbol{0}; \boldsymbol{y}) . \tag{54}$$

4.2.2 Identify Basic KP-Screw Groups

The basic KP-screw group reciprocal to the above two restraining-screws is directly given as

$$\boldsymbol{S}_{i1} = (\boldsymbol{0}; \boldsymbol{x})^{\mathrm{T}}, \boldsymbol{S}_{i2} = (\boldsymbol{0}; \boldsymbol{y})^{\mathrm{T}}, \boldsymbol{S}_{i3} = (\boldsymbol{0}; \boldsymbol{z})^{\mathrm{T}}, \boldsymbol{S}_{i4} = (\boldsymbol{z}; \boldsymbol{0})^{\mathrm{T}} . \tag{55}$$

They represent three fundamental prismatic joints and one revolute joint about z-axis, i.e., a PPPR serial limb. The basic KP-screw group reciprocal to one of the above restraining-screw can be obtained as following

$$\boldsymbol{S}_{i1} = (\boldsymbol{0}; \boldsymbol{x})^{\mathrm{T}}, \boldsymbol{S}_{i2} = (\boldsymbol{0}; \boldsymbol{y})^{\mathrm{T}}, \boldsymbol{S}_{i3} = (\boldsymbol{0}; \boldsymbol{z})^{\mathrm{T}}, \boldsymbol{S}_{i4} = (\boldsymbol{z}; \boldsymbol{0})^{\mathrm{T}}, \boldsymbol{S}_{i5} = (\boldsymbol{y}; \boldsymbol{0})^{\mathrm{T}}, \tag{56}$$

and

$$\boldsymbol{S}_{i1} = (\boldsymbol{0}; \boldsymbol{x})^{\mathrm{T}}, \boldsymbol{S}_{i2} = (\boldsymbol{0}; \boldsymbol{y})^{\mathrm{T}}, \boldsymbol{S}_{i3} = (\boldsymbol{0}; \boldsymbol{z})^{\mathrm{T}}, \boldsymbol{S}_{i4} = (\boldsymbol{z}; \boldsymbol{0})^{\mathrm{T}}, \boldsymbol{S}_{i5} = (\boldsymbol{x}; \boldsymbol{0})^{\mathrm{T}}. \tag{57}$$

The above two KP-screw group can uniformly described with

$$\boldsymbol{S}_{i1} = (\boldsymbol{0}; \boldsymbol{x})^{\mathrm{T}}, \boldsymbol{S}_{i2} = (\boldsymbol{0}; \boldsymbol{y})^{\mathrm{T}}, \boldsymbol{S}_{i3} = (\boldsymbol{0}; \boldsymbol{z})^{\mathrm{T}}, \boldsymbol{S}_{i4} = (\boldsymbol{z}; \boldsymbol{0})^{\mathrm{T}}, \boldsymbol{S}_{i5} = (\boldsymbol{u}; \boldsymbol{0})^{\mathrm{T}}, \tag{58}$$

where, $\boldsymbol{u}=l_u\boldsymbol{x}+m_u\boldsymbol{y}=(l_u, m_u, 0)$ is a vector orthogonal to vector $\boldsymbol{z}$, can be , This KP-screw group represents a PPPRR serial limb.

4.2.3 Construct Equivalent Serial Limbs

The serial limbs, PPPR and PPPRR, both can be used to compose the required parallel mechanism. However, the parallel mechanisms resulted are obviously unsatisfactory. With the principle of equivalent KP-screw groups, many equivalent KP-screw groups can be constructed from the two basic KP-screw groups. For instances, from equation (55), let $\boldsymbol{S}'_{i1}=\boldsymbol{S}_{i3}$, $\boldsymbol{S}'_{i2}=\boldsymbol{S}_{i4}$, $\boldsymbol{S}'_{i3}=\boldsymbol{S}_{i4}+p_{i3}\boldsymbol{S}_{i1}+q_{i3}\boldsymbol{S}_{i2}$, and $\boldsymbol{S}'_{i4}=\boldsymbol{S}_{i4}+p_{i4}\boldsymbol{S}_{i1}+q_{i4}\boldsymbol{S}_{i2}$, a KP-screw group is obtained

$$\boldsymbol{S}'_{i1}=(\mathbf{0};z),\boldsymbol{S}'_{i2}=(z;\mathbf{0}),\boldsymbol{S}'_{i3}=(z;p_{i3}\boldsymbol{x}+q_{i3}\boldsymbol{y}),\boldsymbol{S}'_{i4}=(z;p_{i4}\boldsymbol{x}+q_{i4}\boldsymbol{y}), \tag{59}$$

where p_{ij}, q_{ij} (j=3,4) are the functions of structural and configuration parameters. Similarly, let $\boldsymbol{S}''_{i1}=\boldsymbol{S}_{i3}$, $\boldsymbol{S}''_{i2}=\boldsymbol{S}_{i4}$, $\boldsymbol{S}''_{i3}=l_{i3}\boldsymbol{S}_{i1}+m_{i3}\boldsymbol{S}_{i2}+n_{i3}\boldsymbol{S}_{i3}$, and $\boldsymbol{S}''_{i4}=\boldsymbol{S}_{i4}+p_{i4}\boldsymbol{S}_{i1}+q_{i4}\boldsymbol{S}_{i2}$, another KP-screw group is obtained

$$\boldsymbol{S}''_{i1}=(\mathbf{0};z),\boldsymbol{S}''_{i2}=(z;\mathbf{0}),\boldsymbol{S}''_{i3}=(\mathbf{0};l_{i3}\boldsymbol{x}+m_{i3}\boldsymbol{y}+n_{i3}\boldsymbol{z}),\boldsymbol{S}''_{i4}=(z;l_{i4}\boldsymbol{x}+m_{i4}\boldsymbol{y}), \tag{60}$$

where l_{i3}, m_{i3} and n_{i3} are the direction cosines of center-line $\boldsymbol{s}''_{i3}$ of prismatic joint $\boldsymbol{S}''_{i3}$. They are variables with respect to the input parameters. The KP-screw group in equation (59) represents a PRRR limb with one prismatic joint (*P*) along *z*-axis and three revolute joints *RRR* about *z*-axis. A CRR limb in figure. 1a can be produced when replacing both a prismatic joint and a revolute joint with a cylindrical joint (*C*). The KP-screw group in equation (60) replaces the middle revolute joint in equation (59) with a prismatic joint, and subsequently produces a PRPR or CPR limb in figure 11b.

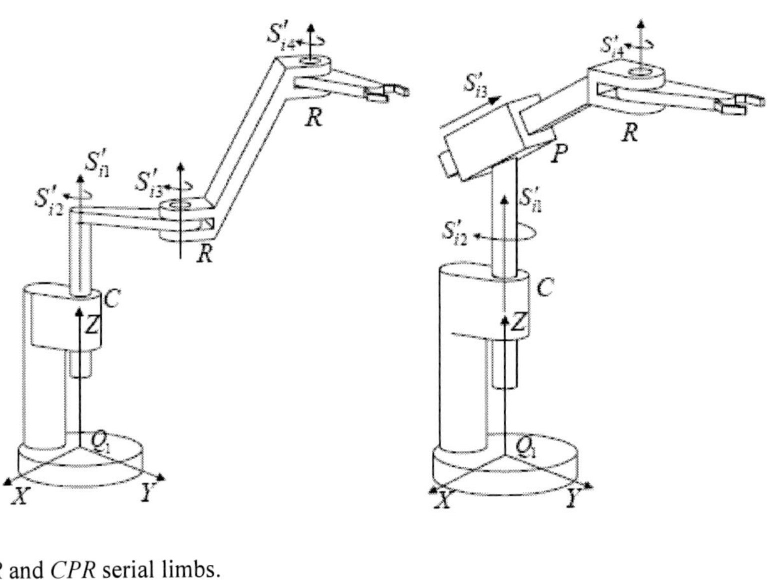

Figure 11. *CRR* and *CPR* serial limbs.

Similarly, the equivalent PK-screw groups can be constructed from equation (58) as

$$S'_{i1}=(z;0),\quad S'_{i2}=(u;0),\quad S'_{i3}=(u;r_{i3}\times u),\quad S'_{i4}=(u;r_{i4}\times u),\quad S'_{i5}=(z;r_{i4}\times z) \quad (61)$$

and

$$S''_{i1}=(z;0),\quad S''_{i2}=(u;0),\quad S''_{i3}=(0;s_{i3}),S''_{i4}=(u;r_{i4}\times u),\quad S''_{i5}=(z;r_{i4}\times z) \quad (62)$$

where r_{ij} (j=3,4) are position vectors of corresponding screws, and are the functions of structural and configuration parameters, s_{i3} is the direction vector of screw S''_{i3}.

KP-screw group in equation (61) represents two revolute joints with their axes along *z*-axis and three revolute joints with their axes parallel to *x-y* plane, i.e., a RRRRR kinematic chain. Since the axes of S'_{i1} and S'_{i2} are orthogonal each other and the axes of S'_{i4} and S'_{i5} are orthogonal each other, the KP-screw group can be represented as a URU kinematic chain in figure. 12a. KP-screw group in equation (62) replaces one revolute joint with one prismatic joint and represents a UPU kinematic limb in figure 12b. With analogous analysis, the KP-screw group also represents a CRU (PUU) kinematic limb in figure 12c.

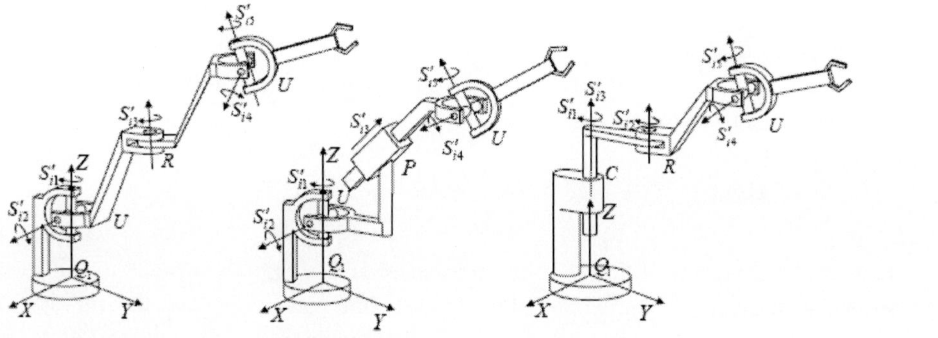

Figure 12. *URU*, *UPU* and *CRU* serial limbs.

4.3 Construction of Four-DOF Parallel Manipulators

4.3.1 Construction of 2-CRR/2-PSS Mechanism

All of the above limbs can be used to construct a parallel manipulator mechanism. In this section, two CRR serial limbs in figure 12 are used to constitute a parallel manipulator mechanism from a symmetry view. To mount actuators on the different limbs, two six-DOF PSS serial limbs are employed to complement. The parallel manipulator mechanism resulted is shown in figure 13. The inputs can be three prismatic joints and one revolute joint.

In the mechanism, each CRR limb contributes two constraints to the moving platform. They are $S^r_{11}=(0;x)$, $S^r_{12}=(0;y)$, $S^r_{31}=(0;x)$ and $S^r_{32}=(0;y)$. It is evident that the four restraining-screws are linearly dependent. There exist only two linearly independent screws. The wrenches on the two screws restrain the rotations of the moving platform around *x*- and *y*-axis. The wrenches on the other two restraining-screws exert repeated or redundant constraints on the moving platform, are called overconstraints. So the output degrees-of-freedom of the manipulator mechanism equals 6-2=4.

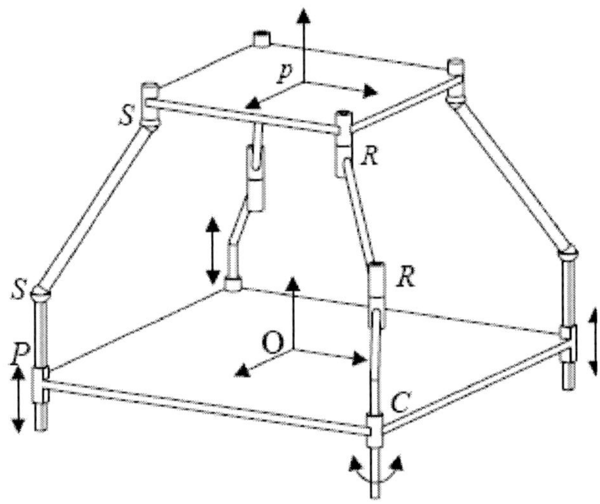

Figure 13. 2-*CRR*/2-*PSS* parallel mechanism.

If a mechanism exists overconstraints, the mobility calculation with equation (2) needs to add the number of overconstraints to the result. Hence, the mobility of a 2-CRR/2-PSS parallel mechanism can be calculated as

$$M = 6\times(10-12-1)+1\times6+2\times2+3\times4+2=6.$$

However, it has been noted that the link between two spherical joints in a PSS limb can rotate around the central connection line of two spherical joints, and the rotation does not affect the motion of the other members in the mechanism. For a parallel manipulator mechanism, the rotation can be dealt with an idle degree of freedom. Hence, because of two PSS serial limbs, the mobility calculation of the parallel manipulator requires the subtraction of two for the reason of the idle degree of freedom, i.e.

$$M = 6\times(10-12-1)+1\times6+2\times2+3\times4+2-2=4.$$

Therefore, the mobility of the parallel manipulator is four, and four inputs are needed for producing desired mechanism outputs. They correspond to the four degrees of freedom of the moving platform. A direct investigation into the mechanics performance shows that the revolute joints in CRR limbs would be loaded forces along the axes, which is undesirable. An improvement is given in the following section.

4.3.2 Construction of 2-CPR/2-SPS Mechanism

With the CPR kinematic limb in figure 11b, a parallel manipulator mechanism can be constructed with two CPR limbs and two SPS limbs in figure 14, where the two CPR limbs restrain the rotations of the moving platform about the x-, y-axes, and four prismatic joints can be chosen as inputs. The mobility of the mechanism and the degree of freedom of the moving platform can be analyzed similarly to that above.

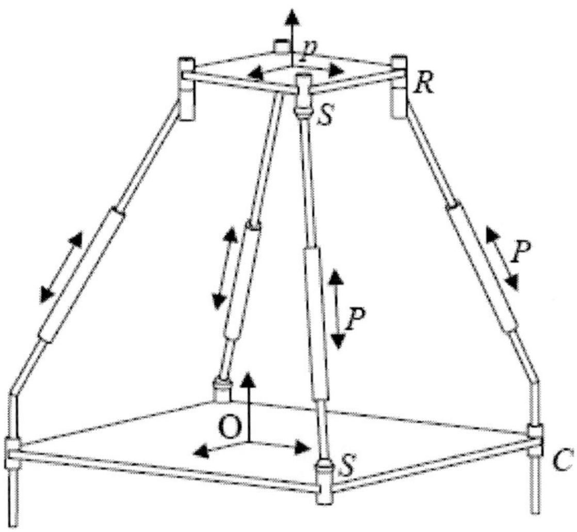

Figure 14. 2-CPR/2-SPS parallel mechanism.

The 2-CPR/2-SPS parallel mechanisms resulted can perform required motion and have better mechanics characteristics than the 2-CRR/2-PSS parallel mechanisms. However, the base platform has to provide spaces for the cylindrical joints, which could result in a small workspace and inconvenience in applications. In the following two subsections, the serial limbs with five joints obtained in section 4.2.2 will be employed to construct more desirable mechanisms.

4.3.3 Construction of 3-URU/SPS Mechanism

A URU serial limb in figure 2a contributes a restraint couple to the moving platform. At least two such limbs are needed for a parallel mechanism with four degrees-of-freedom. To allocate four actuators to different limbs, two six-DOF serial limbs are needed for complements. So a 2-URU/2-SPS parallel mechanism is formed. However, the parallel mechanism will happen singularity when the axes of middle revolute joints of two URU limbs are parallel each other. To solve the problem, three URU limbs and one SPS limb can be employed to constitute required the parallel mechanism. Figure 15 shows a 3-URU/SPS parallel mechanism with four degrees of freedom. The middle revolute joints in the three URU limbs and the prismatic joint in the SPS limb are chosen as four inputs. In the mechanism, three URU limbs contribute three constraints for the moving platform, i.e., $\boldsymbol{S}_i^r = \left(\boldsymbol{0},\ \boldsymbol{s}_i^r\right)$, ($i$=1,2,3), where $\boldsymbol{s}_i^r = \left(m_{iu}, -l_{iu}, 0\right)$ is the vector orthogonal to vector $\boldsymbol{u}_i$ and axis z. Since the rank of matrix $\begin{bmatrix} \boldsymbol{0},\ \boldsymbol{s}_1^r \\ \boldsymbol{0},\ \boldsymbol{s}_2^r \\ \boldsymbol{0},\ \boldsymbol{s}_3^r \end{bmatrix}$ is two, that is the order of the three restraining-screws is two, one overconstraint exists on the moving platform. The three restraining-screws $\boldsymbol{S}_i^r = \left(\boldsymbol{0},\ \boldsymbol{s}_i^r\right)$, ($i$=1,2,3) represent restraint couples whose axes are parallel the moving platform plane. When three URU limbs have geometric allocation as those in figure 15, the axes of the three restraining-screws cannot parallel to the same line. Hence, the wrenches on the three restraining-screws restrain the rotations of the moving platform about the $\boldsymbol{x}$-, y-axes. In addition, since an idle degree-of-freedom exists in the SPS limb, the mobility calculation requires the addition of one due to the overconstraint and subtraction of one due to the idle degree-of-freedom. This is expressed as, $M = 6\times(16-18-1)+1\times16+3\times2+1-1=4$.

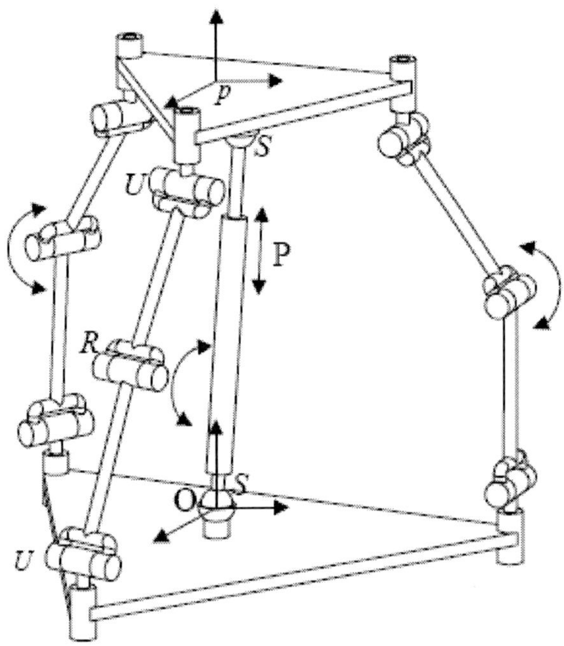

Figure 15. 3-*URU/SPS* parallel mechanism.

When four URU limbs are used to constitute the parallel mechanism, four restraining-screws are exerted on the moving platform. With analogous analysis, two overconstraints on the moving platform can be identified. Such a 4-URU parallel mechanism has 18 links including the moving and base platforms, 20 revolute joints. Hence, the mobility of the parallel mechanism is calculated as $M = 6\times(18-20-1)+1\times20+2=4$.

A 3-URU/SPS parallel mechanism shown in figure 15 is of symmetry and good mechanics performance. However, three motors mounted at the middle revolute joints are inconvenient for some applications. A novel parallel mechanism with four degrees-of-freedom will be constituted in following subsection.

4.3.4 Construction of 4-UPU Mechanism

With the *UPU* limb shown in figure 12b, a 4-UPU parallel mechanism with four degrees of freedom can be constructed in figure 16. The four prismatic joints are then chosen as the inputs. The four UPU limbs contribute four restraint wrenches for the moving platform. The four wrenches restrain together the rotations of the platform about the *x*-, *y*-axes, with two overconstraints. The manipulator mechanism has output-freedom equal to four. The mobility can be expressed by $M = 6 \times (18 - 20 - 1) + 1 \times 20 + 2 = 4$. A direct analysis shows that the 4-UPU mechanisms are of better symmetry, mechanics performance and a bigger workspace than others developed in this chapter.

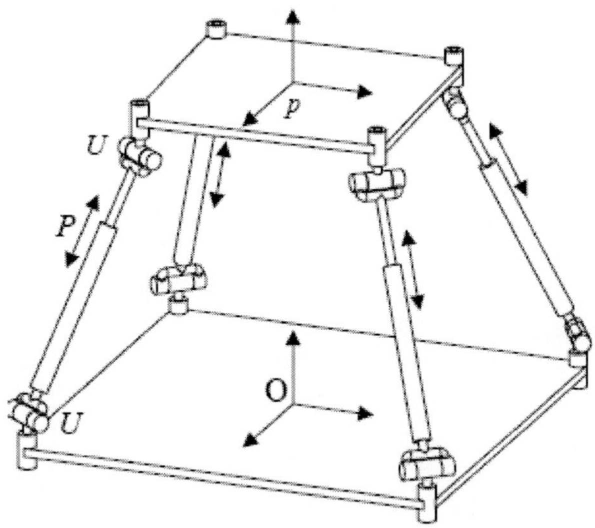

Figure 16. 4-*UPU* parallel mechanism.

Similarly, a 4-CRU (or 4-PUU) parallel mechanism with four degrees of freedom can be constructed in figure 17. The four prismatic joints are still chosen as the inputs. The characteristic of the platform mechanism is that its actuators can be fixed at the base platform.

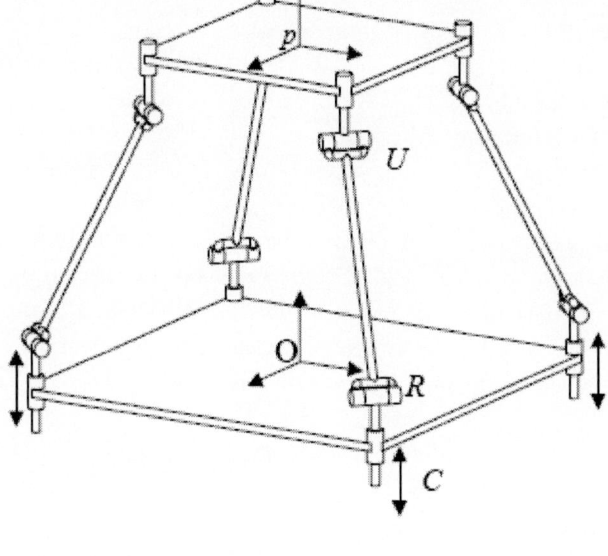

Figure 17. 4-*CRU* parallel mechanism.

REFERENCES

[1] Hunt, K. H., Structural Kinematics of In-Parallel-Actuated Robot Arms, *ASME J. of Mech. Trans. and Aut. in Des.*, Vol.105, (1983), p. 705-712.

[2] Clavel R. Delta, a fast robot with parallel geometry, *Proc. of the Int. Symp. on Industrial Robot,* Switzerland, (1988), p. 91-100.

[3] Miller, K., Modeling of Dynamics and Model-Based Control of DELTA Direct-Drive Parallel Robot, *Journal of Robotics and Mechatronics*, Vol. 17, No. 4, (1995), p. 344-352.

[4] Tsai L.W., Multi-degree-of-freedom mechanisms for machine tools and the like, *U.S.* (1997), *Patent* No. 5656905.

[5] Lee, K. M., and Arjuman, S., A 3-DOF micromotion in-parallel actuated manipulator, *IEEE Trans. Robot. Automat.* Vol.7(5), 1991, p. 634-640.

[6] Bonnet, G.T., A new mechanism, *Engineering.* Vol.76, (1903), p. 777-778.

[7] Goldberg, M., New five-bar and six-bar linkages in three dimensions, *Transaction s of ASME,* Vol.65, (1943), p. 649-661.

[8] Waldron, K. J., Hybrid overconstrained linkages, *J. of Mechanisms,* Vol.3, (1968), p. 73-78.

[9] Dimentberg, F. M., and Yoslovich I. V., A spatial four bar mechanism having two prismatic pairs, *Transactions of Third USSR Conference on The Basic Problems in Theory of Machines and Mechanisms*, Volume on Analysis and Synthesis of Mechanisms, Moscow, (1963), p. 144-150.

[10] Waldron, K. J., A family of overconstrained linkages, *J. of Mechanisms,* Vol.2, (1967), p. 201-211.

[11] Waldron, K. J., Symmetric overconstrained linkages, Transactions of ASME, *J. of Engineering for Industry,* (1969), p. 158-162.

[12] Baker, J. E., On 5-revolute linkages with parallel adjacent joint axes, *Mechanism and Machine Theory*, Vol.19(6), (1984), p. 467-475.

[13] Mavriodis, C. and Roth, B., Analysis of overconstrained mechanisms, *ASME J. Mech. Des.,* Vol.117, (1995), p. 69-74.

[14] Mavriodis, C. and Roth, B., new and Revised overconstrained mechanisms, *ASME J. Mech. Des.,* Vol.117, (1995), p. 75-82.

[15] Waldron, K. J., Raghavan, M., Roth. B., Kinematics of A Hybrid Series-Parallel Manipulation System, *ASME J. of Mechanisms, Trans. And Auto. In design,* Vol.111, (1989), p. 211-221.

[16] Tsai L.W. and Stamper R. E. A parallel mechanism with only translational degree of freedom, *Proc. Of ASME 1996 Desin Engineering Technical Conferences and Computers in Engineering Conference,* 96-DETC/MECH-1152.

[17] Huang Z. and Fang Y.F., Kinematic Characteristics Analysis of 3-DOF In-Parallel Actuated Pyramid Mechanisms, *Mechanism and Machine Theory,* Vol.31(8), (1996), p. 1009~1018.

[18] Huang Z. Tao W.S. and Fang Y.F., Study on the Kinematic Characteristics of 3-DOF In-Parallel Actuated Platform Mechanisms, *Mechanism and Machine Theory,* Vol.31(8), (1996), p. 999~1007.

[19] Gosselin C.M. and Hamel J.F., The agile eye: a high-performance 3-DOF camera-orienting device, *Proc. of the IEEE Int. Conf. on Robotics and Automation*, Vol.3, (1994), p. 781~786.

[20] Gosselin, C.M. and Eric St-Pierre, Development and Experimentation of a Fast 3-DOF Camera-Orienting Device, *Int. J. of Robotics Research,* Vol.16(5), (1997), p. 619-630.

[21] Dai, J.S., Holland, and Kerr, D.R., Task-oriented direct synthesis of serial manipulator using moment invariants, 24th ASME Biennial Mechanisms Conference, ASME: DETC96/MECH-1004, Irvine, California, (1996), 19-22.

[22] Dai, J.S., and Kerr, D.R., A six-component contact force measurement device based on a parallel platform, J. of Mechanical Engineering Science, Vol.214, No.c5, (2000), p.687-697.

[23] Tsai L.W., and Joshi S., Kinematics and optimization of a spatial 3-UPU parallel manipulator, *ASME J. of Mechanical Design*, Vol.122, (2000), p. 439-446.

[24] Zhao, T. S., Analysis and synthesis of parallel mechanisms with less than six degrees of freedom, PhD thesis (in Chinese), Yanshan University, China, (2000).

[25] Zhao, T. S. and Huang, Z., A novel three-dof translational platform mechanism and its kinematics, (2000), ASME DETC2000/MECH-14101.

[26] Dai J.S. and Rees Jones J., Interrelationship between screw systems and their reciprocal system and applications, *Mechanism and Machine Theory*, Vol.37(3), (2001), p. 633-651.

[27] Hunt, K.H., *Kinematic Geometry of Mechanism*, Oxford University Press (1978).

[28] Phillips, J., *Freedom in Machinery,* Volume 1, Introducing screw theory, England: Cambridge University Press (1984).

[29] Dimenberg, F.M., *The screw calculus and its applications in mechanics*, (in Russian) Izad, Nauka, Moscow, English Translated Foreign Technology Division, WP-APB, Ohio, (1968).

[30] Ball R.S., *The Theory of Screws*, England: Cambridge University Press, (1900).

[31] Huang, Z., Fang, Y.F. and Kong, L.F., *Theory and Control of Parallel Robotic Mechanisms*, Publisher of Mechanical Engineering, (in Chinese), Beijing, (1997).

[32] Lee, K. M., and Shah, D. K., Kinematic Analysis of a Three-Degree-of-Freedom In-parallel Actuated Manipulators, Proc. IEEE Int.Conf. Robotics and Automation, 1987, Vol.1, pp. 345-350.

[33] Waldron, K. J., Raghavan, M. and Roth, B., Kinematics of A Hybrid Series-Parallel Manipulation System, *ASME J. of Mechanisms,* Trans. And Auto. In design, 1989, Vol.111, pp. 211-221.

[34] Pierrrot, Fast Parallel Robots, *J. of Robot System*, 1991, Vol.8(6), pp. 829-840.

[35] Dai, J.S., Holland, and Kerr, D.R., Task-oriented direct synthesis of serial manipulator using moment invariants, 1996, 24th ASME Biennial Mechanisms Conference, ASME: DETC96/MECH-1004, Irvine, California, 19-22.

[36] Wang, J. and Gosselin, C.M., Kinematic Analysis of Singularity Loci of Spatial Four-Degree-of-Freedom Parallel Manipulators Using a Vector Formulation, *ASME J. Mech. Des.*, 1998, Vol.120, pp. 555-558.

[37] Tsai, L.W., The enumeration of a class of three-dof parallel manipulators, Proc. of the 10th World Congress on the theory of Machine and Mechanisms, Oulu, Finiand, 1999, pp. 1121-1126.

[38] Carretero, J.A., Podhorodeski, R.P. and Gosselin, C.M., Kinematic Analysis and Optimization of a New Three Degree-of Freedom Spatial Parallel manipulator, *ASME J. Mech. Des.,* 2000, Vol.112, pp. 17-24.

[39] Dimenberg, F.M., The screw calculus and its applications in mechanics, (in Russian) *Izad, Nauka,* Moscow, English Translated Foreign Technology Division, WP-APB, Ohio, 1968.

[40] Murray, R. M., Li, Z. and Shankar Sastry, S., *A Mathematical Introduction to Robotic Manipulator,* Boca Raton, FL: CRC, 1993.

In: Parallel Robotics: Recent Advances in Research… ISBN: 978-1-60456-859-2
Editors: Jinsong Wang and Xin-Jun Liu, pp. 33-52

Chapter 2

TYPE SYNTHESIS OF 2R2T 4-DOF PARALLEL MECHANISMS USING CONSTRAINT SYNTHESIS METHOD

Qinchuan Li, Chuanyu Wu and Xudong Hu*
Mechatronics Institute, Zhejiang Sci-Tech University
Hangzhou, Zhejiang Province, 310018, China

1.1. INTRODUCTION

A parallel mechanism (PM) with less than six degrees of freedoms (DOFs) is called a lower-mobility PM. A well-optimized lower-mobility parallel mechanism (PM) may replace a 6-DOF PM in many tasks which do not require all the six DOFs with reduced cost in design process, manufacturing process, control system, actuation system, Delta robot (Clavel 1988) is one typical example of successful application of lower-mobility PM. Hence, lower-mobility PMs have attracted much attention in recent years.

Mobility analysis and type synthesis are two basic and closely-related topics in research on lower-mobility PMs. A PM comprises a fixed base, a moving platform, and several limbs connecting the base and the moving platform. The limb can be a serial concatenation of kinematic pairs and composite joints like parallelograms. The mobility analysis aims to specify the number of independent variable parameters needed to determine the position and orientation of the moving platform. Often, one also expects to know the motion pattern of the moving platform, namely, rotational or translational DOF, or a combination of both. For example, the moving platform can rotate about the x and y axis, and translate along the x and y axis. The type synthesis aims to invent new architectures based on pre-specified DOFs and motion patterns, which is a reverse process of mobility analysis.

According to the motion pattern, that is, rotational (R) or translational (T), the lower-mobility PMs can be classified. The 5-DOF PMs can be sorted into two subcategories: 3R2T and 2R3T. The 4-DOF PMs can be sorted into three subcategories: 3R1T, 1R3T, and 2R2T.

* lqchuan@zstu.edu.cn, cywu@zstu.edu.cn, xdhu@zstu.edu.cn

The 3-DOF PMs can be sorted into four subcategories: 1R2T, 2R1T, 3T, and 3R. The 2-DOF PMs can be sorted into three subcategories: 2R, 2T, and 1R1T.

The research on type synthesis of lower-mobility PMs can be dated back to 1983, when Hunt (1983) proposed some architectures, including some planar PMs, 3-DOF PMs, 6-DOF PMs and some asymmetrical 4- and 5-DOF PMs. Clavel (1988) invented the famous successful 3-DOF translational Delta Robot. Hervé and Sparacino (1991) proposed the family of 3-DOF translational PMs with 4-DOF limbs. Tsai (1996) proposed the 3-UPU translational PM, which attracted much attention in following years because of its special motion properties. Pierrot and Company (1999) invented the 4-DOF H4 parallel robot, using parallelograms to eliminate rotations. Zhao and Huang (2000) proposed the 4-URU PM with three translational DOFs and one rotational DOF along the norm of the moving platform. Zlatov and Gosselin (2001) proposed a 4-DOF PM with three rotational DOFs and one translational DOF. Huang and Li (2001) invented symmetrical 5-DOF PMs which had been recognized as being vacant for a long time.

Other important progress in inventing new types of lower-mobility PMs include the linear translational PMs (Kong and Gosselin, 2002), 3-URC wrist (Di Gregorio, 2001), singularity-free fully-isotropic translational PMs (Carricato and Parenti-Castelli 2002), fully isotropic 4-DOF parallel mechanisms for Schoenflies motion (Carricato 2005), 3-DOF fully PMs with high rotational capability (Liu, Wang and Pritschow 2005), 3-DOF translational PM with doubly planar limbs (Lee and Hervé 2006) and two-mode 3-DOF PMs (Refaat et.al 2007).

One traditional method for type synthesis of PMs is based on the general Grübler-Kutzbach mobility criterion. Tsai (1999) applied this method to the type synthesis of a kind of 3-DOF PMs. However, this method failed to describe the geometrical arrangement of kinematic pairs, hence invalid results were often obtained (Merlet 2002).

Another method, group-based synthesis method, is based on the pioneering works on the algebraic properties of displacement subgroup theory (Hervé 1978,). The group-based synthesis method is applicable to all kinds of lower-mobility PMs (Hervé and Sparacino1991, Hervé 1999, Li, Huang, and Hervé 2004, Huynh and Hervé 2005a, Karouia and Hervé 2002, Karouia and Hervé 2005, Lee and Hervé 2006). Other researchers (Angeles 1982, 2002, Rico *et al* 2006, Kong and Gosselin 2007, Meng, Liu and Li 2007) also made important contributions related with the group-based synthesis method. Because the displacement subgroup deals with finite motion, PMs synthesized by the group-based method are inherently with full-cycle mobility. Hence, the group-based synthesis method is at the level of finite motion.

The constraint-synthesis method is based on screw theory, particularly, the reciprocal screw theory. Because a screw is inherently instantaneous, the constraint-synthesis method is at the level of instantaneous motion. Hence, one indispensable step of the constraint-synthesis method is to identify if the synthesized mechanism is instantaneous or not, which sometimes is difficult for some complicated architectures. In most cases, finite motion conditions can be established by checking the geometrical conditions of the PM. The constraint-synthesis method has been applied to all kinds of lower-mobility PMs (Huang, Kong and Fang 1997, Frisoli *et al.* 2000, Huang and Li 2002, 2003, Kong and Gosselin 2004a, 2004b, Fang and Tsai 2002, 2004). Numerous novel PMs have been invented using this method, including the symmetrical 5-DOF PMs.

In addition, Jian and Yang (2004) proposed a type synthesis method based single-unit open chains and synthesized some novel mechanisms. Gao, Zhang and Li (2005) proposed the synthesis method based on the Gf coordinates. Gogu (2004, 2007) applied a synthesis method based on linear transform theory. Carricato and Parenti-Castelli (2002) synthesized a family of Singularity-free fully isotropic translational PMs. One common feature of these methods is the use of velocity variables. Hence, we call these methods as motion-synthesis approach for convenience, though not very accuratcly.

However, there are still some gaps in our knowledge of type synthesis of 4-DOF 2R2T PMs with identical limbs. Little has been addressed on the systematical type synthesis of 4-DOF 2R2T PMs (Li and Huang 2003). This paper aims to apply the constraint-synthesis method to the type synthesis of 4-DOF 2R2T PMs systematically. In section 1.2, we review the fundamentals of constraint-synthesis method briefly. In section 1.3 we develop the mechanism and limb constraint-wrench system and identify the structural geometrical conditions under which the combined effect of limb constraint-wrench systems is equal to the mechanism constraint-wrench system. In section 1.4, limb kinematic chains are generated by linear combination of twists reciprocal to the limb constraint-wrench system. On the basis of structural geometrical conditions, numerous 2R2T 4-DOF parallel mechanisms are constructed with limb kinematic chains obtained above. In section 1.5, 2R2T 4-DOF parallel mechanisms with full-cycle mobility are identified based on the integrity of structural geometrical conditions.

1.2. Fundamentals of Constraint Synthesis Method

1.2.1. Twist and Wrench

In screw theory (Hunt 1978), a screw is a line vector associated with a pitch,

$$\boldsymbol{\$} = (\boldsymbol{S};\ \boldsymbol{S}_0) = (\boldsymbol{S};\ \boldsymbol{r} \times \boldsymbol{S} + h\boldsymbol{S}) = (l\ \ m\ \ n;\ \ p\ \ q\ \ r) \tag{1}$$

where $\boldsymbol{S}$ is a unit vector pointing along the line, $\boldsymbol{r}$ is the position vector of any point on the line,

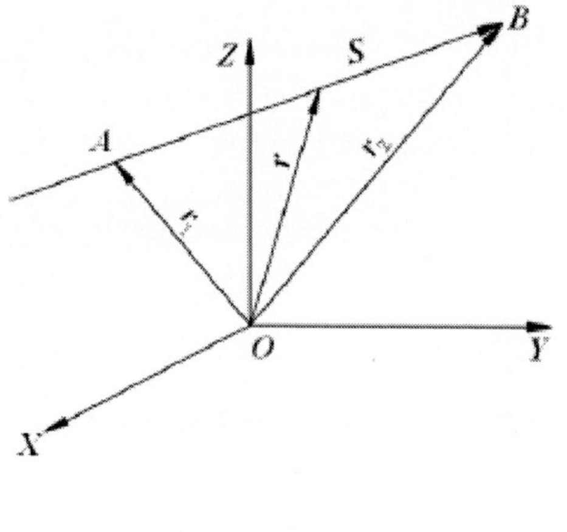

Figure.1. A Straight Line.

and h is the pitch of the screw.

The pitch of a screw is given by

$$h = \boldsymbol{S} \cdot \boldsymbol{S}_0 / \boldsymbol{S} \cdot \boldsymbol{S} \tag{2}$$

When representing a force and moment, a screw is often called a wrench. When representing an instantaneous linear velocity and angular velocity of a rigid body, a screw is often called a twist. Evidently, a screw is a powerful geometrical and mathematical presentation in kinematics and statics.

If the pitch of a screw is equal to zero, the screw is reduced to a line vector. One can describe a line vector by its Plücker coordinates $(\boldsymbol{S};\ \boldsymbol{r} \times \boldsymbol{S})$ which consists of six components. A line vector or a unit screw of zero-pitch ($h = 0$) can be used to represent a revolute joint or a unit force along the screw axis in space.

If the pitch of a screw goes to infinity, the screw is defined as

$$\boldsymbol{\$} = (0;\ \boldsymbol{S}) = \begin{pmatrix} 0 & 0 & 0; & l & m & n \end{pmatrix} \tag{3}$$

A unit screw of infinite pitch ($h = \infty$) can be used to represent a prismatic pair or a couple along the screw axis in space.

A multi-DOF lower kinematic pairs such as universal, cylindrical, or spherical joint can be represented by a group of screws because of its kinematic equivalency to a combination of revolute and prismatic pairs. Note that the geometrical arrangement of joint axes in a mechanism determines the mobility of a mechanism. Hence, it will be useful for mobility analysis to calculate the ranks of screws under different geometrical conditions.

1.2.2 Reciprocal Screws

The reciprocal product of two screws, say $\boldsymbol{\$} = (\boldsymbol{S};\boldsymbol{S}_0)$ and $\boldsymbol{\$}^r = (\boldsymbol{S}^r;\boldsymbol{S}_0^r)$, is defined as

$$\boldsymbol{\$} \circ \boldsymbol{\$}^r = \boldsymbol{S} \cdot \boldsymbol{S}_0^r + \boldsymbol{S}^r \cdot \boldsymbol{S}_0 \tag{4}$$

where symbol "$\circ$" denotes the reciprocal product of two screws. The reciprocal product of two screws represents the work produced by a wrench acting on a rigid body undergoing an infinitesimal twist.

Two screws are said to be reciprocal if their reciprocal product is zero

$$\boldsymbol{\$} \circ \boldsymbol{\$}^r = lp^r + mq^r + nr^r + pl^r + qm^r + rn^r = 0 \tag{5}$$

Equation (5) shows that the wrench $\boldsymbol{\$}^r$ acting on a rigid body undergoing an infinitesimal twist $\boldsymbol{\$}$ yields no work. The wrench $\boldsymbol{\$}^r$ denotes a constraint force when its pitch is zero, or a constraint couple when its pitch is infinite. The former restricts a translational freedom of the rigid body along the direction of the force, and the latter restricts a rotational freedom around the axis of the couple.

1.2.3. Constraint, Mobility and Structural Geometrical Condition

The reduction of DOFs in a lower-mobility PMs is caused by constraint acting on the moving platform exerted by all limbs. One can express a limb constraint by a wrench, which may be a force or a couple. When a constraint force is acting on a rigid body, the translation of the rigid body along the direction being parallel to the force is restricted. When a constraint couple is acting on a rigid body, the rotation of the rigid body around the axis being parallel to the couple is restricted.

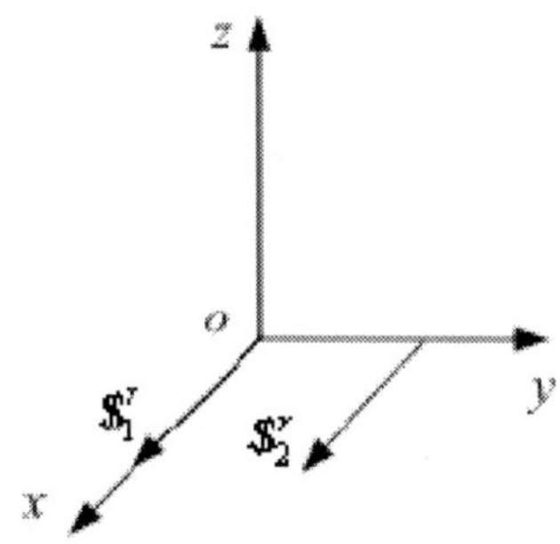

Figure 2. Two parallel constraint forces.

For example, consider two constraint forces, $\$_1^r$ and $\$_2^r$, which are parallel to each other as shown in figure 2. $\$_1^r$ is coincident with the x axis and $\$_2^r$ is in the xy plane. The $\$_1^r$ and $\$_2^r$ is given in terms of screw as follows,

$$\begin{aligned} \$_1^r &= (1\,0\,0\,;0\,0\,0) \\ \$_2^r &= (1\,0\,0\,;0\,0\,c_2) \end{aligned} \tag{6}$$

Note that any two parallel constraint forces in the xy plane can be obtained by linear combination of $\$_1^r$ and $\$_2^r$.

The screws reciprocal to $\$_1^r$ and $\$_2^r$ are

$$\begin{aligned} \$_1 &= (1\,0\,0\,;0\,0\,0) \\ \$_2 &= (0\,1\,0\,;0\,0\,0) \\ \$_3 &= (0\,0\,0\,;0\,0\,1) \\ \$_4 &= (0\,0\,0\,;0\,1\,0) \end{aligned} \tag{7}$$

Equation (7) shows that the instantaneous motions permitted by the two constraint forces, $\$_1^r$ and $\$_2^r$, can be represented by four twists, $\$_1$, $\$_2$, $\$_3$, and $\$_4$. $\$_1$ denotes a rotation about the x axis; $\boldsymbol{\$}_2$ denotes a rotation about the y axis; $\boldsymbol{\$}_3$ denotes a translation along the y axis; $\boldsymbol{\$}_4$ denotes a translation along the z axis. Hence, the two parallel constraint forces restricted the translation parallel to the constraint force and rotation around the normal of the plane determined by the two constraint forces.

Note that all the limb constraint wrenches form a screw system, called limb constraint system. The union of all the limb constraints forms a screw system, called mechanism constraint system, and determines the restricted DOFs of a lower-mobility PM.

The linear dependency of screws is determined by their relative geometrical conditions. For example, consider a chain consists of three successive parallel revolute pairs. The three twists associated with the three pairs are linearly independent provided that the three revolute axes are not coplanar. Using structural geometrical condition to judge the linear dependency of screws is more quick and straightforward than using coordinate transformation which is usually complicated concerning a PM.

1.2.4. Procedure of Constraint Synthesis Method

The detailed constraint-synthesis method is available in an earlier paper (Huang and Li 2002b). For readers' convenience, a brief description is given as follows.

Step 1) Describe the desired DOF of the target PM using twists and form a mechanism twist system.

Step 2) Find the screw system reciprocal to the mechanism twist system obtained in Step 1, which is termed the mechanism constraint system.

Step 3) Find the limb constraint system based on the linear dependence of wrenches under different geometrical conditions.

Step 4) Obtain a base of the desired limb twist system using the reciprocal relation between twist and wrench.

Step 5) Generate the desired limb chain by linear combination of the twists in the base obtained in Step 4. and construct the PM with these limb chains while obeying the geometrical conditions.

Step 6) Identify the synthesized PM is instantaneous or not.

1.2.5. Nomenclature

In the following sections, some symbols are used to describe the limb kinematic chain. Generally, the left lowercase superscript is used to denote the direction of the axis of revolute or prismatic pair. The right capital subscript is used to denote the point on the pair axis if necessary.

The left superscript z denotes that the axis of the pair is perpendicular to the base plane;

The left superscript x or y denotes that the axis of the pair is parallel to the base plane.

The left superscripts i, j and k denote three linearly independent direction vectors. For example, $({}^{i}R{}^{j}R{}^{k}R)_N$ denotes three successive revolute pairs whose axes intersect at a common point N, namely, a 3R spherical subchain or a spherical pair.

Another example is ${}^{ui}U_N$, where the superscript u and i denote the first revolute pair axis and the second revolute pair axis of the universal joint; the subscript N indicates that the second revolute pair axis of the universal joint passes point N.

For convenience, $\boldsymbol{\$}_{ij}$ is used to represent the unit twist associated with the jth kinematic pair in the ith limb; $\boldsymbol{\$}^r_{ij}$ the jth unit wrench exerted by the ith limb; $\boldsymbol{\$}_{mj}$ the jth unit twist in the mechanism twist system; $\boldsymbol{\$}^r_{mj}$ the jth unit wrench in the mechanism constraint system.

We set XY plane of the global frame, O-XYZ, parallel to the base plane, thus Z axis is perpendicular to the base plane and upward. We set z_i axis of the ith limb frame, $o_i - x_i y_i z_i$, upward and perpendicular to the base plane. The limb twist and constraint system are expressed in limb frame while the mechanism twist and constraint systems are expressed in the global frame.

1.3. Constraint-Wrench Systems of 2R2T 4-DOF Parallel Mechanisms

1.3.1. Mechanism Constraint-Wrench System

The moving platform of a 2R2T PM loses one rotational DOF and one translational DOF because a constraint couple and a constraint force act on the moving platform. Obviously, the mechanism constraint system of a 2R2T PM consists of a constraint couple and a constraint force, or two parallel constraint forces.

Without loss of generality, we focus on the $2R^{zy}2T^{xy}$ 4-DOF PMs, where $2R^{zy}$ denotes two rotations around the Z and Y axis, and $2T^{xy}$ denotes two translations along the X and Y axis, respectively. The moving platform is constrained by a force along the Z axis and a couple along the X axis. Hence, the mechanism constraint-wrench system of the $2R^{zy}2T^{xy}$ 4-DOF PM is given by

$$\begin{aligned} \boldsymbol{\$}^r_{m1} &= (\,0 \;\; 0 \;\; 1 \; ; 0 \;\; 0 \;\; 0\,) \\ \boldsymbol{\$}^r_{m2} &= (\,0 \;\; 0 \;\; 0 \; ; 1 \;\; 0 \;\; 0\,) \end{aligned} \tag{8}$$

1.3.2. Limb Constraint-Wrench System and Structural Geometrical Condition

Suppose that the $2R^{zy}2T^{xy}$ 4-DOF PM is symmetrical and consists of four identical limb kinematic chains. One can conclude that the constraint must be a force. Note that these limbs exert same constraints on the moving platform. When the limb kinematic chain has five DOFs, it exerts one constraint on the moving platform. Because the combination of four couples yields only couples and restricts no translations, the constraint must be a force.

Consider the example of the two parallel forces shown in figure 2. If each limb exerts a constraint force on the moving platform, there are total four constraint forces acting on the moving platform. We can arrange the four limbs following some structural geometrical conditions which classify the four forces into two groups, and the two forces in each group are coaxial. Consequently, we can obtain two parallel forces which supply the desired constraint effect. Hence, the limb constraint-wrench system of a $2R^{zy}2T^{xy}$ 4-DOF PM is given by

$$\$^r_{i1} = (0\,0\,1;0\,0\,0) \tag{9}$$

which is a constraint force being coaxial with the z_i axis.

1.3.3. Limb Twist System and Structural Geometrical Condition

Using the reciprocity between twist and wrench, we can obtain the limb twist system which is reciprocal to the $\$^r_{i1}$ as follows

$$\begin{aligned} \$_{i1} &= (1\,0\,0;0\,0\,0) \\ \$_{i2} &= (0\,1\,0;0\,0\,0) \\ \$_{i3} &= (0\,0\,1;0\,0\,0) \\ \$_{i4} &= (0\,0\,0;1\,0\,0) \\ \$_{i5} &= (0\,0\,0;0\,1\,0) \end{aligned} \tag{10}$$

Linear combination of the above five twists can yield different structures of a limb chain. It is noted that the linear combination of the five twists must keep the linear independency among them. With further inspection, one can find more structural geometrical conditions of the limb chain.

A. Existence of 2R or 3R spherical subchain and limb central point

It can be concluded that the limb kinematic chain of the $2R^{zy}2T^{xy}$ 4-DOF PM must contains a 2R spherical subchain or a 3R spherical subchain. To avoid the presence of helical pairs, $\$_{i1}$ can be linearly combined with $\$_{i2}$ and $\$_{i3}$. The linear combination yields

$$\$'_{i1} = (l_{i1}\ m_{i1}\ n_{i1};0\,0\,0) \tag{11}$$

Similarly, $\$_{i2}$ can be linearly combined with $\$_{i1}$ and $\$_{i3}$. The linear combination yields

$$\$'_{i2} = (l_{i2}\ m_{i2}\ n_{i2};0\,0\,0) \tag{12}$$

Note that $\$'_{i1}$ and $\$'_{i2}$ represents two revolute pairs with axes intersecting at a common point. The two pairs form a 2R spherical subchain, dentoed by $({}^iR^jR)_N$, as shown in figure 3. The subscript N denotes the common point, which is termed as limb central point in later sections. $\$_{i3}$ can also be transformed into a revolute pair whose axis passing the limb central point by linear combination with $\$_{i1}$ and $\$_{i2}$, namely,

$$\$'_{i3} = (l_{i3}\ m_{i3}\ n_{i3}\ ; 0\ 0\ 0) \tag{13}$$

Note that the group of $\$'_{i1} - \$'_{i2} - \$'_{i3}$ represents a 3R spherical subchain, $({}^{i}R\,{}^{j}R\,{}^{k}R)_N$, as shown in figure.3.

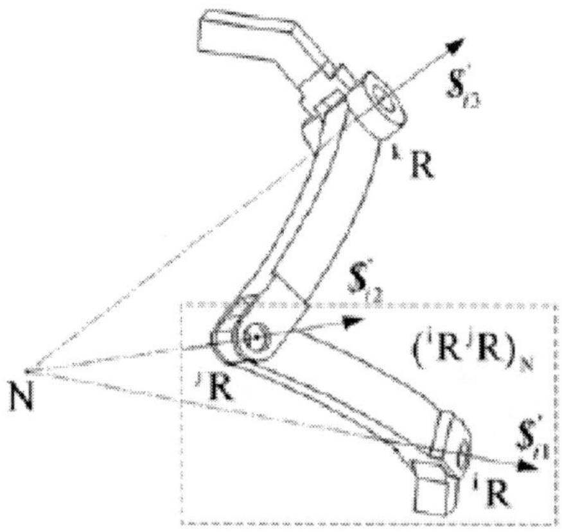

Figure 3. 2R and 3R spherical subchain.

B. Arrangement of Prismatic pairs

The prismatic pair in the $2R^{zy}2T^{xy}$ 4-DOF PM must be parallel to the XY plane. $\$_{i4}$ denotes a prismatic pair along the x_i axis and $\$_{i5}$ a prismatic pair along the y_i axis. The linear combination of $\$_{i4}$ and $\$_{i5}$ produces a twist in the form of $(0\,0\,0\,; l_i\ m_i\ 0)$, which represents a prismatic pair parallel to the XY plane.

C. Arrangement of Revolute pairs not included in 2R or 3R subchain

By appropriate linear combinations, a prismatic pair can be transformed into a revolute pair. For instance, the prismatic pair $\$_{i4}$ can be transformed into a revolute pair $\$'_{i4}$ by linear combination with $\$_{i3}$ and $\$_{i5}$,

$$\$'_{i4} = (0\,0\,1\,; a_{i4}\ b_{i4}\ 0) \tag{14}$$

which represents a revolute pair whose axis is parallel to the z_i axis and passes through any point. The prismatic pair $\$_{i5}$ can be transformed into a revolute pair $\$'_{i5}$ by linear combination with $\$_{i3}$ and $\$_{i4}$,

$$\$'_{i5} = (0\,0\,1\,; a'_{i5}\ b'_{i5}\ 0) \tag{15}$$

which represents a revolute pair whose axis is parallel to the z_i axis and passes through any point.

Though $\$_{i3}$ represents a revolute pair whose axis is parallel to the z_i axis and passes through the origin of the limb frame, one can transform it into a revolute pair whose axis passes through any point by linear combination with $\$_{i4}$ and $\$_{i5}$,

$$\$''_{i3} = (0\,0\,1\,;\, a_{i3}\; b_{i3}\; 0) \tag{16}$$

It is noted that the axes of the revolute pairs not included in the 2R or 3R spherical subchain must be perpendicular to the base. For example, if we put $\$'_{i4}$ next to $\$'_{i5}$, they form a 2R parallel subchain $^{z}R^{z}R$; if we put $\$''_{i3}$, $\$'_{i4}$, and $\$'_{i5}$ successively, they form a 3R parallel subchain $^{z}R^{z}R^{z}R$.

In table 1, we list the constraint and structural features of the $2R^{zy}2T^{xy}$ 4-DOF PM.

Table 1. Constraint features and structural geometrical conditions of the $2R^{zy}2T^{xy}$ 4-DOF PM

Constraint features	Structural geometrical conditions
Each limb exerts a constraint force in the z axis. All the limb constraint forces must be coplanar and parallel to each other.	The limb chain must contain a 2R or 3R spherical subchain The prismatic pairs must be parallel to the base The axes of revolute pairs not included in the 2R or 3R spherical subchain must be perpendicular to the base There are two central points in the mechanism All the 2R or 3R spherical subchains must connect the base or the moving platform

1.4. Type Synthesis of 2R2T 4-DOF Parallel Mechanisms

1.4.1. Generation of Limb Kinematic Chain

__A. Limb Chains only with R and P Pairs__

On the basis of the above analysis, one can enumerate the limb kinematic chains of the $2R^{zy}2T^{xy}$ 4-DOF PM by linear combination of the five twists in equation (10). For example, we put the five pairs in the sequence of $\$_{i4} - \$_{i5} - \$'_{i1} - \$'_{i2} - \$'_{i3}$ and connect them by links. Then we obtain a $^{x_1}P^{x_2}P(^{i}R^{j}R^{k}R)_N$ limb chain as shown in figure.4 (a).

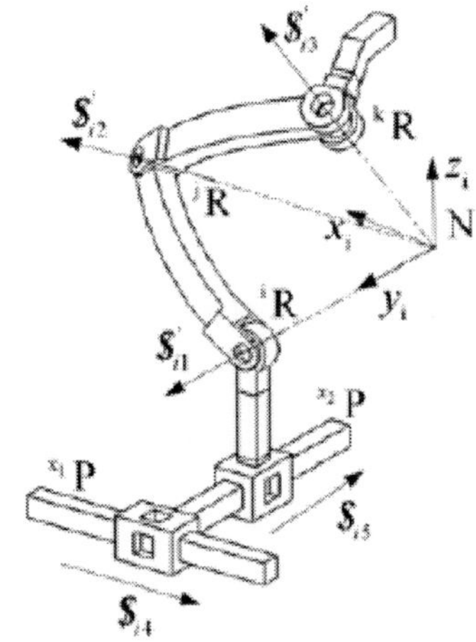

(a) ${}^{x_1}P^{x_2}P({}^iR^jR^kR)_N$

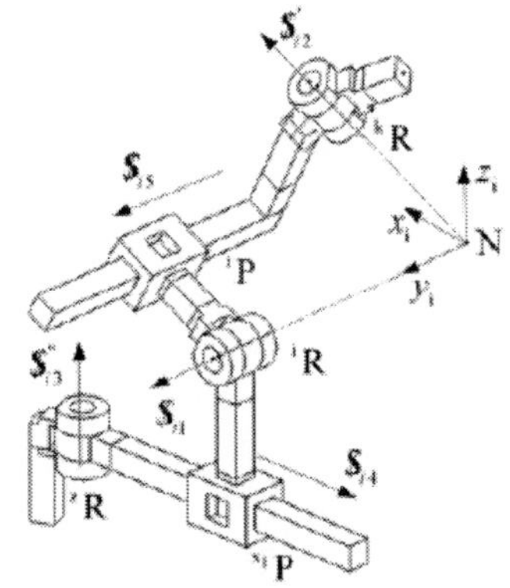

(b) ${}^zR^{x_1}P({}^{x_2}R^{x_2}P^kR)_N$

Figure 4. Two limb chains only with R and P pairs.

Another example is the ${}^zR^{x_1}P({}^{x_2}R^{x_2}P^kR)_N$ limb chain as shown in figure 4 (b). First, $\$_{i3}$ can be transformed into a revolute pair $\$''_{i3}$ by linear combination with $\$_{i4}$ and $\$_{i5}$, as described by equation (15), whose axis passes through any point. Then, we put the five pairs in the sequence of $\$''_{i3}-\$_{i4}-\$_{i1}-\$_{i5}-\$'_{i2}$ and connect them by links.

B. Limb Chains with U Joints

Two-DOF kinematic joints, such as cylindrical pairs and universal joints, can also be employed by some appropriate combinations. If in a serial arrangement of two revolute pairs and the axes are intersecting at point with a right angle, then the sequence of two R pairs can be named a universal joint.

For example, any two of the three revolute pairs $\$_{i1}$, $\$_{i2}$, and $\$_{i3}$ can yields a U joint and the central point of the U joint is also the limb central joint. However, the limb central joint can not coincide with each other which disobeys the structural condition of the $2R^{zy}2T^{xy}$ 4-DOF PM. Hence, such a generation of U joints can not be used for constructing a $2R^{zy}2T^{xy}$ 4-DOF PM.

In addition, note that $\$_{i4}$ and $\$_{i5}$ can be transformed into revolute pairs $\$'_{i4}$ and $\$'_{i5}$ which are perpendicular to the base. Thus, we can use $\$_{i1}$ and $\$'_{i4}$ to form a U joint. Further, $\$'_{i2}$ and $\$'_{i3}$ can form a 2R spherical subchain. Arranging the five pairs in the sequence of $\$'_{i5}-\$'_{i4}-\$_{i1}-\$'_{i2}-\$'_{i3}$, we obtain a ${}^zR^zU^x_N({}^jR^kR)_N$ limb chain.

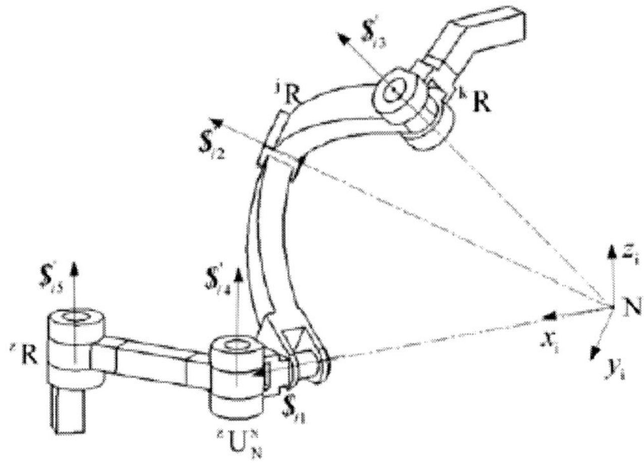

Figure 5. $^{z}R^{z}U^{x}{}_{N}(^{j}R^{k}R)_{N}$ limb chain.

C. Limb Chains with C Pair

A cylindrical pair is a kinematic equalivence to a revolute pair and a prismatic pair which is parallel to the revolute axis. Because the prismatic pair in a $2R^{zy}2T^{xy}$ 4-DOF PM. must be parallel to the base, the C pair must also be parallel to the base. For instance, we can use $\$'_{i2}$ and $\$'_{i3}$ to form a 2R spherical subchain. Then, we can use $\$_{i1}$ and $\$_{i4}$ to construct a C pair. Arranging the five pairs in the sequence of $\$'_{i5}-\$_{i4}-\$_{i1}-\$'_{i2}-\$'_{i3}$, we obtain a $^{z}R^{x}C_{N}(^{j}R^{k}R)_{N}$ limb chain as shown in figure 6 .

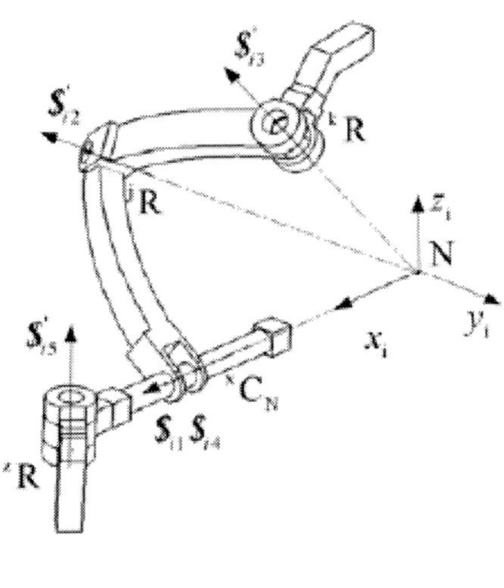

Figure 6. $^{z}R^{x}C_{N}(^{j}R^{k}R)_{N}$ limb chain.

Based on the above analysis, one can enumerate 30 limb kinematic chains by linear combination of the five twists in equation (10), which are listed in table 2.

Table 2. Limb kinematic chains for $2R^{zy}2T^{xy}$ 4-DOF PM

Type	Limb kinematic chains			
1 Containing only R and P pair	${}^{y}P{}^{x}P({}^{i}R{}^{j}R{}^{k}R)_N$ ${}^{y}P{}^{x}P{}^{z}R({}^{j}R{}^{k}R)_N$ ${}^{z}R{}^{y}P{}^{x}P({}^{j}R{}^{k}R)_N$ ${}^{y}P{}^{z}R{}^{x}P({}^{j}R{}^{k}R)_N$	${}^{y}P{}^{y}R_N{}^{x}P({}^{j}R{}^{k}R)_N$ ${}^{z}R{}^{y}P{}^{x}R_N{}^{x}P{}^{k}R_N$ ${}^{y}P{}^{z}R{}^{x}R_N{}^{x}P{}^{k}R_N$ ${}^{z}R{}^{z}R{}^{y}P({}^{j}R{}^{k}R)_N$	${}^{z}R{}^{y}P{}^{z}R({}^{j}R{}^{k}R)_N$ ${}^{y}P{}^{z}R{}^{z}R({}^{j}R{}^{k}R)_N$ ${}^{z}R{}^{x}P({}^{x}R{}^{j}R{}^{k}R)_N$ ${}^{y}P{}^{z}R({}^{x}R{}^{j}R{}^{k}R)_N$	${}^{z}R{}^{x}R_N{}^{x}P({}^{j}R{}^{k}R)_N$ ${}^{z}R{}^{z}R{}^{x}R_N{}^{x}P{}^{k}R_N$ ${}^{z}R{}^{z}R{}^{z}R({}^{x}R{}^{k}R)_N$ ${}^{z}R{}^{z}R({}^{x}R{}^{j}R{}^{k}R)_N$
2 Containing U joint	${}^{z}R{}^{z}U^{x}{}_{N}({}^{j}R{}^{k}R)_N$ ${}^{z}R{}^{z}R{}^{z}U^{x}{}_{N}{}^{k}R_N$ ${}^{z}R{}^{y}P{}^{z}U^{x}{}_{N}{}^{k}R_N$	${}^{y}P{}^{z}R{}^{z}U^{x}{}_{N}{}^{k}R_N$ ${}^{y}P{}^{z}U^{x}{}_{N}({}^{j}R{}^{k}R)_N$ ${}^{x}P{}^{y}P{}^{z}U^{x}{}_{N}{}^{k}R_N$	${}^{z}R{}^{z}U^{x}{}_{N}{}^{x}P{}^{k}R_N$ ${}^{y}P{}^{z}U^{x}{}_{N}{}^{x}P{}^{k}R_N$ ${}^{z}U^{xy}P({}^{j}R{}^{k}R)_N$	
3 Containing C joint	${}^{y}P{}^{x}C_N({}^{j}R{}^{k}R)_N$ ${}^{z}R{}^{x}C_N({}^{j}R{}^{k}R)_N$	${}^{z}R{}^{y}P({}^{x}C{}^{k}R)_N$ ${}^{z}R{}^{z}R({}^{x}C{}^{k}R)_N$	${}^{y}P{}^{z}R({}^{x}C{}^{k}R)_N$	

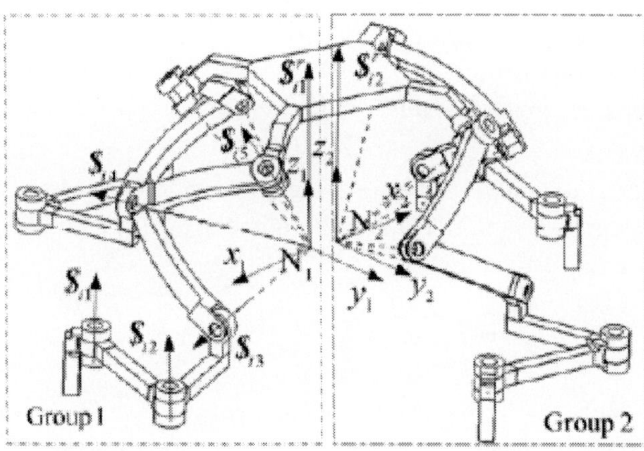

Figure 7. 4- ${}^{z}R{}^{z}R$ $({}^{i}R{}^{j}R{}^{k}R)_N$ PM.

1.4.2. Generation of 2R2T 4-DOF Parallel Mechanisms

Following the structural condition presented in table 1, one can use the limb kineamtic chains listed in table 2 to construct various $2R^{zy}2T^{xy}$ 4-DOF PMs. For example, we can use four ${}^{z}R{}^{z}R$ $({}^{i}R{}^{j}R{}^{k}R)_N$ limbs to construct a $2R^{zy}2T^{xy}$ 4-DOF PM. We divide the four ${}^{z}R{}^{z}R$ $({}^{i}R{}^{j}R{}^{k}R)_N$ limbs into two groups. In each group, the two limb central points are coincident with each other. Note that when we connect the limbs to the base and moving platform, the two central points of the two groups of limb chains must not be coincident with each other. The 4-$^{z}R^{z}R(^{i}R^{j}R^{k}R)_{N}$ $2R^{zy}2T^{xy}$ 4-DOF PM is depicted in figure 7.

Thirty symmetrical $2R^{zy}2T^{xy}$ 4-DOF PMs are depicted in Appendix. It is worth noting that non-symmetrical $2R^{zy}2T^{xy}$ 4-DOF PMs can also be constructed by using different limb kinematic chains listed in table 2.

1.5. Identification of 2R2T 4-DOF Parallel Mechanisms with Full-Cycle Mobility

1.5.1 The Modified Grübler-Kutzbach Criterion

It has long be recongnized that the general Grübler-Kutzbach criterion may lead to wrong results when being applied to mobility analysis of lower-mobility PMs, particularly, to those containing closed-loop in limb kineamtic chain. Using the constrtaint analysis approach, Huang and Li (2002, 2003) proposed a modified Grübler-Kutzbach criterion applicable to all kinds of lower-mobility PMs. Using this modified Grübler-Kutzbach criterion, one can obtain the correct results of mobility analysis with quickness and convenience.

The modified Grübler-Kutzbach criterion is given by (Huang and Li, 2002, 2003)

$$M = d(n-g-1) + \sum_{i=1}^{g} f_i + v, \tag{17}$$

where M denotes the mobility of the mechanism, d the order of the mechanism, n the number of links, g the number of kinematic pairs, f_i the freedom of the ith pair and v the number of the redundant constraint.

The major obstacle in mobility analysis of the lower-mobility PM is the calculation of the order of the mechanism d (Hunt 1978) and the number of the redundant constraint v. The order of a mechanism is given by

$$d = 6 - \lambda \tag{18}$$

where λ is the number of the independent common constraints in the mechanism.

On the basis of screw theory, an independent common constraint can be defined as a screw reciprocal to the unit twists associated with all kinematic pairs in a lower-mobility PM. Let us consider an M-DOF PM, which has p identical limbs. Each limb exerts q structural constraints on the moving platform. Then, the mechanism constraint system is formed by the $p \cdot q$ constraints and its dimension is 6-M in the general non-singular configuration. An geometrical interpretation of the sufficient and necessary conditions for the existence of an independent common constraint is if each limb provides one constraint of the same kind and all the p constraints are coaxial. Using the constraint analysis approach, one can quickly specify the number of the independent common constraints λ.

When a constraint wrench is linearly dependent with other constraints, we call it a redundant constraint. Note that the number of constraints forming the common constraints is

$\lambda \cdot p$, which should not be taken into consideration repeatedly when calculating the redundant constraints. Hence, removing the $\lambda \cdot p$ screws forming the common constraints from the mechanism constraint system, we obtain a new screw system spanned by $p \cdot q - \lambda \cdot p$ constraint screws. Let k be the dimension of the new screw system. The number of the redundant constraint v is given by

$$v = p \cdot q - \lambda \cdot p - k \tag{19}$$

The modified Grübler-Kutzbach criterion can be used to identify the instantaneous PMs. It is noted that only the number of common constraint λ and the number of redundant constraint v in the modified Grübler-Kutzbach criterion are possible to change. If λ and v are constant in any non-singular configuration, the PM is not instantaneous and has a full-cycle mobility.

1.5.2 Geometrical Condition for Finite Motion

The structural geometrical conditions of the $2R^{zy}2T^{xy}$ 4-DOF PM listed in table.2 are also the geometrical conditions for finite motion. $2R^{zy}2T^{xy}$ 4-DOF PM. These geometrical conditions guarantee that the mechanism constraint system remains unchanged after any feasible finite displacement.

1.5.3 Identification of 2R2T 4-DOF Pms with Full-Cycle Mobility

Let us check the limb kinematic chain 4- ${}^{z}R^{z}R\ ({}^{i}R^{j}R^{k}R)_N$ as shown in figure 7. The limb twist system described in a limb frame $N_1 - x_i y_i z_i$ is given by

$$\begin{aligned}
\$_{i1} &= (0\,0\,1\,;a_{i1}\ b_{i1}\ 0)\\
\$_{i2} &= (0\,0\,1\,;a_{i2}\ b_{i2}\ 0)\\
\$_{i3} &= (1\,0\,0\,;0\,0\,0)\\
\$_{i4} &= (l_{i4}\ m_{i4}\ n_{i4}\,;0\,0\,0)\\
\$_{i5} &= (l_{i5}\ m_{i5}\ n_{i5}\,;0\,0\,0)
\end{aligned} \tag{20}$$

When the moving platform is undergoing a finite motion, parameters a_{i2}, b_{i2}, l_{i4}, m_{i4}, n_{i4}, l_{i5}, m_{i5}, and n_{i5} are changing. However, the five twists always are in the form of equation (20) in non-singular configuration, which are determined by the structural geometrical condition of the PM.

Further, the constraint wrench reciprocal to the five twists is given by

$$\$_{i1}^{r} = (0\,0\,1\,;0\,0\,0) \tag{21}$$

$\boldsymbol{\$}_{i1}^{r}$ is a constraint force perpendicular to the base and passing through the limb central point N1.

By similar reasoning, we can obtain the other constraint force $\boldsymbol{\$}_{i1}^{r}$ described in a limb frame $N_2 - x_2 y_2 z_2$.

$$\boldsymbol{\$}_{i2}^{r} = \left(0\,0\,1 ; 0\,0\,0 \right) \tag{22}$$

Clearly, $\boldsymbol{\$}_{i1}^{r}$ and $\boldsymbol{\$}_{i2}^{r}$ contains no changing parameters. One can transform $\boldsymbol{\$}_{i1}^{r}$ and $\boldsymbol{\$}_{i2}^{r}$ into the fixed frame *O-XYZ* to determine their combined effect. Let $\left[\boldsymbol{R}_o^1\right]$ be the rotational transformation matrix from the limb frame $N_1 - x_i y_i z_i$ to the fixed frame *O-XYZ.* Let $\left[\boldsymbol{R}_o^2\right]$ be the rotational transformation matrix from the limb frame $N_2 - x_2 y_2 z_2$ to the fixed frame *O-XYZ.* The position vector of the limb frame $N_1 - x_i y_i z_i$ relative to the fixed frame *O-XYZ.*is $\boldsymbol{ON}_1 = (a_1 \quad a_2 \quad a_3)^{\mathrm{T}}$. The position vector of the limb frame $N_2 - x_2 y_2 z_2$ relative to the fixed frame *O-XYZ.* is $\boldsymbol{ON}_2 = (b_1 \quad b_2 \quad b_3)^{\mathrm{T}}$. Considering the symmetrical arrangement of the four limb chains, we have $a_3 = b_3$.

Then, we have

$$\boldsymbol{\$}_{m1}^{r} = \begin{bmatrix} \left[\boldsymbol{R}_o^1\right] & \boldsymbol{0}_{3\times3} \\ \left[\boldsymbol{ON}_1\right]\left[\boldsymbol{R}_o^1\right] & \left[\boldsymbol{R}_o^1\right] \end{bmatrix} \boldsymbol{\$}_{i1}^{r} = \left(0\,0\,1 ; a_2 \; -a_1 \; 0 \right) \tag{23}$$

where $\left[\boldsymbol{R}_o^1\right] = \begin{bmatrix} p_{11} & p_{12} & 0 \\ p_{21} & p_{22} & 0 \\ 0 & 0 & 1 \end{bmatrix}$ and $\left[\boldsymbol{ON}_1\right] = \begin{bmatrix} 0 & -a_3 & a_2 \\ a_3 & 0 & -a_1 \\ -a_2 & a_1 & 0 \end{bmatrix}$.

Similarly, we have

$$\boldsymbol{\$}_{m2}^{r} = \begin{bmatrix} \left[\boldsymbol{R}_o^2\right] & \boldsymbol{0}_{3\times3} \\ \left[\boldsymbol{ON}_2\right]\left[\boldsymbol{R}_o^2\right] & \left[\boldsymbol{R}_o^2\right] \end{bmatrix} \boldsymbol{\$}_{i2}^{r} = \left(0\,0\,1 ; b_2 \; -b_1 \; 0 \right) \tag{24}$$

where $\left[\boldsymbol{R}_o^2\right] = \begin{bmatrix} q_{11} & q_{12} & 0 \\ q_{21} & q_{22} & 0 \\ 0 & 0 & 1 \end{bmatrix}$ and $\left[\boldsymbol{ON}_2\right] = \begin{bmatrix} 0 & -b_3 & b_2 \\ b_3 & 0 & -b_1 \\ -b_2 & b_1 & 0 \end{bmatrix}$

Hence, the mechanism constraint system is consisted of $\boldsymbol{\$}_{m1}^{r}$ and $\boldsymbol{\$}_{m2}^{r}$. The mechanism twist system reciprocal $\boldsymbol{\$}_{m1}^{r}$ and $\boldsymbol{\$}_{m2}^{r}$ to can be obtained

$$\begin{aligned} \mathbf{\$}_{m1} &= (0\,0\,1;0\,0\,0) \\ \mathbf{\$}_{m2} &= (0\,0\,0;1\,0\,0) \\ \mathbf{\$}_{m3} &= (0\,0\,0;0\,1\,0) \\ \mathbf{\$}_{m4} &= (a_1 - b_1 \;\; a_2 - b_2 \;\; 0;0\,0\,0) \end{aligned} \tag{25}$$

Clearly, $\mathbf{\$}_{m1}$ denotes a rotational DOF about the Z axis, $\mathbf{\$}_{m2}$ and $\mathbf{\$}_{m3}$ denotes a translational DOF along the X and Y axis, respectively. Because $\boldsymbol{N}_2\boldsymbol{N}_1 = (a_1 - b_1 \quad a_2 - b_2 \quad 0)^{\mathrm{T}}$, we know that $\mathbf{\$}_{m4}$ denotes a rotational DOF about the line $\boldsymbol{N}_2\boldsymbol{N}_1$. Thus, the 4- $^{z}R^{z}R$ $(^{i}R^{j}R^{k}R)_N$ PM depicted in figure 7 has two finite rotational DOFs, $2R^{zy}$, and two finite translational DOFs, $2T^{xy}$. It is noted that the rotational axis $\boldsymbol{N}_2\boldsymbol{N}_1$, which is changing in a finite motion, always remains parallel to the XY plane. So we use a superscript y to denote it for convenience.

On the basis of the above analysis, we know that there are no common constraints, namely, $\lambda = 0$. However, the four limb constraints are linearly dependent and span a 2-system. The redundant constraints $v = 4 \cdot 1 - 2 = 2$. Hence, we have

$$M = d(n - g - 1) + \sum_{i=1}^{g} f_i + v = 6(18 - 20 - 1) + 20 + 2 = 4 \tag{26}$$

Conclusions

This paper presents a systematic type synthesis of symmetrical 2R2T 4-DOF PMs with full-cycle mobility. It is shown that the constraint-synthesis based on screw theory is effective in both mobility analysis and type synthesis of 2R2T 4-DOF PMs. The constraint characteristics and structural geometrical conditions of 2R2T 4-DOF PMs are revealed. It is worth pointing out that there are two mechanism central points in the 2R2T 4-DOF PM. The line determined by the two mechanism central points is the axis of one rotational DOF. Thirty limb kinematic chains are enumerated. With these limb kinematic chains, thirty symmetrical 2R2T 4-DOF PMs are disclosed. The results presented in this paper are of importance to the use of 2R2T 4-DOF PMs and contribute to the mechanism theory.

Acknowledgments

The research work reported here is supported by the National Natural Science Foundation of China (Grant No. 50605055) and in part by National High-Tech Research and Development Program of China (863 Program) under Grant 2006AA10Z257.

References

Angeles, J. 1982. Spatial Kinematic Chains: Analysis-Synthesis-Optimization. New York: Springer-Verlag Press.

Angeles, J. 2002. The qualitative synthesis of parallel manipulators. *Proceedings of workshop on Fundamental Issues and Future Research Directions for Parallel Mechanisms and Manipulators*, Quebec, Canada, pp. 160–169.

Carricato, M., and Parenti-Castelli, V. 2002. Singularity-free fully isotropic translational parallel mechanisms. *International Journal of Robotics Research* 21(2):161–174.

Carricato, M. 2005. Fully isotropic four-degrees-of-freedom parallel mechanisms for Schoenflies motion. *International Journal of Robotics Research* 24(5): 397–414.

Clavel, R. Delta, a fast robot with parallel geometry, Proceedings of the Int. Symp. On Industrial Robot, Switzerland, 1988, pp. 91-100.

Di Gregorio, R. 2001. Kinematics of a new spherical parallel manipulator with three equal legs: the 3-urc wrist. *Journal of Robotic Systems* 18(5): 213-219.

Fang, Y. F., and Tsai, L. W. 2002. Structure synthesis of a class of 4-dof and 5-dof parallel manipulators with identical limb structures. *International Journal of Robotics Research* 21(9): 799–810.

Fang, Y. F., and Tsai, L. W. 2004. Structure synthesis of a class of 3-DOF rotational parallel manipulators. *IEEE Transaction on Robotics and Automation* 20(1): 117–121.

Frisoli, A., Checcacci, D., Salsedo, F., and Bergamasco, M. 2000. Synthesis by screw algebra of translating in-parallel actuated mechanisms. *Advances in Robot Kinematics*, Dordrecht: Kluwer Academic. pp. 433–440.

Gao F., Zhang Y., and Li, W. 2005. Type synthesis of 3-dof reducible translational mechanisms. *Robotica* 23(2): 239–245.

Gogu, G. 2004. Structural synthesis of fully-isotropic translational parallel robots via theory of linear transformations. *European Journal of Mechanics A/Solids* 23(6): 1021–1039.

Gogu, G. 2007. Structural synthesis of fully-isotropic parallel robots with Schönflies motions via theory of linear transformations and evolutionary morphology. *European Journal of Mechanics A/Solids* 26(2): 242–269.

Hervé, J. M. Analyse Structurelle Des Mécanismes Par Groupe Des Déplacements. Mechanism and Machine Theory, 1978, 13:437-450

Hervé, J. M. The Lie Group of Rigid Body Displacements, a Fundamental Tool for Mechanism Design. Mechanism and Machine Theory, 1999, 34(5):719-730

Hervé, J. M., Sparacino, F. Structural Synthesis of Parallel Robots Generating Spatial Translation. 5th IEEE Int. Conference on Advanced Robotics, 1991, *Pisa*:808-813

Huang, Z., Li, Q. C. 2001. A decoupled 5-DOF symmetrical parallel mechanism, Patent, China, No. ZL 01122274.3.

Huang, Z., Kong, L. F., and Fang, Y. F. 1997. *Mechanism Theory and Control of Parallel Manipulators.* Beijing: China Machinel Press.

Huang, Z., Li, Q. C. 2002. General methodology for type synthesis of lower-mobility symmetrical parallel manipulators and several novel manipulators. *International Journal of Robotics Research* 21(2): 131–146.

Huang, Z., and Li, Q. C. 2003. Type synthesis of symmetrical lower-mobility parallel mechanisms using constraint-synthesis method. *International Journal of Robotics Research* 22(1): 59–79.

Hunt, K. H. Structural kinematics of in parallel actuated robot arms, *J. of Mechanisms, Transmissions and Automation in Design* 105 (1983) 705-712.

Hunt, K.H. 1978. *Kinematic Geometry of Mechanisms*. Oxford: Oxford University Press.

Huynh, P., and Hervé, J. M. 2005a. Equivalent kinematic chains of 3-dof tripod mechanisms with planar-spherical bonds. *Journal of Mechanical Design.* 127(1): 95–102.

Jin, Q., and Yang, T. L. 2004. Theory for topology synthesis of parallel manipulators and its application to three-dimension-translation parallel manipulators. *Journal of Mechanical Design* 126(1): 625–639.

Karouia, M., Hervé, J. M. A Family of Novel Orientational 3-dof Parallel Robots. Ro. Man. Sy 2002, 14th CISM-IFToMM Symposium, Udine, Italy, 2002:359-368

Karouia, M., Hervé, J. M. 2005b. Asymmetrical 3-dof spherical parallel mechanisms. *European Journal of Mechanics A/Solids*, 24(1): 47–57.

Kong, X. W., Gosselin, C. M. 2002. Type synthesis of linear translational parallel manipulators. *Advances in Robot Kinematics*. Dordrecht: Kluwer Academic, pp. 453–462.

Kong, X. W., Gosselin, C. M. 2007. Type synthesis of 3-DOf linear translational parallel manipulators. IftoMM conference.

Kong, X. W., Gosselin, C. M. 2004a. Type synthesis of 3t1r 4-dof parallel manipulators based on screw theory. *IEEE Transactions on Robotics and Automation* 20(2):181–190.

Kong, X. W., Gosselin, C. M. 2004b. Type synthesis of three-degree-of-freedom spherical parallel manipulators. *International Journal of Robotics Research* 23(3): 237–245.

Lee, C. C., and Hervé, J. M. 2006. Translational parallel manipulators with doubly planar limbs. *Mechanism and Machine Theory* .24(4): 433–435.

Li, Q. C., Huang, Z., and Hervé, J. M. 2004. Type synthesis of 3r2t 5-dof parallel mechanisms using the lie group of displacements. *IEEE Transactions on Robotics and Automation*. 20(2): 173–180.

Liu, X. J., Wang J. S., and Pritschow, G. 2005. A new family of spatial 3-dof fully parallel manipulators with high rotational capability. *Mechanism and Machine Theory* 40(4):475–494.

Meng, J., Liu, G. F., and Li, Z. X. 2007. A geometric theory for synthesis and analysis of sub 6-DOF parallel manipulators. *IEEE Transactions on Robotics.* 23(4): 625–649.

Merlet, J. P. Still a Long Way to Go on the Road for Parallel Mechanisms. ASME Design Engineering Technical Conferences. Keynote speech, Montreal, 2002, http://www-sop.inria.fr/coprin/equipe/merlet/ASME/asme2002.html

Pierrot, F., Company, O. H4: a New Family of 4-dof Parallel Robots. IEEE/ASME International Conference on Advanced Intelligent Mechatronics, Atlanta, 1999:508-513.

Qinchuan Li, Zhen Huang, 2003, Type synthesis of 4-DOF parallel manipulators. In: Proc of 2003 IEEE International Conference on Robotics and Automation, Taiwan, P. R. China, pp. 755-760.

Refaat, S., Hervé, J. M., Nahavandi, S, and Trinh, H. 2007. Two-mode over-constrained three-dofs rotational-translational linear-motor-based parallel-kinematics mechanism for machine tool applications. *Robotica* 25(4): 461–466.

Rico J. M et al. A comprehensive theory of type synthesis of fully parallel platforms. *Proceedings of 2006 ASME Design Engineering Technical conferences,* Philadelphia, USA, DETC2006-99070.

Tsai, L. W. Kinematics of a Three-DoF Platform With Extensible Limbs. J. Lenarčič and V. Parenti-Castelli. Recent Advances in Robot Kinematics, Kluwer Academic Publishers, 1996:401-410

Tsai, L. W. The Enumeration of a Class of Three-Dof Parallel Manipulators. The Tenth World Congress on the Theory of Machines and Mechanisms, Finland, 1999:1123-1126

Zhao, Tieshi, Huang, Zhen. Theory and Application of Selecting Actuating Componenis of Spatial Parallel Mechanisms, *Chinese Journal of Mechanical Engineering*, 2000, 36(10): 81-85

Zlatanov, D., Gosselin, C. M. A New Parallel Architecture with Four Degrees of Freedom. The 2nd workshop on Computational Kinematics, Seoul, 2001:57-66

In: Parallel Robotics: Recent Advances in Research… ISBN: 978-1-60456-859-2
Editors: Jinsong Wang and Xin-Jun Liu, pp. 53-68

Chapter 3

THE DESIGN OF HYBRID MACHINES: A CASE STUDY OF 2-DOF PARALLEL KINEMATIC MECHANISMS

Weizhong Guo
State Key Lab of Mechanical System and Vibration,
Institute of Design and Control Engineering for Heavy Equipments,
Shanghai Jiao Tong University, Shanghai, 200240, P. R. China

ABSTRACT

A hybrid machine is able to be regarded as a special type of parallel kinematic mechanical system which utilizes characteristically different actuators simultaneously to drive a multi-degree-of-freedom (DOF) mechanism for a programmable output motion with a large power. The uncontrollable actuator(s) (e.g. a constant speed motor) is/are expected to provide the main part of the output motion with a large power, while the controllable actuator(s) (e.g. a small servomotor) is/are responsible for the adjustablility/flexibility of the output motion. In this chapter, the design is outlined for this type of machine by taking a new controllable mechanical press as example.

1. INTRODUCTION

A *hybrid machine* is a new type of modern mechanical system which has two or several independent characteristically different driving motors in terms of controllability, where the constant speed motor (CSM) provides a majority of the energy while the variable speed motor (VSM) tunes the performance of the machine during the operation. It was first proposed by L. C. Tokuz and Ree. Jones in 1991 [1, 2] as shown in figure 1.

The hybrid machine is one type of controllable mechanism/controlled mechanism. A controllable mechanism is computer-controlled but different from multi-degree-of-freedom (DOF) robots fully driven by programmed actuators such as servo motors and servo hydraulic systems. The aim of controllable mechanisms is to bridge the gap between traditional or unprogrammable machines and fully flexible/programmable machines such as industrial robots. From a mechanism point of view, controllable mechanisms refer to three types of mechanisms including adjustable mechanism, variable input speed mechanism and hybrid machine. An adjustable mechanism is a mechanism with adjustable link parameters offline. A variable input speed mechanism is a single-dof mechanism with programmable actuator. And a hybrid machine is a two-DOF mechanism with one CSM and one VSM.

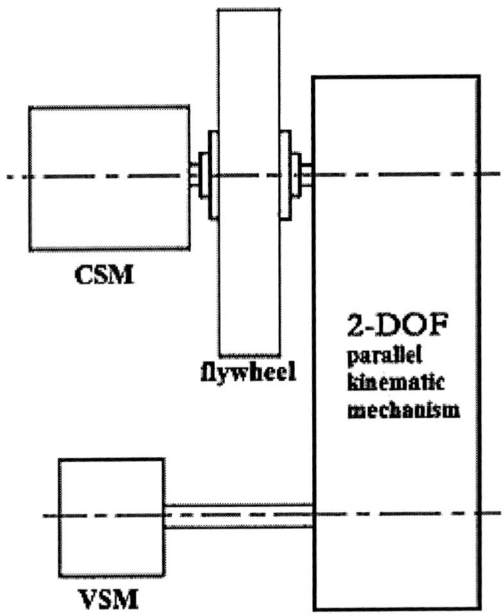

Figure 1. Schematics of a hybrid machine

One potential application of hybrid machines is in metal forming industry. In this chapter, a new design with 2-DOF parallel kinematic mechanism is introduced as an example of hybrid mechines. As well known, metal forming is one of the oldest production processes that can be traced back to the beginning of the Industry Revolution. Today, metal forming press is still one of the most commonly used manufacturing machineries [3]. Everyday, millions of parts are produced by metal forming ranging from battery caps to automotive body panels. Therefore, even a small improvement may add to significant corporative gain.

In the early days, metal forming presses used a simple slide crank mechanism that converted the rotating motion and the energy from the motor to the linear motion of the ram (also called the punch) to form the workpiece. To achieve the desirable performance, many different types of metal forming presses have been developed. In general, they can be divided into two types: mechanical presses and hydraulic presses. The former is fast (high speed presses may reach up to several thousand shots per minute) and energy efficient (the large flywheel eases the impulsive force), but is lack of flexibility. The later is flexible (their motions, including the travel and the velocity, can be programmed), but is expensive to build and to operate. This is because the hydraulic system is expensive and must be constantly pressurized during the operation with or without the metal forming work. Recently, there are programmable mechanical presses driven by servomotors. Unfortunately, they are even more expensive to build because of the special requirement on the servomotor (large torque at low speed). Also, it is known that metal forming process is characterized by large instantaneous force and hence, a very large servomotor is necessary which is difficult to make.

A comprehensive work is done in the handbook by Lange in 1985 [4], which covers practically all aspects of metal forming, including the press design. However, few updates have been found since. To improve machine performance, a nine-bar linkage mechanism is proposed, which is composed of a five-bar linkage, a slider-crank four-bar linkage and a connecting link [5]. A design procedure to optimize the design parameters of the linkage is also included.

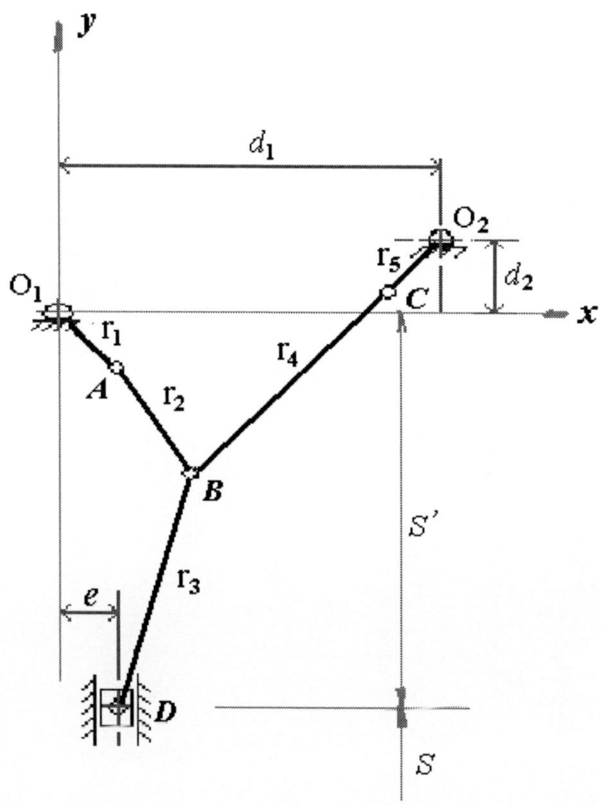

Figure 2. A new design for the mechanical press [Du and Guo, 2003].

A new design [6-9] is presented as shown in figure 2 based on hybrid machine with a 2-DOF pin-jointed seven-bar linkage driven by one CSM and one VSM simultaneously. The computer simulation indicates that the new press has a number of features:

a) It is controllable: its motion trajectory and the punch force can be programmed just like a hydraulic press;
b) It is energy efficient: with a flywheel the large CSM carries 80% of the load and hence, it is energy efficient just like a conventional mechanical press;
c) It is fast: its speed is only limited by the mechanical motion, just like a conventional mechanical press;
d) It is easy to build: it only adds few mechanical joints, a servomotor and a controller, and

e) It is accurate: the feedback control ensures the precision.

However, the design is still far away from the application. According to experiments, the power requirement of the VSM side is of the same numerical level of that of the CSM side due to a flywheel attached to the CSM side. To lower the power requirement of the VSM side, an updated design based on the screw drive is presented in this chapter as shown in figure 3.

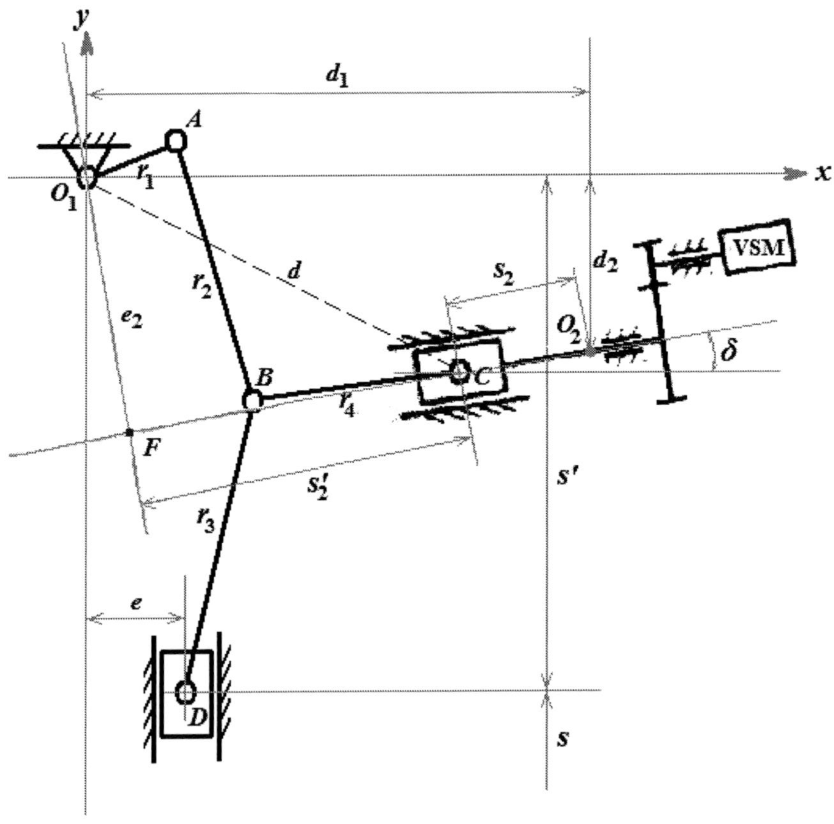

Figure 3. Illustration of the updated design with screw drive.

This chapter consists of four sections. Section 2 presents the design including the configuration, the feasibility conditions, the kinematics model, and a study on the mechanical advantage, torque and power distribution. Section 3 presents an optimized design using Genetic Algorithm (GA). Finally, Section 4 contains the conclusions.

2. The Design

The Configuration

The updated new design is illustrated in figure 3. It is composed of seven links, the driving crank O_1A, Link AB, the driving slide C, Link CB, Link BD, the ram D and the anvil.

The side crank O_1A is driven by a CSM, which is expected to provide a majority of the energy, and the slide C is actuated through a screw drive by a VSM, which tunes the performance of the machine with low power requirement. As shown in figure 3, this design has a total of seven parameters, r_1, r_2, r_3, r_4, e_2, δ and e, where the first four parameters are structure lengths and the last three are assembly parameters with e_2 the distance from O_1 to the guide line of the driving slide C and δ the slide orientation angle from the positive x axis in counterclockwise direction. In order to facilitate the modeling, a coordinate system is setup following the right-hand rule with its origin placed on the center of the driving crank and y axis parallel with the guide line of the ram. A reference point O_2 with (d_1, d_2) is chosen arbitrarily on the guide line of the driving slide for measuring the driving stroke. Obviously, $e_2 = |d_1 \sin\delta - d_2 \cos\delta|$.

Feasibility Conditions

In order to have the new machine move freely, a couple of feasibility conditions must be met. These include the assembly condition and the crank-slide condition. Following the feasibility conditions, the reachable range of the ram can be derived.

(a) Assembly conditions. From a mechanism point of view, as shown in figure 3, the new machine can be divided into two parts: the first part consists of 4R1P-type five-bars: O_1A, AB, the driving slide C, CB and the anvil O_1O_2. The second part is made of Link BD, the ram (the slide), and the ram guidance track, which is referred to as an Assur's RRP group of zero-degree-of-freedom [10]. The first part must satisfy the following assembly condition [11, 12]:

$$e_2 \le r_1 + r_2 + r_4 \tag{1}$$

and

$$e_2 \ge 0 \tag{2}$$

On the other hand, the second part should be able to connect the first part at the Joint B without interference. This converts to the assembly constraint expressed as:

$$r_3 \ge \max(dist) \tag{3}$$

where, *dist* is the distance between Joint B and the central line of the ram guidance track (see figure 3). For simplicity the assembly constraint can be further expressed in sufficient condition as follows:

$$r_3 \ge r_1 + r_2 + e \tag{3'}$$

(b) Crank-slide condition.. In order for the machine to move freely, O_1A must be a crank wherever the driving slide goes. On the other hand, the link *BC* is expected to oscillate. Therefore, the driving slide should be limited to a special motion range and the 4R1P-type five-bar Grashof criteria [11, 12] can be applied. For the five-bar part, the crank-slide conditions are as follows [11, 12]:

$$r_1 = r_{\min} \tag{4.a}$$

$$r_{\min} < r_{\mathrm{mid}} \le r_{\max} \tag{4.b}$$

$$e_2 < r_{\max} + r_{\mathrm{mid}} - r_{\min} \tag{4.c}$$

$$s_{2\min} < s_2 < s_{2\max} \tag{4.d}$$

where, $r_{\min}$, r_{mid}, and $r_{\max}$ refer to the shortest, intermediate and longest links among r_1, r_2, and r_4 respectively, s_2 denotes the algebraic driving travel of the driving slide *C* from O_2 as shown in figure 3, and $s_{2\max}$ as well as $s_{2\min}$ imply the maximum and minimum values of s_2 respectively as follows:

$$s_{2\max} = d_1 \cos\delta + d_2 \sin\delta + \sqrt{(r_{\max} + r_{\mathrm{mid}} - r_{\min})^2 - e_2^2}\text{, if } e_2 > r_{\max} - r_{\mathrm{mid}} + r_{\min}\text{, or}$$

$$s_{2\max} = d_1 \cos\delta + d_2 \sin\delta - \sqrt{(r_{\max} - r_{\mathrm{mid}} + r_{\min})^2 - e_2^2}\text{, if } e_2 \le r_{\max} - r_{\mathrm{mid}} + r_{\min}\text{, and}$$

$$s_{2\min} = d_1 \cos\delta + d_2 \sin\delta - \sqrt{(r_{\max} + r_{\mathrm{mid}} - r_{\min})^2 - e_2^2}\ .$$

The characteristics chart for the five-bar part is depicted in figure 4 [11, 12]. It has six boundaries, including one void boundary A_1, two Grashof boundaries C_2 and C_3, as well as three structure boundaries B_1, B_2 and B_3. The slide *C* cannot arrive at the void region surrounded by A_1 whatever e_2 be. The driving link O_1A is an *unconditioned* crank when the driving slide moves within the green Grashof region between C_2 and C_3. When the driving slide enters the red region between A_1 and C_2 or C_3 and B_3, there will be no crank. Apparently, the green Grashof region should be the best choice for our design.

(c) Reachable range. Due to the two degrees-of-freedom, the new press possesses a workspace of the ram which implies performance flexibility. Note that the ram moves along the guidance track and the reachable range of the ram is the controllable length of the ram's stroke. As shown in figure 3, this reachable range is decided by Point *B*, which is the joint between the RRR group of O_1A and *AB* as well as the RRP group of the driving slide *C* and Link *CB*. Each group can be considered as an open chained binary-bar linkage. Thus, the feasible area of Joint *B* is the area in which the two groups intersect. As shown in figure 5, the feasible region of the RRR group is the ring described by radii $|r_1 - r_2|$ and $(r_1 + r_2)$, and the one of the RRP group is the volume enveloped by the circle of radium r_4 along the guide line

of the driving slide, i.e. between the two boundary lines C_1 and C_1', respectively. The intersection of the two volumes is shown as the shaded areas in the figure. Furthermore, the driving slide should have a driving angle, α less than the permitted pressure angle value, $[\alpha]$. This can be kept by restricting the motion of Joint B within the two boundary lines C_2 and C_2'. Therefore, the latticed part of the shaded area and the corresponding driving slide motion are the good choice.

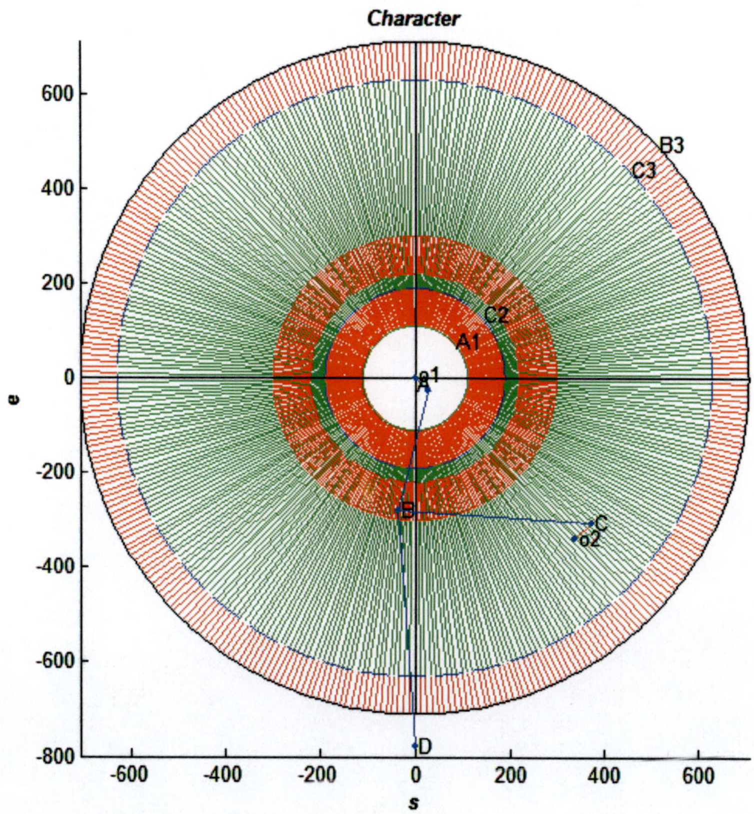

Figure 4. The Characteristics Chart for the Five-Bar Part.

Following the motion of Joint B, the reachable range of ram (i.e., the point D in figure 6) can be determined. Since the length of link BD, r_3, is fixed, we can image that a circle with a radius r_3 is attached to the joint D as it travels. As shown in figure 6, the circle will move along the axis of the ram. Unfortunately, there are no analytical formulas for both the highest intersection point (also called the top dead center, TDC), D_{top}, and the lowest intersection point (also called the bottom dead center, BDC), D_{down}. Nevertheless, they can be found numerically without much difficulty. Furthermore, the driven ram should have a driven angle, β within a reasonable range, i.e. less than the permitted pressure angle value, $[\beta]$. This can be kept by restricting the motion of Joint B within the two boundary lines D_1 and D_1'. Therefore, the latticed part of the shaded area and the corresponding driving slide motion are the good choice.

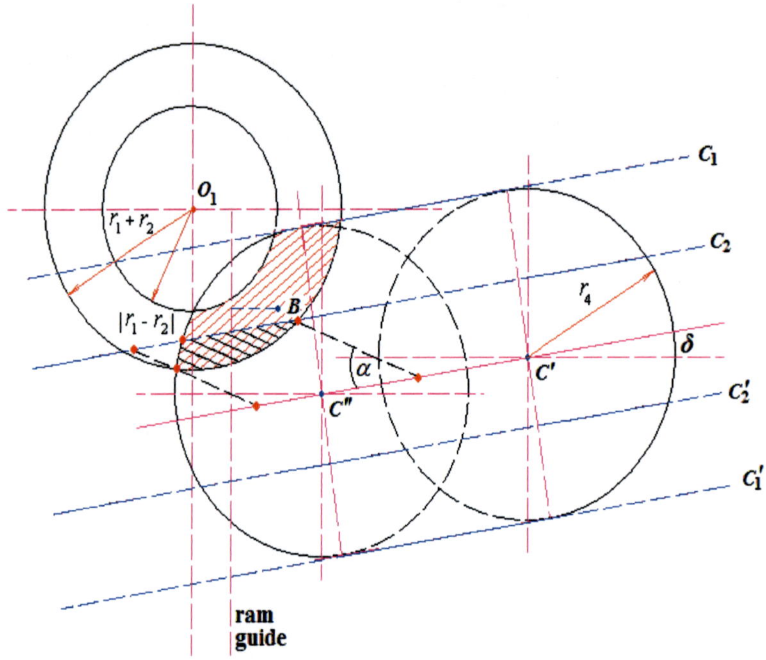

Figure 5. Illustration of the reachable area of Joint B (in shaded area).

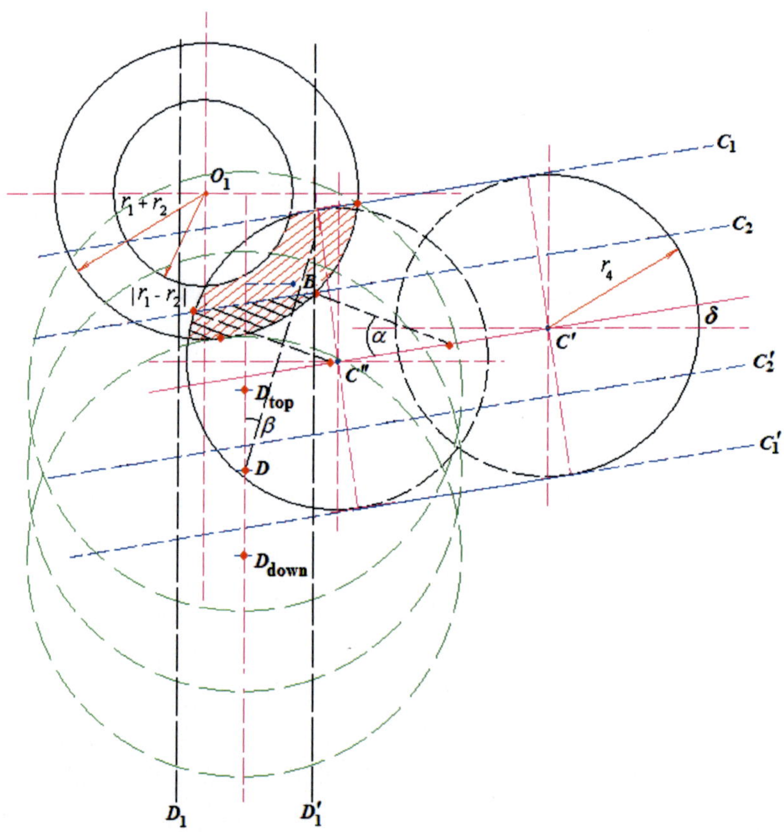

Figure 6. Illustration of the reachable range of the ram (Point D).

Kinematical Model

To derive the kinematical model of the machine, the Assur's Group method [10] is used. Its basic idea is divide-and-conquer. First, the machine is divided into four groups: Group 1 consists of the anvil of the machine and Link O_1A, Group 2 consists of the anvil of the machine and the slide *C*, Group 3 is made of Link *AB* and Link *CB*, and Group 4 is made of Link *BD* and the ram. Next, the kinematical model for each group is derived. Finally, by combining them together, the kinematical model of the machine is found.

(a) Group 1: From figure 3, it is seen that the position of the joint point *A* is:

$$\begin{cases} x_1 = r_1 \cos\theta_1 \\ y_1 = r_1 \sin\theta_1 \end{cases} \tag{5}$$

where, θ_1 is the angular position of the CSM.

(b) Group 2: Similar to Group 1, the position of the joint point *C* is as follows:

$$\begin{cases} x_2 = d_1 + s_2 \cos\delta \\ y_2 = d_2 + s_2 \sin\delta \end{cases} \tag{6}$$

with $s_2 = s_{20} + p\dfrac{\theta_2}{2\pi}$, where s_{20} is the initial displacement and p is the screw lead with $p = \pi D \tan\gamma$.

(b) Group 3: The joint point *B* is the critical point of the machine. From figure 3, it is seen that the position of *B* of the RRR group is determined by:

$$\begin{cases} x_B = x_1 + r_2 \cos\varphi_1 \\ y_B = y_1 + r_2 \sin\varphi_1 \end{cases} \tag{7}$$

where, φ_1 is the angle between the link *AB* and the positive *x* axis and it is defined below:

$$\varphi_1 = 2\tan^{-1}\frac{K \pm \sqrt{K^2 - L^2 + M^2}}{L - M}$$

where, $K = 2r_2(y_2 - y_1)$, $L = 2r_2(x_2 - x_1)$, and $M = r_2^2 + (x_2 - x_1)^2 + (y_2 - y_1)^2 - r_4^2$.

(d) Group 4: This RRP group describes the motion of the ram. Following the configuration in figure 3, assuming that the ram reaches the BDC at $s'_{\min}$, then the ram position is determined by

$$s = s' - s'_{\min} \tag{8}$$

where, $s'_{\min} = \min(s')$, and $s' = y_B - \sqrt{r_3^2 - (e - x_B)^2}$.

Equation (8) defines the displacement of the ram. By single and double differentiations with respect to time respectively, the velocity and acceleration of the ram can be found as well.

The Mechanical Advantage, Torque Distribution and Power Distribution

The new machine has two distinct advantages. First, it is flexible, which has been shown in Section 2. Second, it is energy efficient. This is achieved by distributing the majority of the energy to the CSM, which carries a flywheel to ease the large impulsive metal forming force. The VSM, on the other hand, tunes the motion, velocity and acceleration of the press but takes only a small load. To show this, it is necessary to study the mechanical advantage, torque distribution and power distribution.

(a) Mechanical advantage with respect to the CSM-driven crank. According to the static mechanics, the mechanical advantage of the machine with respect to the CSM-driven crank (i.e. the proportion of the output force F of the machine over the torque M_1 of the CSM-driven crank) can be derived as shown below:

$$\frac{F}{M_1} = \frac{\sin(\pi - \gamma_1)}{\sin(\pi - \gamma_2)} \cdot \frac{\cos \gamma_4}{r_1 \sin(\pi - \gamma_3)} \tag{9}$$

where, γ_1 is the angle between Link BC and Link BA, γ_2 is the angle between Link BD and Link BC, γ_3 is the angle between Link AB and Link AO_1, and γ_4 is the angle between Link DB and the guidance of the ram. Note that they are all known since the instantaneous positions of the links can be found as shown in Section 2.

(b) Mechanical advantage with respect to the VSM-driven slide. The mechanical advantage with respect to the driving slide, i.e. the proportion of the output force of the machine over the driving force F_2 of the driving slide is:

$$\frac{F}{F_2} = \frac{\sin(\pi - \gamma_1)}{\sin(\pi - \gamma_5)} \cdot \frac{\cos \gamma_4}{\cos(\pi - \gamma_6)} \tag{10}$$

where, γ_5 is the angle between the link *BA* and the link *BD*, and γ_6 the one between the slide guide link CO_2 and Link *CB* counterclockwise. They are both known.

The driving slide is driven by a VSM through a gear reducer and screw transmission. The driving torque of the screw axis is as follows [13]:

$$M_2 = iF_2 \tag{11}$$

where, $i = \frac{d_2}{2}\tan(\varphi - \rho')$ denotes the screw transmission ratio with φ the screw angle and ρ' the friction angle.

From Equations 10 and 11 we have

$$\frac{F}{M_2} = \frac{1}{i}\frac{\sin(\pi - \gamma_1)}{\sin(\pi - \gamma_5)} \cdot \frac{\cos\gamma_4}{\cos(\pi - \gamma_6)} \tag{12}$$

(c) Mechanical advantage of Hybrid Press. Integrating Equations (9) and (12), we can get the mechanical advantage for the hybrid press represented as

$$\begin{aligned} \frac{F}{M} &= \frac{F}{M_1 + M_2} \\ &= \frac{\sin(\pi - \gamma_1)\cos\gamma_4}{\sin(\pi - \gamma_2)\cdot r_1 \sin(\pi - \gamma_3) + i\sin(\pi - \gamma_5)\cdot\cos(\pi - \gamma_6)} \end{aligned} \tag{13}$$

From Equation (13), it is seen the mechanical advantage of the machine is a function of its structure parameters as well as the angular positions of the motors.

(d) Moment distributions for CSM and VSM. By rearranging Equations (9) and (12), the moment requirements of the CSM and the VSM can be easily found. Neglecting the influences of the friction, damping, *et al*, it is seen that the moments of both the CSM and VSM are functions of the workload, the structures, and the angular positions of the motors.

(e) Power distribution for CSM. The power distributed to the CSM-driven crank is represented as follows:

$$P_1 = M_1\dot{\theta}_1 \tag{14}$$

where, $\dot{\theta}_1$ is the velocity of the CSM-driven crank and it is constant.

(f) Power distribution for VSM. The power distributed to the VSM-driven screw is expressed by

$$P_2 = M_2\dot{\theta}_2 = M_2\frac{d\theta_2}{d\theta_1}\dot{\theta}_1 \tag{15}$$

where, $\dot{\theta}_2$ is the angular velocity of the VSM-driven crank that is controllable.

3. Parameter Optimization Using Genetic Algorithm

From the simulation results presented in the previous section it is seen that the structure parameters have a significant influence onto the performance of the machine. Therefore, it is necessary to optimize these parameters.

The Optimization Model

The optimization parameters can be denoted as a vector:

$$X = (r_1, r_2, r_3, r_4, e_2, \delta, e) \tag{16}$$

The objective function is defined as follows:

$$F = \varpi_1\frac{\max(M_1)}{\max(M_2)} + \varpi_2(1/\Phi) + \varpi_3(1/\min(SA)) \tag{17}$$

where, ϖ_1, ϖ_2, and ϖ_3 are weighting factors. Obviously, it is desirable to have the moment of the CSM much bigger than that of the VSM. Also, Φ is the maximal rotational angle of the CSM when the ram is close to the dead point D_{down}. This term favors the long dwelling time. Finally, *SA* is the range of the ram travel, which describes the adjustability of the stroke as show in figure 7. In the figure, the top curve represents the highest possible position of the ram, and the bottom curve represents the lowest position of the ram. The optimization constraints include the assembly conditions and the double crank condition given in Section 2.

Clearly, the optimization model is highly nonlinear. Therefore, it is necessary to use the Genetic Algorithm (GA) [6].

Case Studies

The optimization studies were carried out. The first one is to minimize the moments of the VSM without considering the dwelling angle (i.e., $\varpi_2 = 0$) and the ram travel range (i.e.,

$\varpi_3 = 0$). The second one is to minimize the moments of the VSM as well as the dwelling angle and the ram travel range. The optimization results are shown in Table 2.

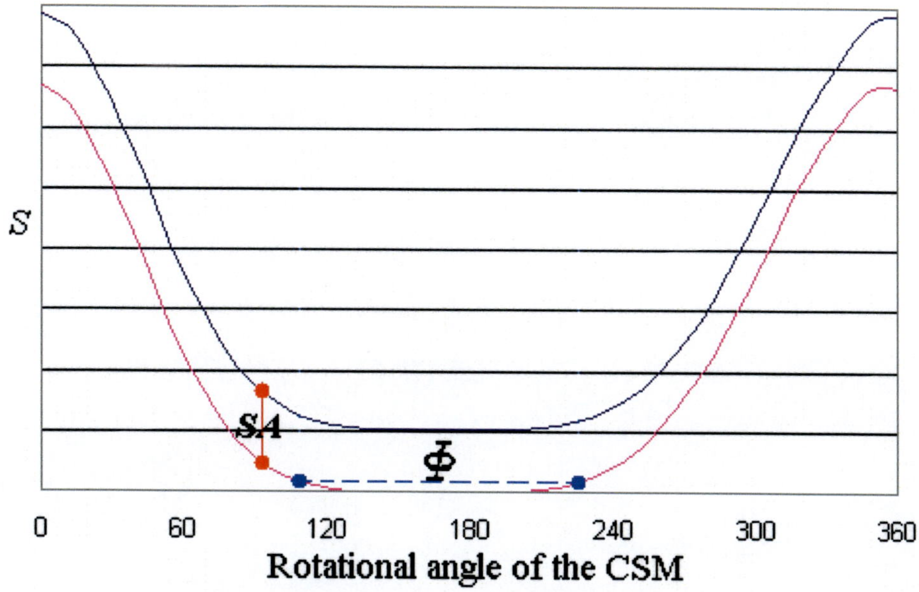

Figure 7. Illustration of the optimization parameters Φ and *SA*.

Table 2. Optimized Design Parameters (in mm)

	r_1	r_2	r_3	r_4	e_2	δ	e
Case 1	85.0	229.8	878.6	103.0	145.9	-20.2	163.5
Case 2	13.3	61.0	513.9	324.8	232.0	-39.6	-84.9

Figures 8(a) and (b) show the simulation results. Note that each figure has three graphs. The one on the left represents the feasible region of the ram travel as a function of the angular positions of the CSM and the VSM. The contour lines are shown in the bottom. The one in the middle shows the feasible ram travel against the angular position of the CSM. In particularly, the bottom curve represents the lowest reach of the ram, while the top curve represents the highest reach of the ram. The graph on the right represents the feasible region of Joint *B*. Note that in these figures, a dark line is used to describe the performance of the machine with a specific set of the CSM and VSM angles. Furthermore, the mark "+" represents the beginning.

From figure 8, it is seen that the ram travel range in Case 2 is much larger than that of Case 1. Hence, in the machine design, it is necessary to balance different factors, which can be done by adjusting the weighting factors in Equation (17).

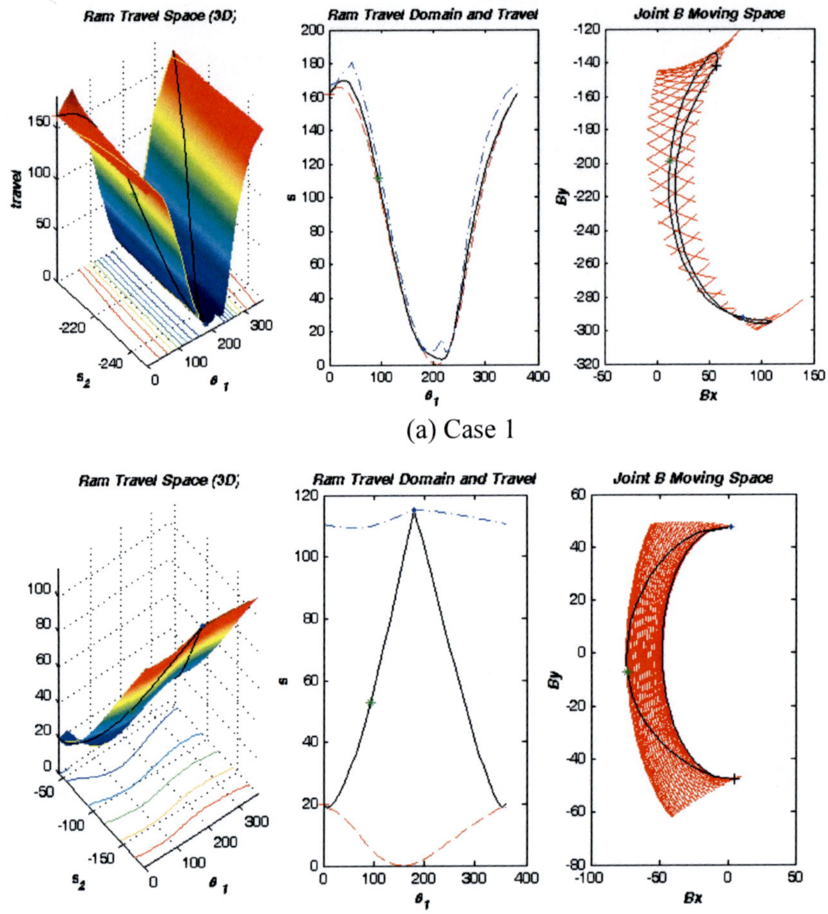

(a) Case 1

(b) Case 2

Figure 8. The travel of the optimized designs.

4. CONCLUSIONS

Hybrid machines are a special type of parallel robotic mechanism with one un-controllable actuator. This chapter discusses the design of such mechanical system by taking a new metal forming press as example. Based on the discussions above, following conclusions can be drawn:

a) The new machine based on hybrid machines can be easily adjusted to produce different stroke. In other words, it can behave like a hydraulic press or a servomotor driven press, though it is much cheaper to build and to operate;
b) Compared with the hydraulic presses or servomotor driven presses, the new machine is very efficient in terms of energy usage which is one potential advantage of hybrid machines. This is because it can use a large constant speed motor with a flywheel to produce majority of the force, and at the same time use a small servomotor to tune the performance;

c) The design shall be optimized, which provides the best performance balancing the kinematical flexibility and the energy consumption.

ACKNOWLEDGMENTS

The author thanks the partial financial supports under the projects from the National Natural Science Foundation of China (Grant No. 50405017), the National High-Tec Research and Development Plan of China (Grant No. 2006AA04Z118), the Young Scientist Fund of Higher Education of Shanghai Education Committee (Grant No. 01QN10), and the National Basic Research Program of China (Grant No. 2006CB705402), as well as a fund from the Chinese University of Hong Kong (Order No. 5100002646).

REFERENCES

[1] Tokuz, L. C. and Jones, J. R., 1991, *"Programmable Modulation of Motion Using Hybrid Machines,"* Proceedings of IMECHE, C413/071, pp. 85-91

[2] Tokuz, L. C., 1992, Hybrid Machine Modeling and Control, Ph.D. Dissertation, Liverpool Polytechnic University.

[3] Wagener, H. -W., 1997, "New Developments in Sheet Metal Forming: Sheet Materials, Tools and Machinery," *Journal of Materials Processing Technology*, Vol. 72, pp. 342-357.

[4] Lange, Kurt, 1985, *Handbook of Metal Forming*, SME Press.

[5] Tso, P. L., Liang, K. C., 2002, "A Nine-bar Linkage for Mechanical Forming Presses," *International Journal of Machine Tools and Manufacture*, Vol. 42, No.1, pp. 139 – 135.

[6] Du, R. and Guo, W. Z., 2003, "The Design of a New Metal Forming Press with Controllable Mechanism," *ASME Journal of Mechanical Design*, Vol. 125, No. 3, pp. 582-592.

[7] W. Z. Guo, K. He, K. Yeung and R. Du. A New Type of Controllable Mechanical Press: Motion Control and Experiment Validation. *ASME Journal of Manufacturing Science and Engineering,* 2005, 127(4): 731-742

[8] Guo, W. Z. and Du, R., 2003, "Trajectory Planning of a New Mechanical Press with Controllable Mechanism," *Proceedings of the IASTED, International Conference on Robotics and Applications*, Salzburg, Austria: 115-121

[9] Yeung, K., Guo, W. Z. and Du, R., 2003, "The Prototype of a New Metal Forming Press with Controllable Mechanism," *CIRP 2nd International Conference on Reconfigurable Manufacturing,* Ann Arbor, USA, 2003.

[10] Liang, C. G. and Ruan, P. S. 1986, *Computer-Aided Design of Linkages*, Machine Press, Beijing. (in Chinese)

[11] Guo, W. Z., Zou, H. J., Han, B., Zhang, Q. 2004, "Mobility of 4R1P-type Five-Bars Using Characteristics Chart", Proceedings of the *ASME International* Design Engineering Technical Conferences and Computers and Information in Engineering Conference, DETC2004-57236, pp. 627-634, Sept. 28 - Oct. 2, 2004, Salt Lake City, Utah, USA

[12] Guo, W. Z., Huang, Q. G., etc., 2005, "Grashof Criteria for 4R1P-type Five-bar Planar Parallel Mechanisms with Active/Passive Prismatic Joint," *Chinese Journal of Mechanical Engineering,* 41(8), pp. 66-73 (in Chinese)

[13] Yang, K. Z. and Chen, G. Y., 1999, Foundation of Machine Design (Fourth Edition), Higher Education Press, Beijing. (in Chinese)

In: Parallel Robotics: Recent Advances in Research... ISBN: 978-1-60456-859-2
Editors: Jinsong Wang and Xin-Jun Liu, pp. 69-94

Chapter 4

A FAMILY OF PARALLEL MANIPULATORS WITH MOBILE PLATFORM ROTATING CONTINUOUSLY

Jianjun Zhang
Parallel Manipulator Research Center, School of Mechanical Engineering,
Hebei University of Technology, P.R. China

1. INTRODUCTION

Parallel manipulators are always welcome in conditions where large payload capability, high stiffness and fine precision are needed. They have been studied intensively for more than a couple of decades and still attract attention from scholars of academe and industry [1-5]. To work in more complicated situations, many parallel manipulators with specified numbers and types of DOF (degree of freedom) have been proposed, and many researchers focused on the type synthesis of parallel manipulators. Several approaches for the type synthesis of parallel manipulators were also developed, such as the method based on the screw theory [6], the method based on the displacement group theory [7] and the special motion coordinate technique [8, 9].

However, the workspace of the parallel manipulators is relatively smaller than the serial manipulators. This results in lower translational or rotational capability and larger equipment fixing space. Many researchers have focused on new methods to avoid these disadvantages. Liu, Wang and Pritschow proposed a new family of spatial 3-DoF fully-parallel manipulators with high rotational capability [10]. Merlet's research showed that the joints' layout could have large influence on the workspace volume for robots [11]. Many serial-parallel manipulators were developed to achieve larger workspace [12, 13]. Also, Kim proposed an Eclipse-II parallel manipulator that used a circular leader rail to achieve the continuous 360-degree spinning [14]. Except the Eclipse-II, parallel manipulators with the mobile platform rotating continuously are seldom.

In this chapter, we will discuss three types of parallel manipulators with special topology structures and special limbs to make their mobile platforms rotate continuously. The first types of parallel manipulators possess redundant actuators and multi-actuators in each limb (at least in one limb). The second types of parallel manipulators possess a special limb, a RUPU (R-revolute joint, U-universal joint, and P-prismatic joint) limb or a RP^+ (P^+-two P joints configured in vertical directions) limb. The third types of parallel manipulators are those where their limbs are embedded together. Mechanisms of these parallel manipulators are described respectively, and some topology mechanisms of these types of parallel manipulators are given. Moreover, their kinematic equations are developed to analyze their input-output characteristics.

2. Parallel Manipulators with Redundant Actuators and Multi-Actuators in Their Limbs

2.1 Structure Description

For a planar four-bar mechanism shown in figure 1, by changing the length of the four bars, we can get several types of mechanisms with different kinematic characteristics, such as double-crank mechanisms, double-rocker mechanisms, crank-rocker mechanisms, and so on. In a double-rocker mechanism, the first rocker driven by a motor and the second passive rocker cannot rotate in a circular domain, as shown in figure 2(a). Theoretically, the link can rotate continuously driven by the first rocker in a peculiar trajectory. However, a dead point, shown in figure 2(b), would appear to prevent the link from rotating smoothly where the link can rotate clockwise or counterclockwise. If the second rocker is also driven by a motor, the rotation direction of the link can be determined. This 1-DOF manipulator is driven by two actuators, so-called redundant driving.

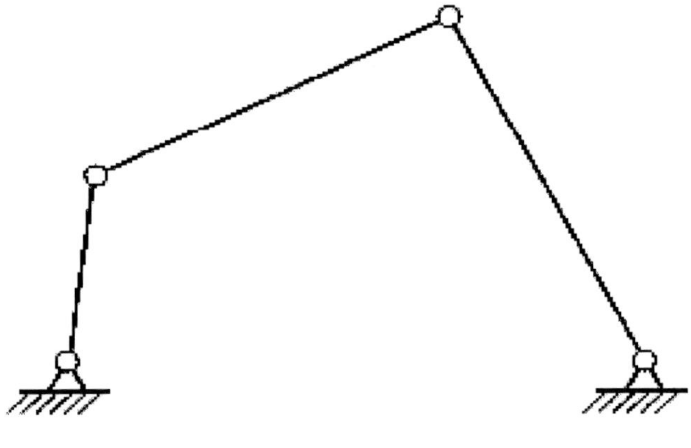

Figure 1. Planar four-bar mechanism.

Accordingly, we would discuss some parallel manipulators that involve the double rocker mechanism with redundant driving. A new type of parallel manipulator that is named as ***JLZN*** by its outline shape in three directions is illustrated in figure 3.

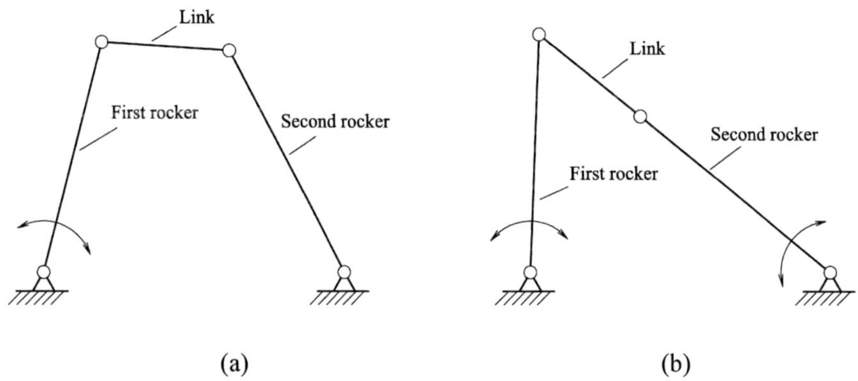

Figure 2. Double-rocker mechanism.

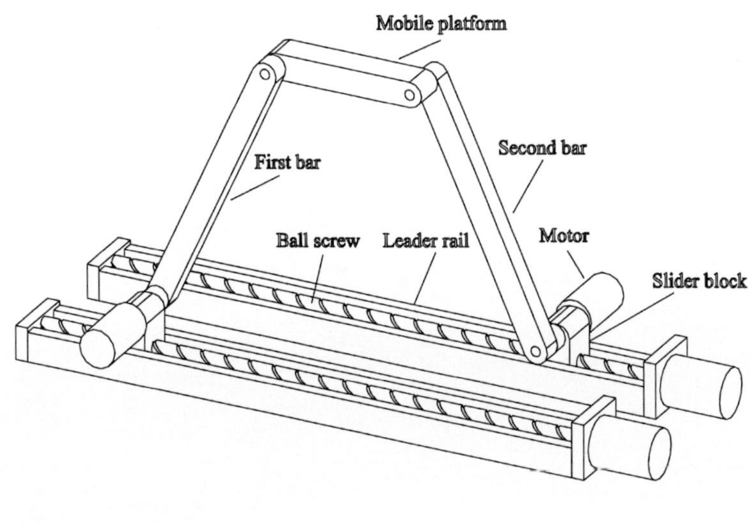

Figure 3. ***JLZN*** parallel manipulator.

The ***JLZN*** parallel manipulator includes a planar four-bar mechanism where the two bars with the same length are driven by two motors, respectively. Driven by two other motors and two ball screws, the two-slider blocks, where the two motors are fixed on, can translate along two leader rails, respectively. Therefore, the distance between the two rotational axes of the two bars is not fixed but varies with the translation of the two slider blocks. At the same time, the link of the original planar mechanism becomes the mobile platform of the ***JLZN*** parallel manipulator.

It is obvious that the ***JLZN*** is a parallel manipulator with two limbs, each of which possesses a P joint and two R joints. It is different from the traditional parallel manipulators that two joints in each limb are actuated. Since this manipulator is a planar 3-DOF mechanism with one rotational and two translational motions, one of the actuation is redundant of the four. The objective of a redundant actuator is to avoid the dead point in some positions or orientations. In a word, the ***JLZN*** parallel manipulator is different from the traditional parallel mechanisms in two aspects that it possesses redundant actuation and its limbs (at least one of all its limbs) possess more than one actuator. The peculiarities of the ***JLZN*** make the kinematics and workspace change tremendously. Thus, its mobile platform can rotate continuously at some positions, and possesses higher rotational ability. Its area of workspace would, also, increase enormously.

2.2. Kinematics of the *JLZN* Parallel Manipulator

2.2.1. Inverse Kinematics

The inverse kinematics problem involves mapping a known pose (position and orientation) of the mobile platform of the manipulator to a set of input joint variables that will achieve that pose. A kinematic model of the manipulator is developed as shown in figure 4. P_i (i=1, 2) represent the positions of the two R joints connected with the mobile platform. B_i (i=1, 2) represent the positions of the two slider blocks. A fixed coordinate system, $O-XY$ is built and B_i (i=1, 2) are located on the X-axis. A mobile coordinate system $O'-X'Y'$ is fixed on the mobile platform, where the origin O' is located at the mid-point of P_1P_2 and the Y'-axis is in the same direction of P_1P_2. The length of the two legs is represents as l, where $B_iP_i = l$ (i=1, 2). The length of the mobile platform is $2a$, i.e. $P_1P_2 = 2a$.

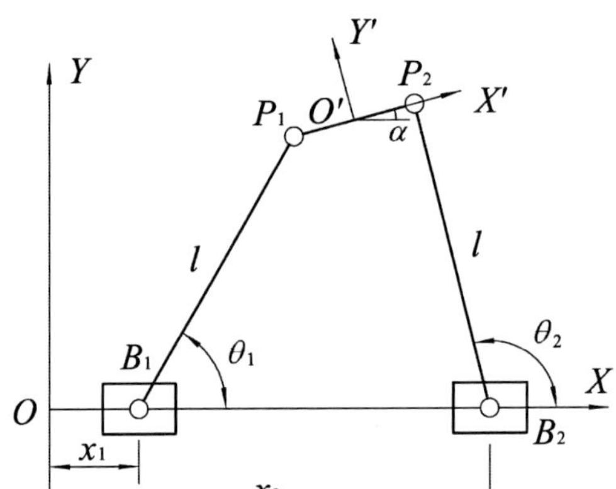

Figure 4. Mechanism sketch of the ***JLZN*** parallel manipulator.

For the inverse kinematics analysis, the pose of the mobile platform is considered as known. The position of the mobile platform is given by a position vector $\boldsymbol{OO'}$, and the orientation of the mobile platform is given by a matrix $\boldsymbol{R}$, as

$$\boldsymbol{OO'} = (x \quad y)^{\mathrm{T}} \tag{1}$$

$$\boldsymbol{R} = \begin{pmatrix} \cos\alpha & -\sin\alpha \\ \sin\alpha & \cos\alpha \end{pmatrix} \tag{2}$$

Here, the angle α is the rotational DOF of the mobile platform. The inverse kinematics of the parallel manipulator can be solved by writing the constraint equation as

$$\| \boldsymbol{OP}_i - \boldsymbol{OB}_i \| = l \quad (i=1, 2) \tag{3}$$

where $\boldsymbol{OP}_i$ and $\boldsymbol{OB}_i$ are position vectors of the points B_i and P_i in O - XY, respectively. Position vector of the point P_i in O' - $X'Y'$ is represented as $(\boldsymbol{O'P}_i)'$, so we can get

$$\boldsymbol{OP}_i = \boldsymbol{OO'} + \boldsymbol{R} \cdot (\boldsymbol{O'P}_i)' \quad (i=1, 2) \tag{4}$$

For a given manipulator and the prescribed values of the position and orientation of the mobile platform, the required actuator inputs can be directly derived from Eq. (3), that is

$$x_1 = \pm\sqrt{l^2 - (y - a\sin\alpha)^2} + x - a\cos\alpha \tag{5}$$

$$x_2 = \pm\sqrt{l^2 - (y + a\sin\alpha)^2} + x + a\cos\alpha \tag{6}$$

where x_i (i=1, 2) represents the position of B_i in the X-axis. In Eqs. (5) and (6), we can see that there are four inverse kinematic solutions for a given pose of the parallel manipulator, as shown in figure 5.

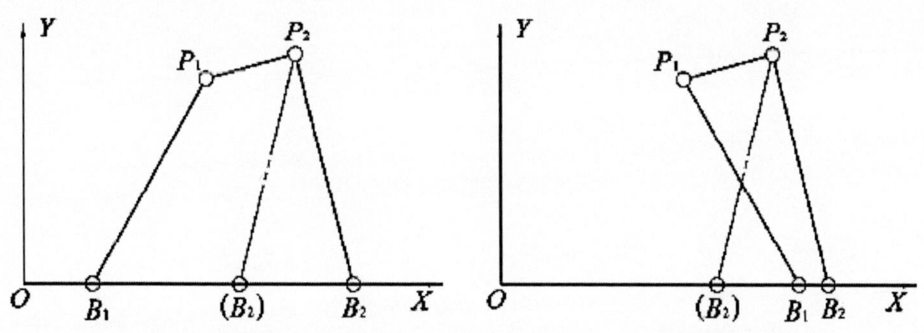

Figure 5. Four inverse kinematic solutions for a given pose of the parallel manipulator.

Based on Eqs. (5) and (6), the two other rotational inputs θ_1 and θ_2 can be solved. Eq. (3) can be rewritten as.

$$\begin{pmatrix} l\cos\theta_1 \\ l\sin\theta_1 \end{pmatrix} = \begin{pmatrix} x \\ y \end{pmatrix} + \boldsymbol{R}\begin{pmatrix} -a \\ 0 \end{pmatrix} - \begin{pmatrix} x_1 \\ 0 \end{pmatrix} \tag{7}$$

$$\begin{pmatrix} l\cos\theta_2 \\ l\sin\theta_2 \end{pmatrix} = \begin{pmatrix} x \\ y \end{pmatrix} + \boldsymbol{R}\begin{pmatrix} a \\ 0 \end{pmatrix} - \begin{pmatrix} x_2 \\ 0 \end{pmatrix} \tag{8}$$

Letting Eq. (7) add Eq. (8), we have

$$\begin{pmatrix} l\cos\theta_2 + l\cos\theta_1 \\ l\sin\theta_2 + l\sin\theta_1 \end{pmatrix} = \begin{pmatrix} 2x - x_1 - x_2 \\ 2y \end{pmatrix} \tag{9}$$

θ_1 and θ_2 can be solved by Eq. (9), and they also have four solutions due to the four inverse kinematic solutions of x_i.

2.2.2. Forward Kinematics

Forward kinematics is to define a mapping from a known set of the actuated inputs to an unknown pose of the mobile platform. For the manipulator in figure 4, the inputs that are considered as known can be denoted as a vector

$$\boldsymbol{q} = \begin{pmatrix} x_1 & x_2 & \theta_1 & \theta_2 \end{pmatrix}^{\mathrm{T}} \tag{10}$$

The pose of the mobile platform is described as a position vector

$$\boldsymbol{X} = \begin{pmatrix} x & y & \alpha \end{pmatrix}^{\mathrm{T}} \tag{11}$$

Equation (9) can be changed to

$$\begin{pmatrix} x_2 + x_1 + l\cos\theta_2 + l\cos\theta_1 \\ l\sin\theta_2 + l\sin\theta_1 \end{pmatrix} = 2\begin{pmatrix} x \\ y \end{pmatrix} \tag{12}$$

Equation (8) minus Eq. (7) leads to

$$\begin{pmatrix} x_2 - x_1 + l\cos\theta_2 - l\cos\theta_1 \\ l\sin\theta_2 - l\sin\theta_1 \end{pmatrix} = 2a\begin{pmatrix} \cos\alpha \\ \sin\alpha \end{pmatrix} \tag{13}$$

It is obvious that there is one group of solutions x, y and α when the values of x_1, x_2, θ_1, θ_2 are given. Since there is a redundant actuator in the manipulator, θ_2 which depends on x_1, x_2 and θ_1 is not an independent input, that is

$$\theta_2 = f(x_1, x_2, \theta_1) \tag{14}$$

In Eq. (13), by squaring and adding the elements of each side, we obtain

$$(x_2 - x_1)^2 + 2l^2[1 - \cos(\theta_1 - \theta_2)] + 2l(x_2 - x_1)(\cos\theta_2 - \cos\theta_1) = 4a^2 \tag{15}$$

Obviously, Eq. (15) is a concealed formula of Eq. (14).

2.2.3 Velocity Equation

Equations (7) and (8) can be differentiated with respect to time to obtain the velocity equations, this leads to

$$\begin{cases} \dot{x}_1 - l\sin\theta_1\dot{\theta}_1 = \dot{x} + a\sin\alpha\dot{\alpha} \\ l\cos\theta_1\dot{\theta}_1 = \dot{y} - a\cos\alpha\dot{\alpha} \\ \dot{x}_2 - l\sin\theta_2\dot{\theta}_2 = \dot{x} - a\sin\alpha\dot{\alpha} \\ l\cos\theta_2\dot{\theta}_2 = \dot{y} + a\cos\alpha\dot{\alpha} \end{cases} \tag{16}$$

Equation (16) can be changed to an equation of vector form as

$$\boldsymbol{A}\dot{\boldsymbol{q}} = \boldsymbol{B}\dot{\boldsymbol{X}} \tag{17}$$

where $\dot{\boldsymbol{q}}$ is a vector of input velocities defined as

$$\dot{\boldsymbol{q}} = \begin{pmatrix} \dot{x}_1 & \dot{x}_2 & \dot{\theta}_1 & \dot{\theta}_2 \end{pmatrix}^{\mathrm{T}} \tag{18}$$

And $\dot{\boldsymbol{X}}$ is a vector of output velocities defined as

$$\dot{\boldsymbol{X}} = \begin{pmatrix} \dot{x} & \dot{y} & \dot{\alpha} \end{pmatrix}^{\mathrm{T}} \tag{19}$$

Matrices $\boldsymbol{A}$ and $\boldsymbol{B}$ can be expressed as

$$\boldsymbol{A} = \begin{pmatrix} 1 & 0 & -l\sin\theta_1 & 0 \\ 0 & 0 & l\cos\theta_1 & 0 \\ 0 & 1 & 0 & -l\sin\theta_2 \\ 0 & 0 & 0 & l\cos\theta_2 \end{pmatrix} \tag{20}$$

$$\boldsymbol{B} = \begin{pmatrix} 1 & 0 & a\sin\alpha \\ 0 & 1 & -a\cos\alpha \\ 1 & 0 & -a\sin\alpha \\ 0 & 1 & a\cos\alpha \end{pmatrix} \tag{21}$$

With the Eq. (20), we have

$$\det(\boldsymbol{A}) = -l^2\cos\theta_1\cos\theta_2 \tag{22}$$

If $\theta_1 \neq 90^\circ$ and $\theta_2 \neq 90^\circ$, i.e. $\det(\boldsymbol{A}) \neq 0$, Eq. (17) can be changed to

$$\dot{q} = G\dot{X} \tag{23}$$

where G is a matrix with four rows and three columns and satisfies $G = A^{-1}B$. Since $\det(B^{\mathrm{T}} B) = 8a^2 \neq 0$, the left pseudo-inverse matrix B^{LM} of B is

$$B^{\mathrm{LM}} = (B^{\mathrm{T}} B)^{-1} B^{\mathrm{T}} \tag{24}$$

Equation (17) can be also changed into

$$\dot{X} = J\dot{q} \tag{25}$$

where J is a velocity Jacobian matrix with three rows and four columns and satisfies $J = B^{\mathrm{LM}} A$. It is obvious that G is the pseudo-inverse matrix of J .

2.3 Position and Orientation Analysis of the *JLZN* Parallel Manipulator with Continuous Rotation

For the ***JLZN*** parallel manipulator, the distance between the points B_1 and B_2 can be changed by driving the two ball screws. With the variety of the distance, the kinematic characteristics of the mobile platform will be different. It seems that the kinematic characteristics of a four-bar mechanism will be changed if the length of all its bars is changed.

Figure 6 demonstrates the rotational course of the mobile platform in one situation which is similar to the double-rocker mechanism in figure 2. The mobile platform rotates when the position and orientation of the two bars change. Also, a dead point appears as shown in figure 6(c), and the rotational direction of the second bar determines rotational direction of the mobile platform.

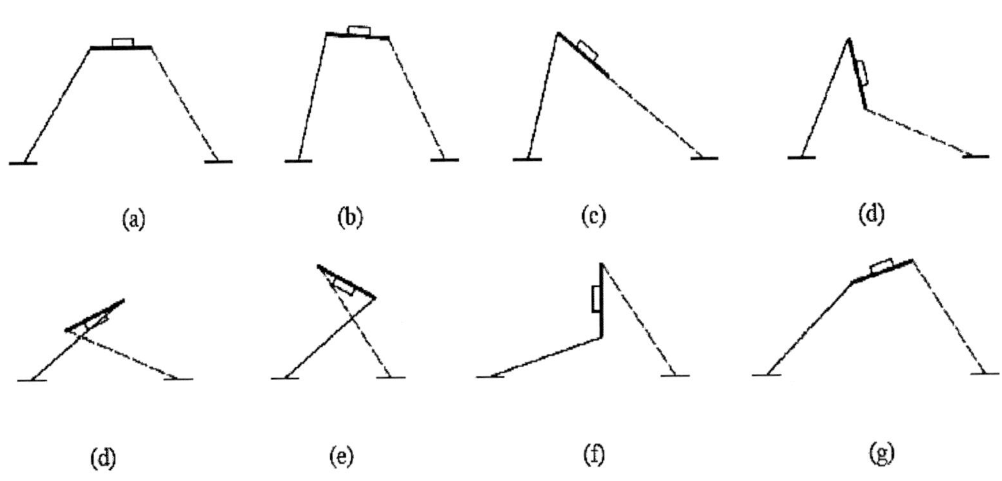

Figure 6. The rotational course of the mobile platform like a double rocker mechanism.

Figure 7 shows another situation that the mobile platform rotates continuously. When the distance between the two slider blocks gets less than the distance of the two R joints in the mobile platform (and $l \geq 2a$ here), shown in figure 7(c), the mechanism turns into a planar double-crank mechanism. The mobile platform can rotate continuously with a longer rotational radius. A special situation happens when the distance between the two-slider blocks gets zero, shown in figure 8. The mobile platform rotates continuously along a circular trace.

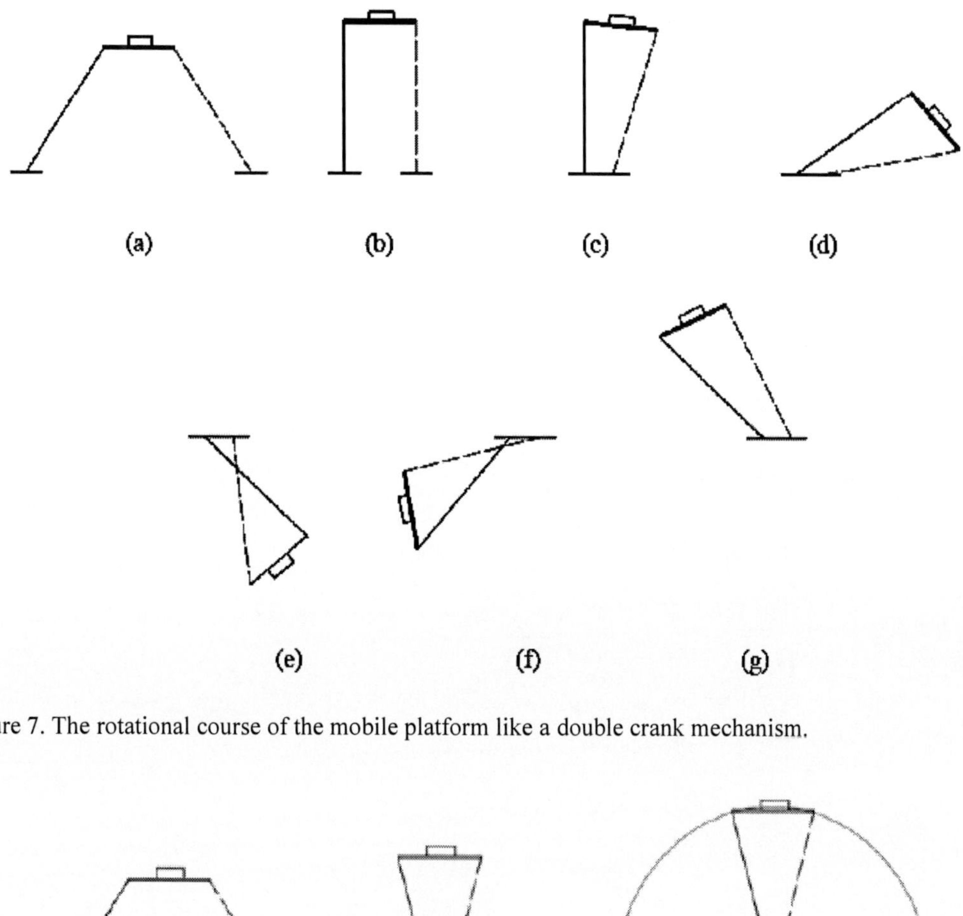

Figure 7. The rotational course of the mobile platform like a double crank mechanism.

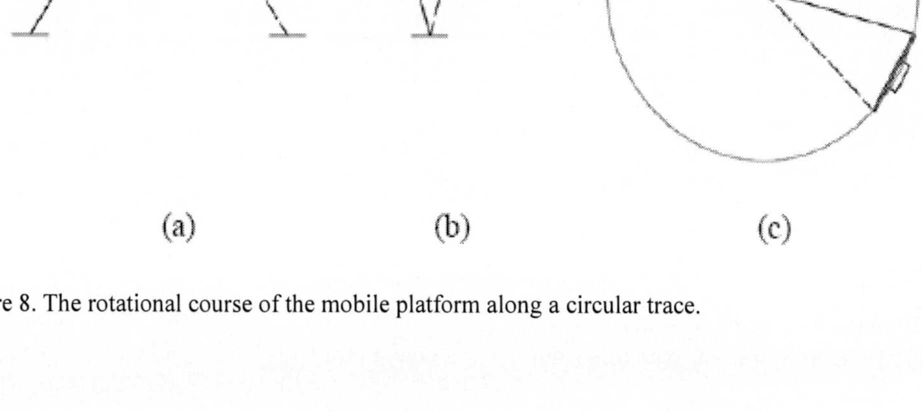

Figure 8. The rotational course of the mobile platform along a circular trace.

The ***JLZN*** parallel manipulator enhances not only its rotational ability, but also its translational ability. In figure 9, when the distance between the two-slider blocks gets equal to the length of the mobile platform, a parallelogram mechanism is derived. The mobile platform can translate along a circular trace as shown in figure 9(c). Also in figure 10, the mobile platform can translate up and down in a large domain.

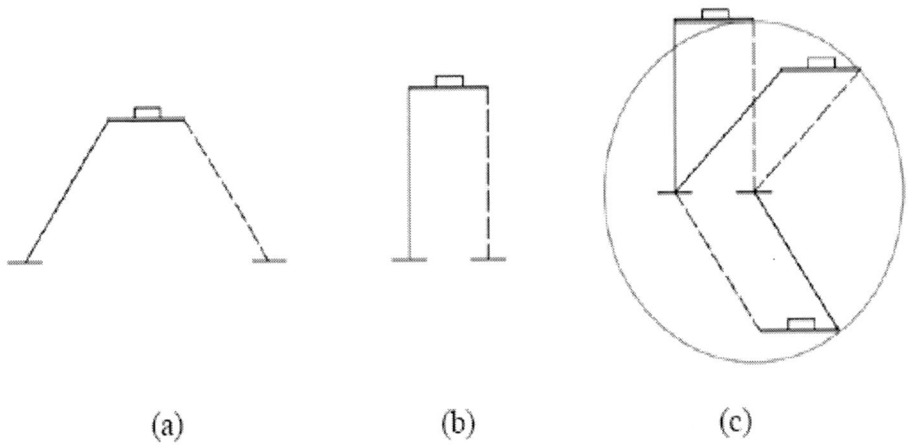

Figure 9. The translational course along a circular trace.

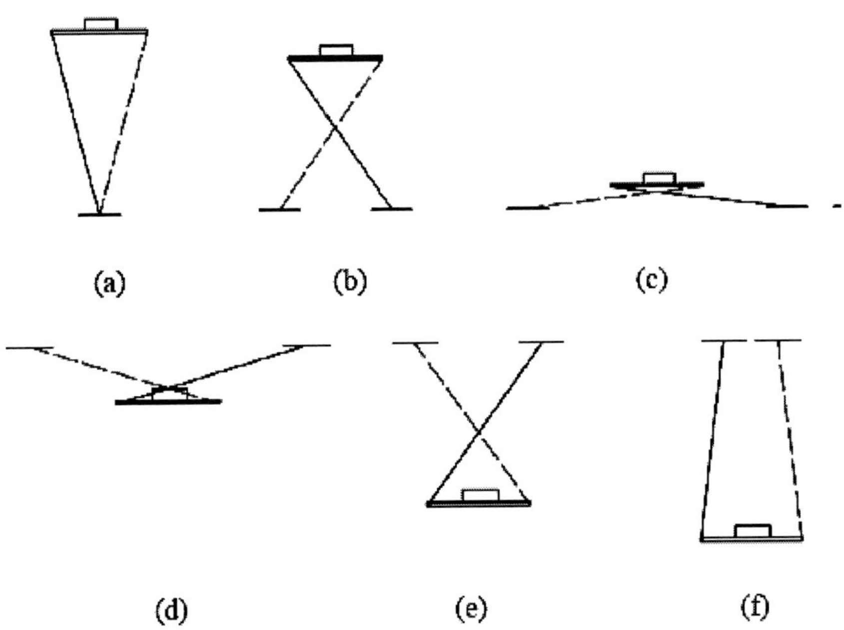

Figure 10. The up and down translational course.

In the course of the continuous rotation, the mobile platform of the ***JLZN*** parallel manipulator is located in some singular configuration. For example, figure 6(c) demonstrates a singular configuration, which is referred to as dead point. If the second bar is not actuated, the mobile platform of the manipulator cannot rotate while the first bar is actuated. It means that the mobile platform loses one rotational DOF instantaneously. Another singular configuration is shown in figure 7(b) or figure 9(b), which is called as extreme displacement singularity. In this position, we know that det($\boldsymbol{A}$) = 0 in Eq. (20) since the input angles $\theta_1 = \theta_2 = 90^\circ$. It means that the certain output velocity of the mobile platform needs the infinity input velocities of the actuated joints. In fact, the manipulator loses one translational DOF along the Y-axis and the rotational DOF instantaneously.

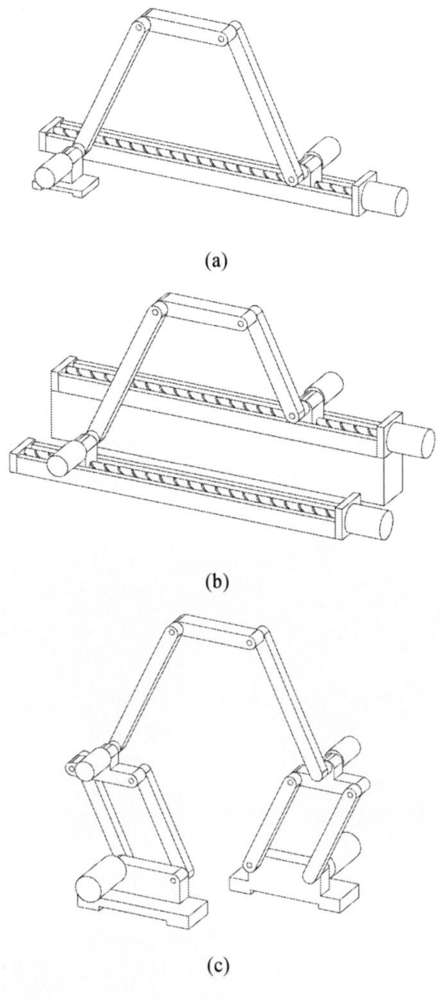

Figure 11. Several parallel manipulators with redundant actuators and multi-actuators in each limb.

2.4 Topology Mechanisms of the *JLZN* Parallel Manipulator

In the ***JLZN*** parallel manipulator, if one slider block is fixed, a 2-DOF parallel manipulator is obtained as shown in figure 11(a). Its kinematic characteristics are similar to those of the ***JLZN***, except that it possesses only one translational DOF that is synthesized by the translations in two directions (up to down, left to right). Another topology mechanism is shown in figure 11(b), the lengths of the two bars are not equal to each other at all and the translational axes of the two slider blocks are not coincident. Moreover, if the two P joints of the ***JLZN*** parallel manipulator are substituted by two parallelogram mechanisms, a new parallel manipulator is obtained, shown in figure 11(c). The kinematics of the last two manipulators would present some different situations and we do not discuss here.

3. Parallel Manipulators with Special Limbs

3.1. Structure Description

3.1.1. Parallel Manipulators with a UPU Limb

A U joint, shown in figure 12(a), is often used as a coupling to transfer rotation and torque between two shafts whose axes are intersectant. Also in parallel manipulators, the U joints pay important roles. And usually, a UPU limb which is composed of two U joints and a P joint is shown in figure 10(b).

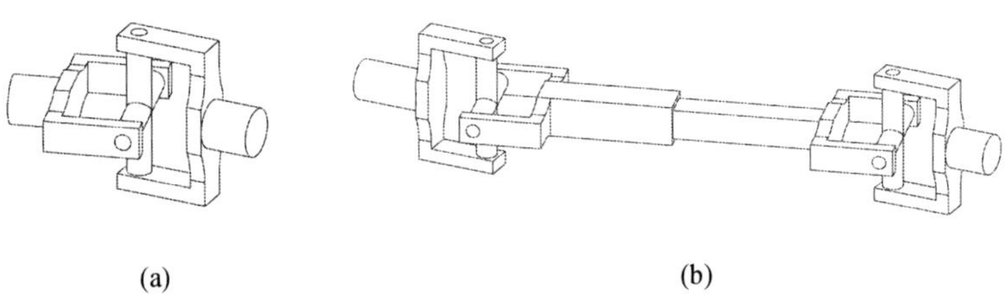

Figure 12. U joint and UPU limb.

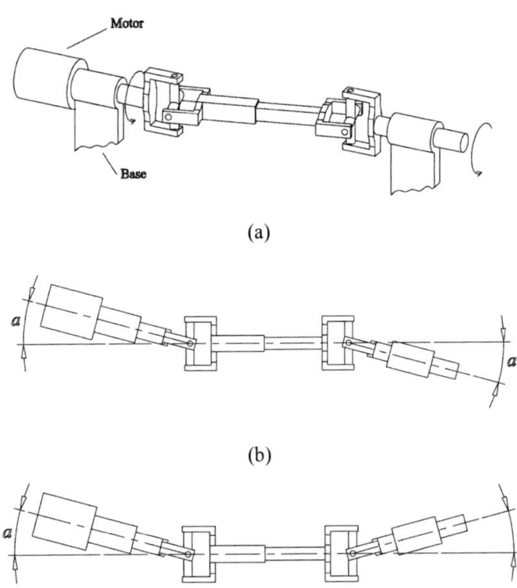

Figure 13. Two shafts connected with a UPU limb.

In figure 13(a), the two shafts connected with a UPU limb are supported on the base. And if one shaft is driven by a motor and rotates continuously, another shaft must rotate continuously whatever the orientation of the two shafts' axes is. Figures. 13(b) and 13(c) show top views of the two situations that the two shafts are parallel and intersectant. In fact, the angle α cannot be too large, otherwise the transfer efficiency would get lower remarkably. So many parallel manipulators can be obtained by aid of the UPU limb.

A new type of 3-DOF parallel manipulator with a UPU limb is given in figure 14. In fact, an R joint is involved in the UPU limb which is really a RUPU limb. The other two same limbs, i.e. PPR chains, are located in two vertical directions and possess one common R joint connected with the mobile platform. The axis of their common R joint is parallel to the axis of the R joint in the RUPU limb. According to the motion description of Gao [8] and Zhang [9], the motion of the 1-RUPU&2-PPR parallel manipulator is described in Table 1, in which T_j (j denotes x, y or z) and R_j (j denotes x, y or z) are motion coordinates to describe the motion performances. T_j takes 1 means that the limbs or the mobile platform have the translational degree of freedom along that direction, while T_j takes 0 means that they have no that degree of freedom. Similarly, R_j takes 1 means that the limbs or the mobile platform have the rotational degree of freedom about that axis, while R_j takes 0 means that they have no that degree of freedom. Moreover, the motion coordinates of the mobile platform is the intersection of the motion coordinates of all the limbs, which means only if T_j (or R_j) of all the limbs take 1, T_j (or R_j) of the mobile platform takes 1.

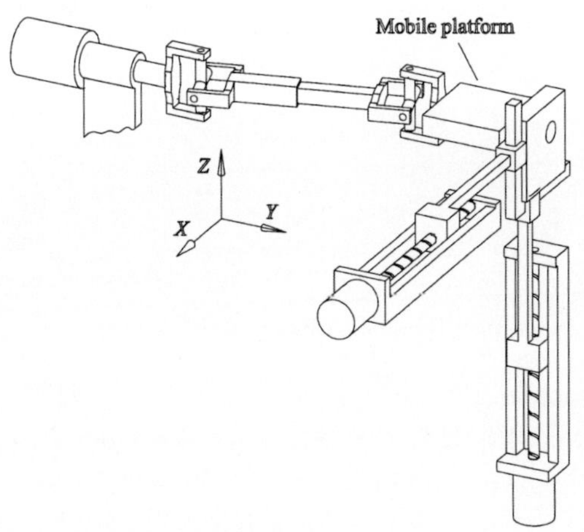

Figure 14. A new type 3-DOF parallel manipulator with a UPU limb.

Table 1. Motion description of the 1-RUPU&2-PPR parallel manipulator

	the instantaneous motion coordinates					
	T_x	T_y	T_z	R_x	R_y	R_z
RUPU limb	1	1	1	1	1	1
PPR limb	1	0	1	0	1	0
PPR limb	1	0	1	0	1	0
Mobile platform	1	0	1	0	1	0

From table 1, we know that the 1-RUPU&2-PPR parallel manipulator has two translational degrees of freedom along the *X*- and *Y*-axes, and one rotational degree of freedom about the *Y*-axis. Theoretically, whatever position the mobile platform translates to, it can rotate about the *Y*-axis continuously. If one limb of the1-RUPU&2-PPR parallel manipulator is got rid of, a 2-DOF parallel manipulator, shown in figure 15, is obtained. The 2-DOF parallel manipulator has one translational degree of freedom along the *X*-axis and one rotational degree of freedom about the *Y*-axis. Above all, the mobile platform can rotate about the *Y*-axis continuously.

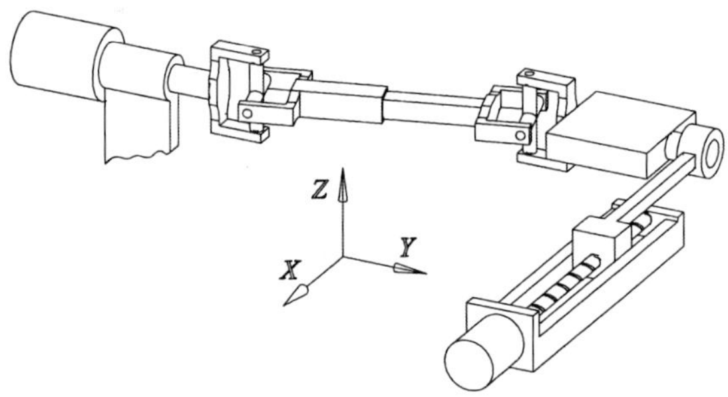

Figure 15. A 2-DOF parallel manipulator with a UPU limb.

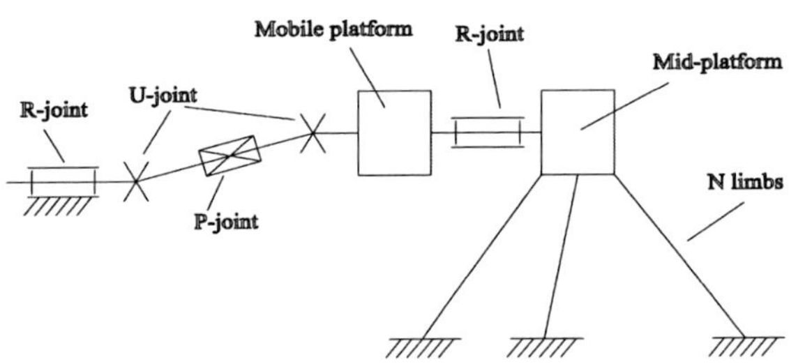

Figure 16. A general structure of parallel mechanism with UPU limb.

Then, a general structure for this type of parallel manipulators is given in figure 16, which denotes a general description. The general structure has N+1 limbs, one of which is a RUPU limb. The N limbs have a common R joint to connect the mobile platform with the mid-platform. In fact the N (when $N \geq 2$) limbs and the mid-platform, where the common R joint is not involved, compose a parallel manipulator. The mid-platform possesses N DOFs depending on the N limbs. So the mobile platform possesses N+1 DOFs, one of which is the rotation about the axis of the common R joint. By aid of the general description, many other parallel manipulators that possess this motion characteristic can be derived. In this type of parallel manipulators, the axes of the two R joints can be parallel (such as a case in figure 14) or intersectant. In figure 17, a case of 2-DOF parallel manipulator in which the two R joints' axes are intersectant is demonstrated. A RUPU limb is involved to make sure that the mobile platform can rotate continuously, and the other limb is a simple one, RR. When the mobile platform changes its position, the axis of the R joint attached to the mobile platform changes its direction and forms a different angle with the axis of the R joint in the RUPU limb.

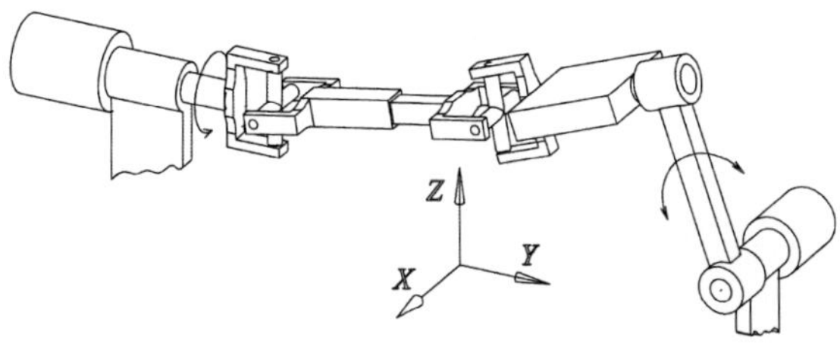

Figure 17. A 2-DOF parallel manipulator of the two R joints with intersectant axes.

3.1.2. Parallel Manipulators with a RP⁺ Limb

A P^+ joint, shown in figure 16(a), is composed of two P-joints configured in two vertical directions. And P^+ joints are used in the parallel manipulators with perpendicular structures [9, 10]. Similar to U-joints, P^+ joints are often used as couplings to transfer rotation and torque between two shafts whose axes are parallel. Figure 16(b) is another type of structure of P^+ joints.

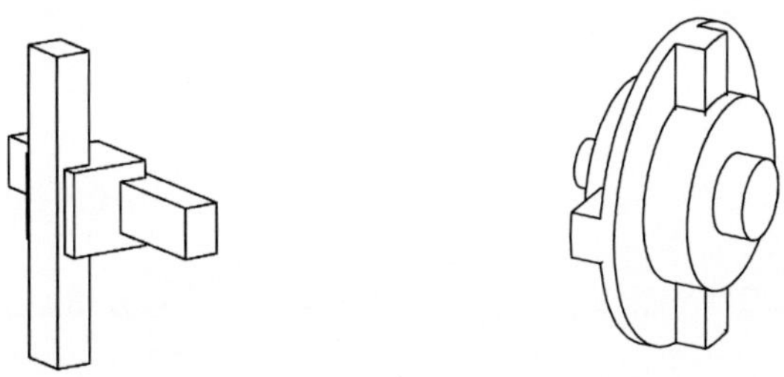

(a) (b)

Figure 18. P^+-joint.

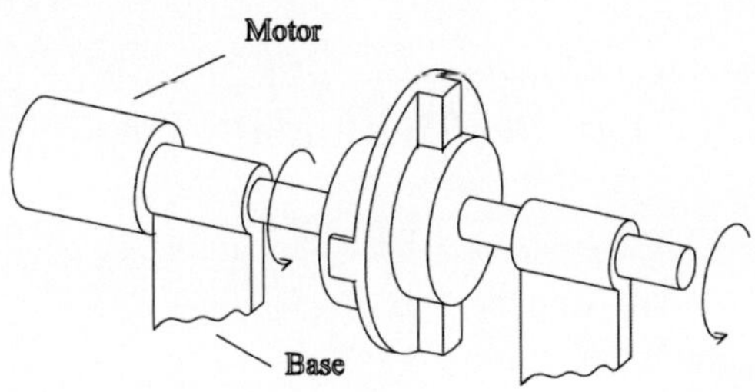

(a)

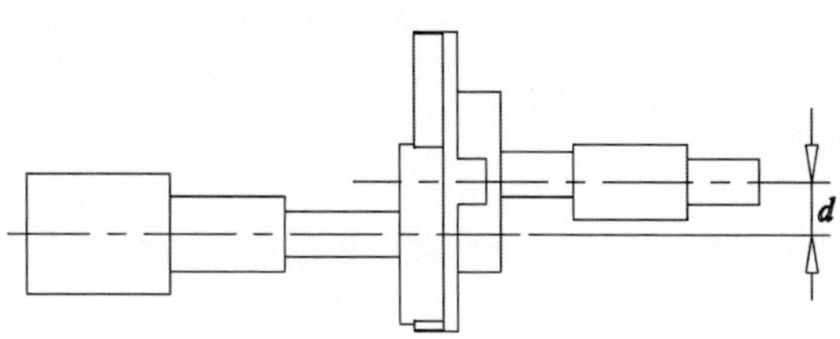

(b)

Figure 19. Two shafts connected with a P^+-joint.

In Figure 19(a), the two shafts attached to the base are connected with a P^+ joint. If one shaft is driven by a motor and rotates continuously, another shaft must rotate continuously although the two shafts' axes are not coincident. But the two shafts' axes must be parallel, which is shown in its planform, figure 19(b). Moreover, the distance of the two shafts' axes, *d* can not be too large, otherwise the transfer efficiency would get lower remarkably. Many parallel manipulators can be obtained by aid of RP^+ limbs.

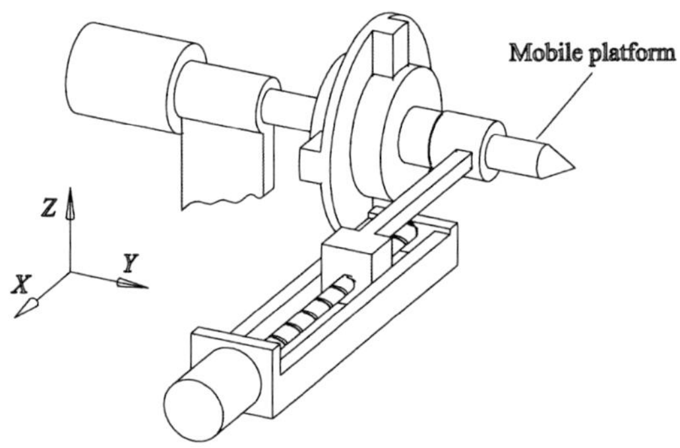

Figure 20. A 1-RP^+&1-PR 2-DOF parallel manipulator.

A new 2-DOF parallel manipulator with a RP^+ limb and a PR limb is given in figure 20, which has two DOFs of one translational and one rotational motion. And the translational direction is vertical to the rotational axis, which makes the axis of the R joint near the mobile platform be parallel to the axis of the R joint fixed on the base forever. Parallel manipulators of this type are similar to those with a UPU limb. We can also give a general structure for this type of parallel manipulators like figure 16. Because the two R joints must be parallel, it becomes simple. Moreover, the parallel manipulator in figure 14 can change to a new one if the RUPU limb is replaced by a RP^+ limb.

3.2. Kinematics of the Parallel Manipulators with Special Limbs

It is just because a RUPU limb or a RP^+ limb is used that the mobile platform of the parallel manipulators can rotate continuously. As a typical case, we only study the kinematics of the 1-RUPU&2-PPR parallel manipulator here.

3.2.1. Kinematics of a RUPUR Manipulator

In figure 16, the mobile platform can rotate continuously because of both the UPU limb and the two R joints. So we solve the kinematics of a RUPUR manipulator in advance.

In figure 14, the distance between the centers of the two U joints changes with the moving of the mobile platform, so a P joint between the two U joints is used to adapt the changed distance. The mechanism sketch of a RUUR manipulator is shown in figure 21, where the P joint is omitted because the distance between point O_1 and O_2 is fixed. Whether a P joint is used can not change the relationship between the rotational input and output. We would solve the relationship between the input angle θ rotating about the axis OO_1 and the output angle α rotating about the axis O_2O'. Here, we only discuss the situation that the input axis OO_1 and the output axis O_2O' are parallel. It is obvious that the two axes form an equal angle ϕ_0 with the middle axis O_1O_2.

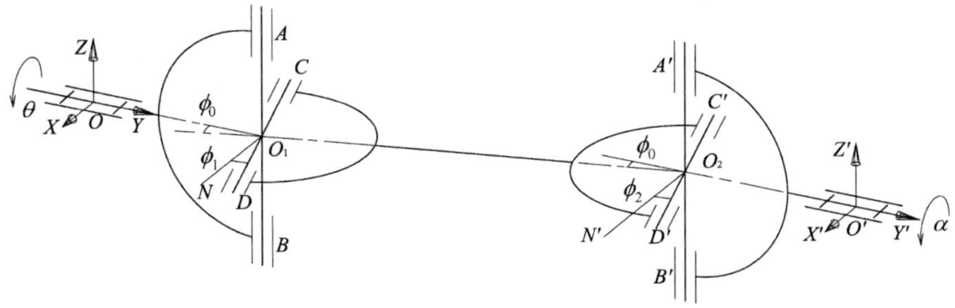

Figure 21. Mechanism sketch of a RUUR manipulator.

At the same time, the configuration of the RUUR manipulator must be the distribution as shown in figure 21, in which the two rotational axes CD and $C'D'$ in the two U joints are parallel. In the left U joint, AB is perpendicular to the plane (named as V_1) determined by CD and OO_1 because it is perpendicular to CD and OO_1 at any pose. A similar condition happens in the right U joint that $A'B'$ is perpendicular to the plane (named as V_2) determined by $C'D'$ and O_2O'. It is obvious that the two planes V_1 and V_2 are parallel according to the given geometry constraint. So AB is parallel to $A'B'$ whatever the RUUR manipulator rotates at any orientation. By the way, O_1N and O_2N' are two normal lines of the left and right half circle planes, respectively.

In the RUUR manipulator as shown in figure 21, a fixed coordinate system O - XYZ and a mobile coordinate system O' - $X'Y'Z'$ which is attached to the output axis O_2O', are constructed. So the orientation of the output shaft, i.e. the axis O_2O', can be described by a rotation transformation matrix

$$\boldsymbol{R} = Rot(Y,\alpha) = \begin{pmatrix} c_\alpha & 0 & s_\alpha \\ 0 & 1 & 0 \\ -s_\alpha & 0 & c_\alpha \end{pmatrix} \tag{26}$$

Here, c_α and s_α are abbreviations for $\cos\alpha$ and $\sin\alpha$, respectively.

Also, the rotational transformation matrix of the output shaft can be obtained by the rotation of all the joints, that is

$$\boldsymbol{R} = Rot(Y,\theta)Rot(Z,\theta_1)Rot(Z,\phi_1)Rot(X,\theta_2)Rot(X,\theta_3)Rot(Z,-\phi_2)Rot(Z,\theta_4) \tag{27}$$

In Eq. (27)

$$Rot(X,\gamma)=\begin{pmatrix}1 & 0 & 0\\ 0 & c_\gamma & -s_\gamma\\ 0 & s_\gamma & c_\gamma\end{pmatrix} \tag{28}$$

$$Rot(Y,\gamma)=\begin{pmatrix}c_\gamma & 0 & s_\gamma\\ 0 & 1 & 0\\ -s_\gamma & 0 & c_\gamma\end{pmatrix} \tag{29}$$

$$Rot(Z,\gamma)=\begin{pmatrix}c_\gamma & -s_\gamma & 0\\ s_\gamma & c_\gamma & 0\\ 0 & 0 & 1\end{pmatrix} \tag{30}$$

where γ denotes the corresponding angle. Also in Eq. (27), θ is the input angle rotating about the axis OO_1, θ_i ($i=1,2,3,4$) denote the angles that the two U joints rotate about the four axes AB, CD, $C'D'$ and $A'B'$, respectively. Two angles ϕ_1 and ϕ_2 are equal to ϕ_0 in the given pose shown in figure 21. From Eqs. (26) and (27), we get

$$Rot(Y,\theta)Rot(Z,\theta_1)Rot(Z,\phi_1)Rot(X,\theta_2)Rot(X,\theta_3)Rot(Z,-\phi_2)Rot(Z,\theta_4)=Rot(Y,\alpha) \tag{31}$$

In Eq. (31), the elements at the same row and column of the left and right matrix are equal. Considering the elements in column 3, row 2 and row 3, we can obtain

$$\begin{cases}\sin\theta=\sin\alpha\\ \cos\theta=\cos\alpha\end{cases} \tag{32}$$

It is obvious that $\theta=\alpha$ or $\dot{\theta}=\dot{\alpha}$. So the RUUR manipulator has the kinematic characteristic that the output rotational angle or velocity is equal to the input rotational angle or velocity.

3.2.2. Kinematics of the 1-RUPU&2-PPR Parallel Manipulator

Figure 22 shows a mechanism sketch of the 1-RUPU&2-PPR parallel manipulator, in which a fixed coordinate system O - XYZ and a mobile coordinate system O' - $X'Y'Z'$ are constructed. The two PPR limbs configured in two vertical directions make the mobile platform translate along the *X*- and *Y*-axes, respectively. With the translation of the mobile platform, the distance between the two U joints in the RUPU limb changes and the P joint translates to adapt the change. However the mobile platform translates, the axes of the two R joints in the two PPR limbs and in the RUPU limb remain parallel. So the rotational output of the mobile platform follows the rotational characteristic of the RUUR manipulator.

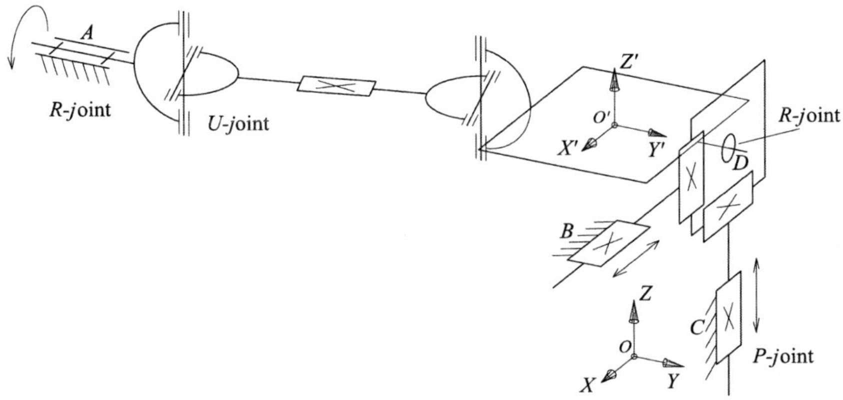

Figure 22. Mechanism sketch of the 1-RUPU&2-PPR parallel manipulator.

It is obvious that the 1-RUPU&2-PPR parallel manipulator possesses a simple kinematic model. Here, the velocity equation is only given. Let $\dot{\boldsymbol{q}}$ denote the vector of the input velocities and be defined as

$$\dot{\boldsymbol{q}} = \begin{pmatrix}\dot{q}_1 & \dot{q}_2 & \dot{\theta}\end{pmatrix}^{\mathrm{T}} \tag{33}$$

where $\dot{q}_1$ and $\dot{q}_2$ are the translational velocities of the two actuated P joints along the X- and Y-axes, respectively, $\dot{\theta}$ is the rotational velocity of the actuated R joint. And let $\dot{\boldsymbol{X}}$ denote the vector of the output velocities which is defined as

$$\dot{\boldsymbol{X}} = \begin{pmatrix}\dot{x} & \dot{y} & \dot{\alpha}\end{pmatrix}^{\mathrm{T}} \tag{34}$$

where $\dot{x}$ and $\dot{y}$ are the translational velocities of the mobile platform along the X- and Y-axes, respectively, $\dot{\alpha}$ is the rotational velocity of the mobile platform about the Z-axis. The input and output velocities have a simple relationship

$$\dot{\boldsymbol{X}} = \boldsymbol{J}\dot{\boldsymbol{q}} \tag{35}$$

where the velocity Jacobian $\boldsymbol{J}$ is a unit diagonal matrix, that is

$$\boldsymbol{J} = \begin{pmatrix}1 & & \\ & 1 & \\ & & 1\end{pmatrix} \tag{36}$$

For the general parallel manipulators with a UPU limb as shown in figure 16, the kinematic characteristics of the mobile platform can be regarded as the combination of the kinematic characteristics of the mid-platform and the RUUR manipulator. For the parallel manipulators with a RP^+ limb, similar conditions happen. In figure 21, if the rotational axes of the two R joints are not parallel, the kinematics would be more complex. Its kinematic equations can be derived by the same method and we don't consider here.

4. PARALLEL MANIPULATORS WITH LIMBS OF EMBEDDING STRUCTURES

4.1. Structure Description

In figure 8(c), the axes of the two actuated R joints in the ***JLZN*** parallel manipulator are coincident. Also in Ref. [15], a 5R mechanism is studied and the axes of the two rotational actuators are coincident. This peculiar structure makes the mobile platform rotate about the coincident axis continuously. Similar condition happens in the following parallel manipulators.

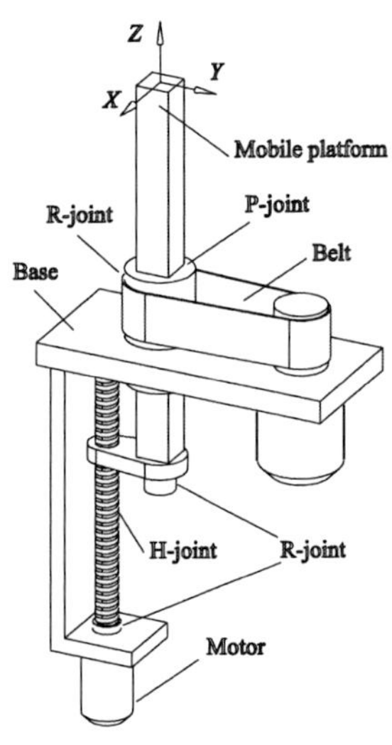

Figure 23. A new 2-DOF parallel manipulator with limbs of embedding structures.

A new 2-DOF parallel manipulator as shown in figure 23 possesses two limbs, i.e. a RHRP (H-helix joint) limb and a RP limb. The peculiarities of its structures are that the axis of the second R joint in the RHRP limb and the axis of the R joint in the RP limb are coincident, and the two limbs have a common P joint. Moreover, the common P joint runs through the R joint in the RP limb whose axis is parallel to the translational direction of the P joint. A motor drives the first R joint in the RHRP limb, which makes the mobile platform translate along the *Z*-axis. And another motor drives the R joint in the RP limb by a belt (or the other driving mode), which makes the mobile platform rotate about the *Z*-axis continuously. Above all, the two motions, translation along the *Z*-axis and rotation about the *Z*-axis, are decoupled. It means that the kinematic equation should be a simple formula.

Obviously, the structures of the new 2-DOF parallel manipulator are different from the traditional parallel manipulators in two aspects. First, one of its limbs embeds another partly. Secondly, the two R joints in the different limbs possess a coincident axis, which is similar to the description in the former paragraph.

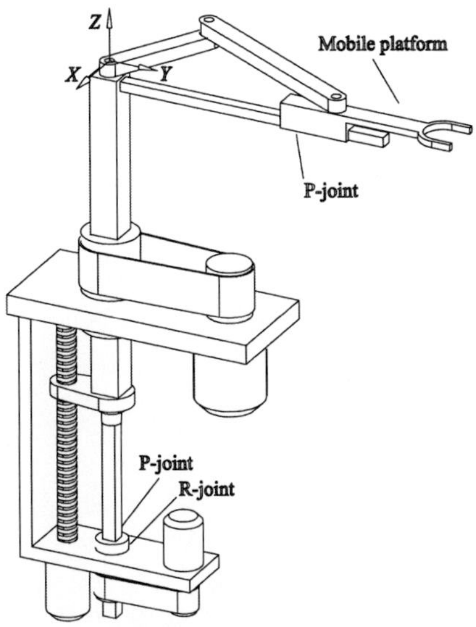

Figure 24. A new 3-DOF parallel manipulator with limbs of embedding structures.

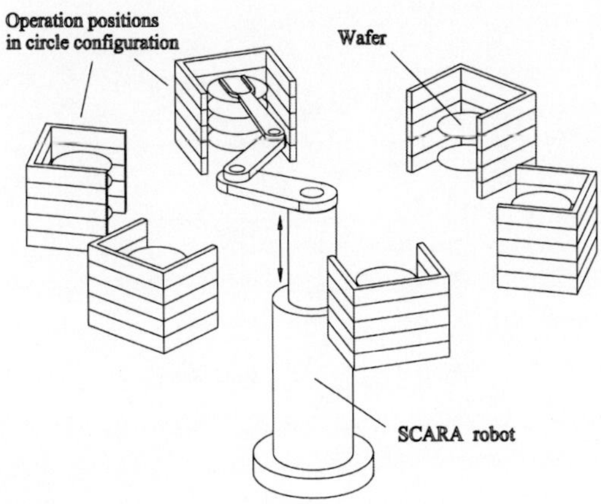

Figure 25. SCARA type wafer-handing robot.

Based on the 2-DOF parallel manipulator with limbs of embedding structures, a new type of 3-DOF parallel manipulator, shown in figure 24, is derived, in which a crank-slider mechanism is involved. The crank, driven by the third motor by a belt (or the other driving mode), rotates about the R joint's axis and makes the slider block (a mobile platform here) translate. According to definition description of parallel manipulators, the new 3-DOF parallel manipulator has three limbs, i.e. a RHRPP limb, a RPP limb and a RPRRR limb.

Compared with the traditional parallel manipulators, the peculiarities are that the three limbs are embedding together and the axes of the three R joints in each limb are coincident. Moreover, the RHRPP and RPP limbs possess two common P joints. Similar to the 2-DOF parallel manipulator, the mobile platform of the 3-DOF one can translate along and about the *Z*-axis continuously. Besides, it can translate along a leader-rail in radial direction.

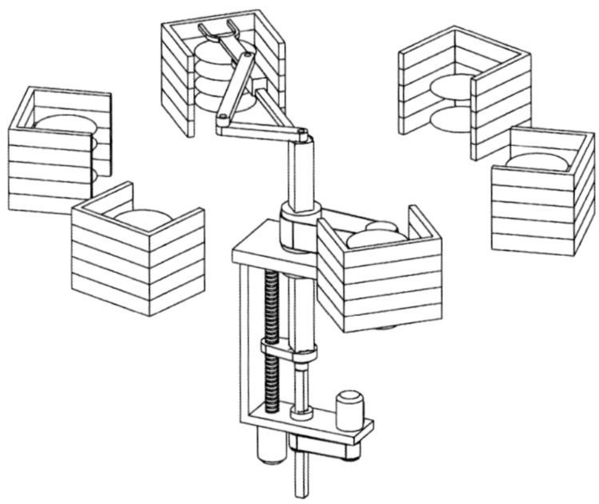

Figure 26. Similar-SCARA type wafer-handing robot.

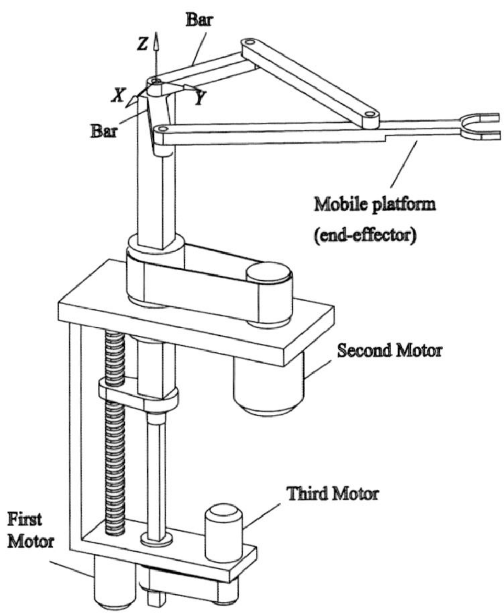

Figure 27. A new 3-DOF parallel manipulator with limbs of embedding structures.

A SCARA robot is a well-known serial manipulator which is used in the conditions of assembling, welding and carrying operation etc. In figure 25, a SCARA type wafer-handing robot is carrying wafers between the operation positions configured in a circle. The robot can operate in a lager area with a smaller equipment fixing space. Generally speaking, it is difficult for the traditional parallel manipulators to finish the wafer-handing operation. The 3-DOF parallel manipulator in figure 24 can do though it possesses one less DOF than the

SCARA. We name it similar-SCARA. Also in figure 26, a similar-SCARA type wafer-handing robot is carrying wafers between the operation positions configured in a circle.

Based on the 2-DOF parallel manipulator in figure 23, some other parallel manipulators with limbs of embedding structures can be derived. If the top P joint in the similar-SCARA is replaced by an R joint, a new 3-DOF parallel manipulator, shown in figure 27, can be obtained. Driven by the second and third motors, respectively, the two bars rotate about a same axis continuously. In fact, its kinematic characteristics are similar to those of the similar-SCARA.

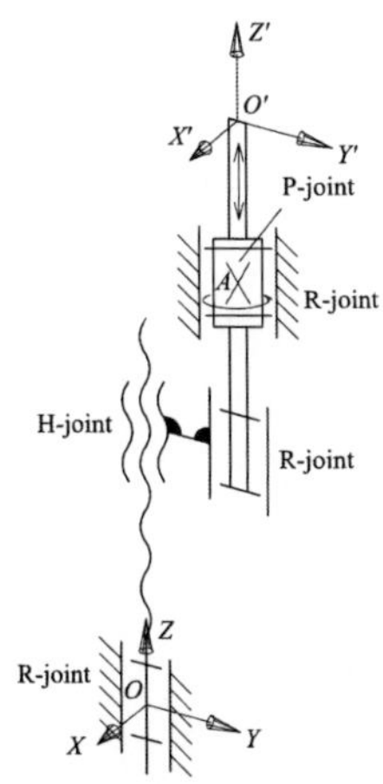

Figure 28. Mechanism sketch of the 1-RHRP&1-RP 2-DOF parallel manipulator.

4.2. Kinematics of the Parallel Manipulators with Limbs of Embedding Structures

Figure 28 is a mechanism sketch of the 1-RHRP&1-RP 2-DOF parallel manipulator. As the two motions, translation along the *Z*-axis and rotation about the *Z*-axis, are decoupled, it is easy to get the kinematic model. The fixed coordinate system $O-XYZ$ and the mobile coordinate system $O'-X'Y'Z'$, attached to the mobile platform, are constructed, as shown in figure 28. And we let the *Z* coordinate z and the rotational angle α about the *Z*-axis denote the output of the manipulator. The inputs are two rotational angles θ_1 of the R joint in point *O*, θ_2 of the R joint in point *A*. It is easy to get the relation between the inputs and outputs

$$\begin{cases} z = z_0 + \dfrac{p}{2\pi}\theta_1 \\ \alpha = \theta_2 \end{cases} \tag{37}$$

Here, z_0 is the original distance in the *Z*-axis direction between points O and O', p is the pitch of the H joint. Eq. (37) is differentiated with respect to time, velocity equations are obtained

$$\begin{cases} \dot{z} = \dfrac{p}{2\pi}\dot{\theta}_1 \\ \dot{\alpha} = \dot{\theta}_2 \end{cases} \tag{38}$$

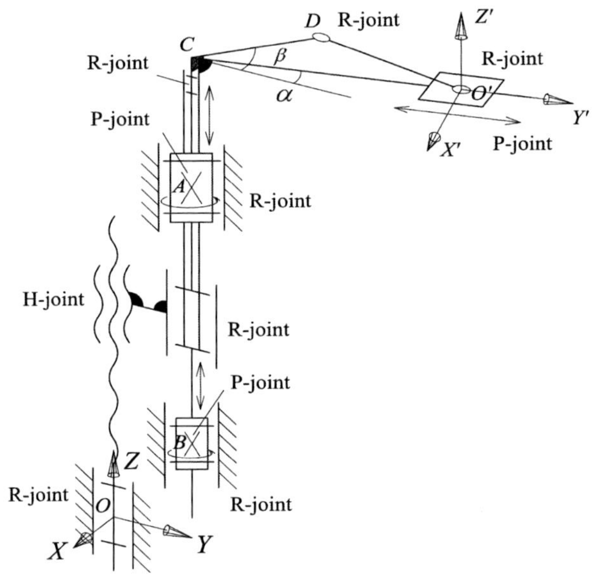

Figure 29. Mechanism sketch of the 3-DOF similar-SCARA parallel manipulator.

For the mechanism of the 3-DOF similar-SCARA parallel manipulator in figure 29, the outputs involve not only the z coordinate and the rotational angle α about Z-axis but also the distance r between points O' and C. So the cylinder coordinate vector $(r \quad \alpha \quad z)^{\mathrm{T}}$ is used to denote the motion of the mobile platform. The inputs are three rotational angles which are θ_1 of the R joint in point O, θ_2 of the R joint in point A and θ_3 of the R joint in point B. The outputs α and z (or $\dot{\alpha}$ and $\dot{z}$) are expressed by the same style in Eq. (37) (or Eq. (38)), and the left output r can be solved in the planar triangle CDO' where we have

$$b^2 = a^2 + r^2 - 2ar\cos(\beta - \alpha) \tag{39}$$

Here, a is the length of CD, b is the length of DO', β is the rotational angle of CD about the Z-axis and satisfies

$$\beta = \theta_3 \tag{40}$$

Differentiating Eq. (39) with respect to time and integrating with Eq. (40), we obtain

$$[2r - 2a\cos(\theta_3 - \theta_2)]\dot{r} = -2ar\sin(\theta_3 - \theta_2)(\dot{\theta}_3 - \dot{\theta}_2) \tag{41}$$

Combining Eqs. (38) and (41), we obtain a velocity equation of the similar-SCARA manipulator in a vector form, that is

$$\boldsymbol{A}\dot{\boldsymbol{q}} = \boldsymbol{B}\dot{\boldsymbol{X}} \tag{42}$$

Here, $\dot{\boldsymbol{X}} = (\dot{r} \quad \dot{\alpha} \quad \dot{z})^{\mathrm{T}}$ is the output velocity vector in which the elements are the translational and rotational velocities, respectively, $\dot{\boldsymbol{q}} = (\dot{\theta}_1 \quad \dot{\theta}_2 \quad \dot{\theta}_3)^{\mathrm{T}}$ is the input velocity vector. Matrices $\boldsymbol{A}$ and $\boldsymbol{B}$ can be expressed as

$$\boldsymbol{A} = \begin{pmatrix} 0 & 2ar\sin(\theta_3 - \theta_2) & -2ar\sin(\theta_3 - \theta_2) \\ 0 & 1 & 0 \\ \dfrac{p}{2\pi} & 0 & 0 \end{pmatrix} \tag{43}$$

$$\boldsymbol{B} = \begin{pmatrix} 2r - 2a\cos(\theta_3 - \theta_2) & 0 & 0 \\ 0 & 1 & 0 \\ 0 & 0 & 1 \end{pmatrix} \tag{44}$$

If $|\boldsymbol{B}| \neq 0$, Eq. (41) can be changed to

$$\dot{\boldsymbol{X}} = \boldsymbol{J}\dot{\boldsymbol{q}} \tag{45}$$

in which $\boldsymbol{J}$ is a velocity Jacobian matrix and satisfies

$$\boldsymbol{J} = \boldsymbol{B}^{-1}\boldsymbol{A} \tag{46}$$

5. Summarization and the Future Work

The parallel manipulators discussed in this chapter change the traditional opinion of parallel mechanisms. Because the mobile platforms can rotate continuously, it is possible that the parallel manipulators possess very large dexterous workspaces, which makes the application area of parallel mechanisms enlarge. But here we discuss only their topology structures that realize the continuous rotation and their kinematics.

As a new conception and style of parallel manipulators, many other aspects would be studied in near future, which are involved in the following.

1) Considering the kinematics performances of these parallel manipulators such as workspace, singularity etc., the mechanism parameters should be optimized.
2) In the course of continuous rotation of the mobile platforms, the static and dynamic performances would be different in different pose. Moreover, the stiffness characteristics of the similar-SCARA parallel manipulator would be peculiar because of the limbs of embedding structures.

REFERENCES

[1] R. Clavel, DELTA: A fast robot with parallel geometry, 18th International Symposium on Industrial Robot, Apr. 1988, pp. 91-100.

[2] H.S. Kim, L.W. Tsai. Design optimization of a Cartesian parallel manipulator. Journal of Mechanical Design Transactions of the *ASME*, 2003, 125(1), pp. 43-51.

[3] X.W. Kong and C.M. Gosselin, Type synthesis of input-output decoupled parallel manipulators, *Transactions of the CSME*, 2004, 28(2A), pp. 185-196.

[4] C.M. Gosselin, St-Pierre, Development and experimentation of a fast 3-DOF camera-orientating device, *The International Journal of Robotics Research*, 1997, 16 (5) , pp.619–630.

[5] T. Huang, X. Zhao, and D. J. Whitehouse, Stiffness estimation of a tripod-based parallel kinematic machine, *IEEE Transactions on Robotics and Automation*, 2002, 18(1), pp. 50–58.

[6] Y. Fang, L.W. Tsai, Structure synthesis of a class of 4-DOF and 5-DOF parallel manipulators with identical limb structures, *International Journal of Robotics Research*, 2002, 21(9), pp.799-810.

[7] J.M. Hervé, The Lie group of rigid body displacements, a fundamental tool for mechanism design, *Mechanism and Machine theory*, 1999, 34(5), pp.719-730.

[8] F. Gao, W.M. Li, X.C. Zhao, etc., New kinematic structures for 2-, 3-, 4- and 5-DOF parallel manipulator designs, *Mechanism and Machine Theory,* 2002, 37 (11), pp. 1395-1411.

[9] J. Zhang, Design Method and key technique of 6-DOF parallel macro/micro-manipulators with perpendicular structures, Dissertation for Ph. D. Degree, Beijing University of Aeronautics and Astronautics, 2004

[10] X.-J. Liu, J. Wang and G. Pritschow, A new family of spatial 3-DoF fully-parallel manipulators with high rotational capability, *Mechanism and Machine Theory*, 2005, 40(4), pp.475-494.

[11] J.P. Merlet, Determination of 6D workspaces of Gough-type parallel manipulator and comparison between different geometries, International *Journal of Robotics Research,* 1999, 18(9), pp.902-916.

[12] J. Russakow, O. Khatib, S.M. Rock, Extended operational space formulation for serial-to-parallel chain (branching) manipulators, Proceedings of IEEE International Conference on Robotics and Automation Nagoya, Japan, 1995, pp.1056-1061.

[13] L. Romdhane, Design and analysis of a hybrid serial-parallel manipulator. Mechanism and Machine Theory, 1999, 34(7), pp. 1037-1055.

[14] J. Kim, J. Hwang, J. Kim, C.C. Iurascu, Eclipse-II: A new parallel mechanism enabling continuous 360-degree spinning plus three-axis translational motions, *IEEE Transactions on Robotics and Automation,* 2002, 18(3), pp. 367-373.

[15] X.-J. Liu, J. Wang, and H. Zheng, Optimum design of the 5R symmetrical parallel manipulator with a surrounded and good-conditioning workspace, *Robotics and Autonomous Systems,* 2006, 54(3), pp 221-233.

[16] J. Zhang, X. Wang, W. Li, F. Gao, Study on kinematics decoupling for parallel manipulator with perpendicular structures, *IEEE/RSJ proceeding of IROS*, Oct. 9-14, 2006, Beijing China, pp.748-753.

In: Parallel Robotics: Recent Advances in Research... ISBN: 978-1-60456-859-2
Editors: Jinsong Wang and Xin-Jun Liu, pp. 95-121

Chapter 5

A GENERAL METHOD FOR FORWARD DYNAMICS OF PARALLEL MANIPULATORS BASED ON NEWTON-EULER FORMULATION WITH GENERALIZED COORDINATES

Hao Wang*, Genliang Chen and Yong Zhao
Institute of Automobile Engineering
Shanghai Jiaotong University, Shanghai, 200240, P. R. China

1. INTRODUCTION

The dynamic analysis is a very important aspect on design, analysis and control of the parallel manipulators and parallel kinematic machine tools. The dynamics of any parallel manipulators can also be divided into the inverse problem and the forward (or direct) problem as its kinematics does. The inverse dynamic analysis gives the driving forces and driving moments of all active joints according to the desired trajectory of the movable platform, which provides a prerequisite for the design of control strategy. The forward dynamic analysis calculates the real dynamic response of the movable platform and all legs according to the forces and moments applied to the system. The goal of forward dynamic modeling is to simulate the velocity fluctuation, vibration and other dynamic behavior and to investigate their effect on positioning or machining quality when the parallel manipulator performs a certain task such as machining or picking up and placing.

However, contrast to the vast literature on the synthesis and kinematics of the parallel manipulators, studies on its dynamics, especially the forward dynamics, are relatively few. The forward dynamic modeling and simulation of the parallel manipulators are more complex than other mechanisms because of two main factors: 1) any parallel manipulator has closed loop topology, the modeling of such a topology is harder than that of a tree or chain topology when the mechanism is considered as a multi-body system (MBS); 2) calculation of the kinematics of the mechanism is needed at each step in dynamic analysis, however, not all the

*E-mail: wanghao@sjtu.edu.cn

parallel manipulators have a closed-form solution for forward kinematics; moreover, the assembly problem caused by numerical fault is inevitable when solutions of different legs have errors resulting from a certain numerical integration algorithm. Considering these characteristics of the parallel manipulators, this chapter puts the emphasis on the forward dynamics of parallel manipulators and improving the efficiency of analysis.

A general modeling and simulation method for the forward dynamics of parallel manipulators is proposed in this chapter, which is based on the contracted Newton-Euler formulation with generalized coordinates. The chapter consists of six sections: section one is an introduction part; in section two, the state of the art will be presented on research progress on dynamics of parallel manipulators; the aforementioned general method for the forward dynamics of parallel manipulators will be proposed and depicted in detail in section three; in section four, a complete analysis of the forward and inverse dynamics of a 3-DOF (degree of freedom) parallel manipulator will be presented; and in section five, numerical solution will be given as a case study to show the feasibility and application of the method; section six will give the conclusion of the whole chapter.

2. State of the Art

During the past two decades, a lot of research work has been done on the dynamic analysis of the parallel manipulators. Among all this work, three main principle applied to the parallel manipulators could be summarized as principle of virtual work, Lagrange's equations of the second kind in analytic mechanics, and Newton-Euler equations in vectorial mechanics.

The principle of virtual work and the general equation of dynamics are the rationale for the analysis of constraint rigid body systems in which the applied forces and reaction forces are distinguished clearly. With the assumption that all the joints are ideal constraint, all reaction forces could be eliminated by the principle of virtual work. Such elimination simplifies the items in dynamic equations compared with the Newton-Euler method. The principle of virtual work is applied to the inverse dynamics of some parallel manipulator, e.g., the Gough-Stewart style parallel manipulator[1,2], the DELTA style mechanism[3], 3-RPS[4] and 3-HSS[5] mechanism. This principle is also employed to compare the method to form dynamic model of parallel manipulator with full mobility[6], and to study inverse dynamics of parallel manipulator combined with screw theory[7]. In the statics of parallel manipulator, a mapping could be easily established between the driving force on active joint and applied forces summarized on the end effecter by the principle of virtual work, assuming that the gravity of the movable platform and all legs are neglected or balanced beforehand. Such a mapping is considered as a dual part of the kinematic analysis, for the matrices in static equation are transpositions of the Jacobian in kinematics. The limitation of modeling by the principle of virtual work is that the constraint forces, which provides a boundary condition for the stress analysis of the structure, can not be solved directly. Besides, although the friction forces in the joint could be traded as applied forces, the clearance and elasticity of the joints are difficult to model because of the ideal assumption of all the joints.

The Lagrange's equations of second kind lead to a minimal set of ordinary differential equations (ODE). These equations are also often used in the dynamics of parallel manipulator in recent researches, e.g. Stewart-Gough parallel manipulators[8], the 3-RPS parallel mechansims[9], and Hexapod mechanism[10]. Since the kinetic energy must be calculated before the equations of motion are obtained, the influences of inertial parameters of the mechanism are easier to investigate, for instance, in [10]. The computational cost of the Lagrange's equations of the second kind will increase vastly when the number of bodies in a multi-body system increases. Thus, such equations are much applicable in a parallel manipulator with fewer bodies than the one with complex chains.

The Newton-Euler equations are classical method in vectorial mechanics. In the application of the equations, there are nearly no restrictions, e.g., no assumption of ideal constraints, no matter holonomic or nonholonomic systems. The dynamic equations of the Stewart platform have been derived by the Newton-Euler approach by some researchers with different precision. In the beginning the mass of legs was neglected and legs were simplified as ideal two-force rods[11]. Then, the inverse dynamic model of Stewart type parallel manipulators has been established considering the influence of the limb inertia[12,13] and the friction of joints are also taken into account in the inverse and forward dynamic model of 6-UPS and 6-SPS mechanisms[14,15,16]. The dynamic model of Stewart type parallel manipulators by the Newton-Euler method can also be found in [17-23]. There are also some report on the dynamic modeling of 3-DOF parallel manipulator by the Newton-Euler equation, e.g., CaPaMan[24]. When performing the forward dynamic analysis, the dynamic model of parallel manipulator by Newton-Euler method is a set of differential-algebra equations (DAE) with a maximum number of coordinates, which consists of the Newton-Euler equations of all bodies and constraint equations of all joints. The dimension of this set of DAE for solution will be huge, for instance, the number of DAE of a typical 6-UPS mechanism is 150 (78 Newton-Euler equations and 72 constraint equations). To solve these equations will be a tedious work for any numerical integration algorithm.

Beside these three main methods, other approaches such as Kane's equation[25], Hamilton's equation[26], screw theory[7,12], and differential manifold[27] have also been reported.

Among all these aforementioned work, most of the researches are within inverse dynamic analysis. A few of them present the modeling and formulation of forward dynamics[14,17-19,25,26]. However, only [14, 25, 26] give the simulation result as dynamic responds from the forces applied on the parallel manipulator. Comparing with the achievements on inverse dynamics, the researches on forward dynamics are relatively few. Contrast to the purely algebra equations of inverse dynamics, the equation of forward dynamic analysis are DAEs or ODEs which lead to several challenges in modeling and simulation. It seems that the precision of the model and efficiency of computation have conflicts. On one hand, the number of integration variables may be minimal by principle of virtual work or Lagrange's equations of second kind, whereas the behavior of joint, e.g., elasticity of ball joint, can not be taken into account. On the other hand, the integration of all Cartesian coordinates in the DAE generated from Newton-Euler methods will be hard work. Since the dynamic analysis on parallel manipulator is still incomplete, further researches are required on efficient simulation by proper modeling and simulation method, and on precise modeling, e.g. considering the elasticity of linkages and joints.

From point of view on dynamic modeling, all the mechanisms are typical multi-body systems (MBS). The theories and methods of multi-body system dynamics provide a foundation for modeling the dynamics of parallel manipulators. By now, there are two main modeling strategies in the multi-body system: use of relative coordinates and absolute coordinates. Relative coordinates methods consider a pair of adjacent rigid body as an unit, one of which is set to be reference body, while the configuration of the other could be depicted by the generalized coordinate defined on the joints. The configuration of the system could be completely determined by the generalized coordinate of all joints in a system with tree structure. As a result, a set of ODE is found as the equation of motion of the system. The typical relative coordinates methods, the Roberson-Wittenburg method[28], gives an elegant way to solve the dynamics of MBS with tree topology. However, circuit matrix has to be introduced for modeling the mechanisms with closed loop topology. Since the parallel manipulators have many closed loops, the computation cost of such method is relatively high. Relative coordinates method is quite efficient for serial manipulators, but may not be the best way to model parallel manipulators. By the strategy of absolute coordinates, each body of the system is considered as a unit. The configuration of the system is defined relative to a common coordinate system. The general form of system dynamic model could be written as a set of differential-algebra equations. The absolute coordinates are more suitable for the parallel manipulator for modeling the system with loop topology is no longer distinct.

3. A General Method for Forward Dynamics of the Parallel Manipulator

A general modeling and simulation approach for the forward dynamics of the parallel manipulator is proposed in this section. The approach is based on the method proposed by Schiehlen in 1980s[29,30], who derived the equation of motion of MBS by Newton-Euler equation in terms of generalized coordinates. This method, also known as Schiehlen's method, uses both the Newton-Euler equations and the principle of virtual work. The equations of motion of the system are expressed as a set of ODE and the number of equations equals to the system DOF. This set of equations is easer to be solved than the traditional Newton-Euler methods because the integration variables are reduced to the minimal set. The following section will also show that the inverse dynamic analysis can also be implemented by this method effectively.

The general 6-DOF parallel manipulators could be depicted as a MBS with n=31 bodies and 36 revolute or prismatic joints[31] which introduce s=180 constraints. The degrees of freedom of the mechanism could be obtained according to the following equation

$$f = 6n - s \tag{1}$$

As shown in figure 1, the reference frame $O_0x_0y_0z_0$ is fixed on geometric center of the base. The body-fixed frame $O_ix_iy_iz_i$ (i=1, 2, …31) are originated from the mass center of each body. The mass center of movable platform, body 31, is assumed to be located in its geometric center. Then, the configuration of each body could be expressed by the 3×1 translation vector of mass center C_i relative to the reference frame and the 3×3 rotational tensor as

$$\mathbf{r}_i = [x_i, y_i, z_i]^T \tag{2}$$

$$\mathbf{S}_i = \mathbf{S}_i(\alpha_i, \beta_i, \gamma_i) \tag{3}$$

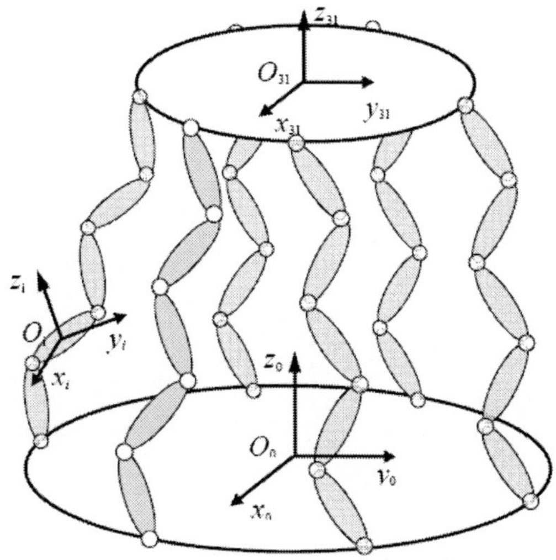

Figure 1. Coordinate system definition of the general 6-DOF parallel manipulators.

The rotational tensor $\mathbf{S}_i$ depends on the Cardan angle α_i, β_i, γ_i and corresponds with the direction cosine matrix relating the reference frame $O_0x_0y_0z_0$ and body-fixed frame $O_ix_iy_iz_i$.

Although the generalized coordinates could be selected arbitrarily in any MBS, computational efficiency will be quite different. There are two options for the selection of generalized coordinates for a parallel manipulator: the configuration of movable platform and the input variables in active joints. Since the inverse kinematic analysis is more efficient in general than the forward kinematic analysis, it is recommended that the generalized coordinates be selected on the movable platform rather than on the active joints even in forward dynamic analysis. Then, the generalized coordinates of this 6-DOF general parallel manipulator are defined by the Cartesian coordinate of the platform mass center relative to the reference frame and the Cardan angle of the movable platform.

$$\mathbf{q} = [q_1, q_2, q_3, q_4, q_5, q_6]^T = [x_{31}, y_{31}, z_{31}, \alpha_{31}, \beta_{31}, \gamma_{31}]^T \tag{4}$$

By this definition, one step forward kinematic calculation may still be needed to determine the initial value of the generalized coordinates by the given displacements and velocities of all active joints.

Then, the 3×1 translation vector of the mass centers and 3×3 rotation tensor for each body could be represented in terms of the generalized coordinates as[30]

$$\mathbf{r}_i = \mathbf{r}_i(\mathbf{q}, t) \in \mathbb{R}^3 \tag{5}$$

$$\mathbf{S}_i = \mathbf{S}_i(\mathbf{q}, t) \in \mathbb{R}^{3\times 3} \tag{6}$$

And the translational and rotational velocities could be derived by differentiating Eq.(5) and (6) with respect to the inertial frame as

$$\mathbf{v}_i = \dot{\mathbf{r}}_i = \frac{\partial \mathbf{r}_i}{\partial \mathbf{q}}\dot{\mathbf{q}} + \frac{\partial \mathbf{r}_i}{\partial t} = \mathbf{J}_{T_i}(\mathbf{q},t)\dot{\mathbf{q}} + \overline{\mathbf{v}}_i(\mathbf{q},t) \tag{7}$$

$$\boldsymbol{\omega}_i = \dot{\mathbf{s}}_i = \frac{\partial \mathbf{s}_i}{\partial \mathbf{q}}\dot{\mathbf{q}} + \frac{\partial \mathbf{s}_i}{\partial t} = \mathbf{J}_{R_i}(\mathbf{q},t)\dot{\mathbf{q}} + \overline{\boldsymbol{\omega}}_i(\mathbf{q},t) \tag{8}$$

where the $3\times f$ Jacobian matrices $\mathbf{J}_{T_i}$ and $\mathbf{J}_{R_i}$ defined in Eq.(7) and (8) could be derived from the inverse kinematic analysis of the general parallel manipulator and they characterize the virtual translational and rotational displacement of the system respectively. And $\mathbf{s}_i$ denotes the infinitesimal 3×1 rotation vector follows analytically from the corresponding infinitesimal skew-symmetrical 3×3 rotation tensor.

The acceleration for each body could be obtained by a second-order differentiation of Eq.(5) and (6) with respect to the inertial frame

$$\ddot{\mathbf{r}}_i = \mathbf{J}_{T_i}(\mathbf{q},t)\ddot{\mathbf{q}} + \frac{\partial \dot{\mathbf{r}}_i}{\partial \mathbf{q}^T}\dot{\mathbf{q}} + \frac{\partial \dot{\mathbf{r}}_i}{\partial t} \tag{9}$$

$$\dot{\boldsymbol{\omega}}_i = \mathbf{J}_{R_i}(\mathbf{q},t)\ddot{\mathbf{q}} + \frac{\partial \boldsymbol{\omega}_i}{\partial \mathbf{q}^T}\dot{\mathbf{q}} + \frac{\partial \boldsymbol{\omega}_i}{\partial t} \tag{10}$$

Rewrite the acceleration equations as follows[32]

$$\ddot{\mathbf{r}}_i = \boldsymbol{\alpha}_i\ddot{\mathbf{q}} + \boldsymbol{\mu}_i \tag{11}$$

$$\dot{\boldsymbol{\omega}}_i = \boldsymbol{\beta}_i\ddot{\mathbf{q}} + \boldsymbol{\sigma}_i \tag{12}$$

where

$$\boldsymbol{\alpha}_i = \mathbf{J}_{T_i}(\mathbf{q},t) \in \mathbb{R}^{3\times f},\ \boldsymbol{\mu}_i = \frac{\partial \dot{\mathbf{r}}_i}{\partial \mathbf{q}^T}\dot{\mathbf{q}} + \frac{\partial \dot{\mathbf{r}}_i}{\partial t} \in \mathbb{R}^3,$$

$$\boldsymbol{\beta}_i = \mathbf{J}_{R_i}(\mathbf{q},t) \in \mathbb{R}^{3\times f},\ \boldsymbol{\sigma}_i = \frac{\partial \boldsymbol{\omega}_i}{\partial \mathbf{q}^T}\dot{\mathbf{q}} + \frac{\partial \boldsymbol{\omega}_i}{\partial t} \in \mathbb{R}^3$$

Thus, the translational and rotational accelerations for each body have been expressed in terms of the generalized coordinates. The acceleration equations of all the bodies in the system could be combined together in a matrix form as

$$\ddot{\mathbf{r}} = \boldsymbol{\alpha}\ddot{\mathbf{q}} + \boldsymbol{\mu} \tag{13}$$

$$\dot{\boldsymbol{\omega}} = \boldsymbol{\beta}\ddot{\mathbf{q}} + \boldsymbol{\sigma} \tag{14}$$

where

$$\ddot{\mathbf{r}} = [\ddot{\mathbf{r}}_1^{\mathrm{T}}\ \ddot{\mathbf{r}}_2^{\mathrm{T}} \cdots \ddot{\mathbf{r}}_n^{\mathrm{T}}]^{\mathrm{T}} \in \mathbb{R}^{3n\times 1}, \dot{\boldsymbol{\omega}} = [\dot{\boldsymbol{\omega}}_1^{\mathrm{T}}\ \dot{\boldsymbol{\omega}}_2^{\mathrm{T}} \cdots \dot{\boldsymbol{\omega}}_n^{\mathrm{T}}]^{\mathrm{T}} \in \mathbb{R}^{3n\times 1}$$

$$\boldsymbol{\alpha} = [\boldsymbol{\alpha}_1^{\mathrm{T}}\ \boldsymbol{\alpha}_2^{\mathrm{T}} \cdots \boldsymbol{\alpha}_n^{\mathrm{T}}]^{\mathrm{T}} \in \mathbb{R}^{3n\times f}, \boldsymbol{\beta} = [\boldsymbol{\beta}_1^{\mathrm{T}}\ \boldsymbol{\beta}_2^{\mathrm{T}} \cdots \boldsymbol{\beta}_n^{\mathrm{T}}]^{\mathrm{T}} \in \mathbb{R}^{3n\times f}$$

$$\boldsymbol{\mu} = [\boldsymbol{\mu}_1^{\mathrm{T}}\ \boldsymbol{\mu}_2^{\mathrm{T}} \cdots \boldsymbol{\mu}_n^{\mathrm{T}}]^{\mathrm{T}} \in \mathbb{R}^{3n\times 1}, \boldsymbol{\sigma} = [\boldsymbol{\sigma}_1^{\mathrm{T}}\ \boldsymbol{\sigma}_2^{\mathrm{T}} \cdots \boldsymbol{\sigma}_n^{\mathrm{T}}]^{\mathrm{T}} \in \mathbb{R}^{3n\times 1}$$

The Newton-Euler equation for each body in this parallel manipulator could be written as

$$m_i \ddot{\mathbf{r}}_i = \overline{\mathbf{F}}_i^a + \overline{\mathbf{F}}_i^n \tag{15}$$

$$\mathbf{J}_i \cdot \dot{\boldsymbol{\omega}}_i + \boldsymbol{\varepsilon}_i = \overline{\mathbf{M}}_i^a + \overline{\mathbf{M}}_i^n \tag{16}$$

where m_i denotes the mass of body i, and $\mathbf{J}_i$ represents the 3×3 inertia tensor with respect to the center of mass. $\overline{\mathbf{F}}_i^n$, $\overline{\mathbf{M}}_i^n$ are the vectors of constraint forces and moments in the ideal joints, $\overline{\mathbf{F}}_i^a$, $\overline{\mathbf{M}}_i^a$ are the equivalent applied forces and moments summarized to the center of mass (including the external applied forces and moments, gravity, and actuating forces in the active joints), and $\boldsymbol{\varepsilon}_i = \boldsymbol{\omega}_i \times (\mathbf{J}_i \cdot \boldsymbol{\omega}_i)$ represents the Coriolis and centrifugal forces and moments with respect to the mass center.

The Newton-Euler equations of the whole mechanism could also be expressed by assembling the above ones in a matrix form as

$$\mathbf{m}\ddot{\mathbf{r}} = \overline{\mathbf{F}}^a + \overline{\mathbf{F}}^n \tag{17}$$

$$\mathbf{J} \cdot \dot{\boldsymbol{\omega}} + \boldsymbol{\varepsilon} = \overline{\mathbf{M}}^a + \overline{\mathbf{M}}^n \tag{18}$$

where $\mathbf{m}$ and $\mathbf{J}$ are the system mass and inertia tensor, $\overline{\mathbf{F}}^a$ and $\overline{\mathbf{M}}^a$ are vectors of total applied forces and moments; while $\overline{\mathbf{F}}^n$ and $\overline{\mathbf{M}}^n$ are the vectors of total constraint forces and moments in the ideal joints, where

$$\mathbf{m} = diag(m_1\mathbf{I}_3, m_2\mathbf{I}_3, \cdots, m_n\mathbf{I}_3),\ \mathbf{J} = diag(\mathbf{J}_1, \mathbf{J}_2, \cdots, \mathbf{J}_n)$$

$$\overline{\mathbf{F}}^a = \begin{bmatrix} \overline{\mathbf{F}}_1^a \\ \overline{\mathbf{F}}_2^a \\ \vdots \\ \overline{\mathbf{F}}_n^a \end{bmatrix}, \overline{\mathbf{M}}^a = \begin{bmatrix} \overline{\mathbf{M}}_1^a \\ \overline{\mathbf{M}}_2^a \\ \vdots \\ \overline{\mathbf{M}}_n^a \end{bmatrix}, \overline{\mathbf{F}}^n = \begin{bmatrix} \overline{\mathbf{F}}_1^n \\ \overline{\mathbf{F}}_2^n \\ \vdots \\ \overline{\mathbf{F}}_n^n \end{bmatrix}, \overline{\mathbf{M}}^n = \begin{bmatrix} \overline{\mathbf{M}}_1^n \\ \overline{\mathbf{M}}_2^n \\ \vdots \\ \overline{\mathbf{M}}_n^n \end{bmatrix}, \boldsymbol{\varepsilon} = \begin{bmatrix} \boldsymbol{\varepsilon}_1 \\ \boldsymbol{\varepsilon}_2 \\ \vdots \\ \boldsymbol{\varepsilon}_n \end{bmatrix}$$

Substituting Eq.(13) and Eq.(14) into Eq.(17) and Eq.(18), and then combining them together, the system dynamic equations of the general parallel manipulator could be given as Newton-Euler formulations with the generalized coordinates $\mathbf{q}$ defined on moveable platform as

$$\mathbf{\Phi} \cdot \mathbf{H}\ddot{\mathbf{q}} + \mathbf{K}(\dot{\mathbf{q}}, \mathbf{q}) = \hat{\mathbf{F}}^{a} + \hat{\mathbf{F}}^{n} \tag{19}$$

where $\hat{\mathbf{F}}^{a}$ are the vectors of applied forces and moments, $\hat{\mathbf{F}}^{n}$ are the vectors of constraint forces and moments; $\mathbf{K}(\dot{\mathbf{q}}, \mathbf{q})$ is the bias term of the forces and moments including the Coriolis forces and centrifugal forces. These coefficient matrices could be determined as follows[30,32]

$$\mathbf{\Phi}_{6n\times 6n} = \begin{bmatrix} \mathbf{m} & \\ & \mathbf{J} \end{bmatrix}, \mathbf{H}_{6n\times f} = \begin{bmatrix} \boldsymbol{\alpha} \\ \boldsymbol{\beta} \end{bmatrix}, \mathbf{K}_{6n\times 1} = \begin{bmatrix} \mathbf{m}\boldsymbol{\mu} \\ \mathbf{J}\boldsymbol{\sigma} + \boldsymbol{\varepsilon} \end{bmatrix}, \hat{\mathbf{F}}^{a}_{6n\times 1} = \begin{bmatrix} \bar{\mathbf{F}}^{a} \\ \bar{\mathbf{M}}^{a} \end{bmatrix}, \hat{\mathbf{F}}^{n}_{6n\times 1} = \begin{bmatrix} \bar{\mathbf{F}}^{n} \\ \bar{\mathbf{M}}^{n} \end{bmatrix} \tag{20}$$

Equation (19) is a set of $6n$ differentiable-algebraic equations which have been derived as a function of the generalized coordinates defined on movable platform.

The $6n\times 1$ vector of constraint forces and moments $\hat{\mathbf{F}}^{n}$ could be expressed as

$$\hat{\mathbf{F}}^{n} = \bar{\mathbf{Q}}\boldsymbol{\lambda} \tag{21}$$

where $(6n\text{-}f)\times 1$ vector $\boldsymbol{\lambda}$ denotes the generalized constraints forces known as the Lagrange multiplier, and $\bar{\mathbf{Q}}$ denotes the $6n\times(6n\text{-}f)$ distribution matrix. So Eq.(19) could be rewritten as

$$\mathbf{\Phi} \cdot \mathbf{H}\ddot{\mathbf{q}} + \mathbf{K}(\dot{\mathbf{q}}, \mathbf{q}) = \hat{\mathbf{F}}^{a} + \bar{\mathbf{Q}}\boldsymbol{\lambda} \tag{22}$$

Equation (22) leads to the Lagrange equation of the first kind. The constraint forces in Eq.(22) could be eliminated by utilizing the principle of virtual work if all joint are assumed to be ideal or the friction in the joint is considered as applied forces. Left-multiplying Eq.(22) with the transposition of $\mathbf{H}$, we get

$$\mathbf{H}^{T} \cdot \mathbf{\Phi} \cdot \mathbf{H}\ddot{\mathbf{q}} + \mathbf{H}^{T} \cdot \mathbf{K}(\dot{\mathbf{q}}, \mathbf{q}) = \mathbf{H}^{T} \cdot \hat{\mathbf{F}}^{a} + \mathbf{H}^{T} \cdot \bar{\mathbf{Q}}\boldsymbol{\lambda} \tag{23}$$

According to the principle of virtual work, the term containing the constraint forces can be eliminated, for

$$\mathbf{H}^{T} \cdot \bar{\mathbf{Q}}\boldsymbol{\lambda} = 0 \tag{24}$$

So the equation of motion of this general parallel manipulator could be derived by rewriting the above equation as

$$\mathbf{H}^{T} \cdot \mathbf{\Phi} \cdot \mathbf{H}\ddot{\mathbf{q}} = \mathbf{H}^{T} \cdot \hat{\mathbf{F}}^{a} - \mathbf{H}^{T} \cdot \mathbf{K}(\dot{\mathbf{q}}, \mathbf{q}) \tag{25}$$

Equation (25) is a set of *f* ordinary differential equations with the minimal number of degree of freedom. Both the forward and inverse dynamic analysis equation could be derived from this equation.

For the forward dynamic analysis, Eq.(25) is the equation of motion of the system. The trajectory of the movable platform could be obtained by numerical integration when the generalized applied forces are given. The constraint forces could be also obtained based on this method. By left-multiplying Equation (22) with $\bar{\mathbf{Q}}^T \cdot \mathbf{\Phi}^{-1}$, the term containing the second-order differentiation of the generalized coordinates will also be eliminated according to the principle of virtual work. Then, the equation of reaction could be obtained as

$$\bar{\mathbf{Q}}^T \cdot \mathbf{\Phi}^{-1} \cdot \bar{\mathbf{Q}}\boldsymbol{\lambda} = \bar{\mathbf{Q}}^T \cdot \mathbf{\Phi}^{-1} \cdot \mathbf{K} - \bar{\mathbf{Q}}^T \cdot \mathbf{\Phi}^{-1} \cdot \hat{\mathbf{F}}^{\mathbf{a}} \tag{26}$$

Equation (26) is a set of $6n$-f purely algebraic equations, and the generalized constraint forces on all the ideal joints could be easily calculated out.

For the inverse dynamic problem, the required actuating forces could be obtained without solving the constraint forces. Eq.(25) could be rewritten as

$$\mathbf{H}^T \cdot \hat{\mathbf{F}}^{\mathbf{a}} = \mathbf{H}^T \cdot \mathbf{\Phi} \cdot \mathbf{H}\ddot{\mathbf{q}} + \mathbf{H}^T \cdot \mathbf{K}(\dot{\mathbf{q}}, \mathbf{q}) \tag{27}$$

Equation (27) turns out to be a set of *f* purely algebraic equations. When the configuration, velocity and acceleration of the movable platform are given, and the item of applied forces could be solved out, from which the actuating forces in the active joints can be obtained.

4. Case Study of a 3-DOF Parallel Manipulator

In this section, the forward dynamics of a 3-DOF parallel manipulator proposed by X.J.Liu[33] is presented as the case study of the method. Numerical solutions of the trajectory and velocity of the platform are given as the response of driving force on actuated joints. The inverse dynamic analysis of this mechanism, which has been reported on [34], is also implemented by the proposed methods in this chapter.

As shown in figure 2 (a), the mechanism[33] is composed of a movable platform connected to a fixed base by three legs, two of which have the identical structure, which consists of a prismatic joint connecting the leg to the fixed base, a universal joint connecting the leg to the platform and a revolute joint connecting the two links of the leg; the third leg consists of a four-bar parallelogram and two revolute joints which connects the leg with the platform and a slider. This leg is attached to the fixed base by a prismatic joint. According to the kinematics analysis, the movable platform has two degrees of translational freedom and one degree of rotational freedom.

Since all the vectors and tensors are expressed by the matrixs form in a reference frame in this section, the vectors and tensors will not be express in bold in the equations as ther are expressed in section 3.

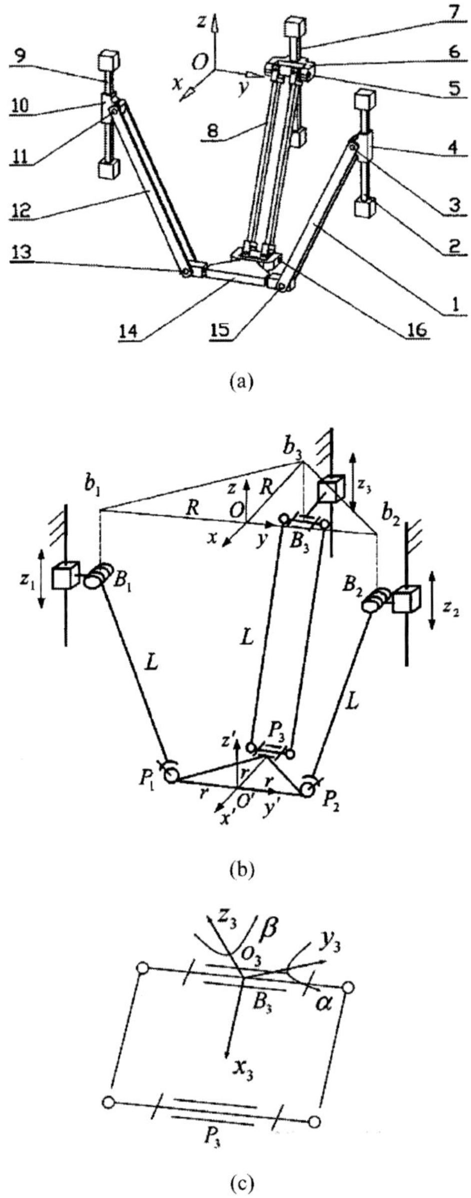

Figure 2. Structure and geometric parameters of spatial 3-DOF parallel manipulator[33]

As shown in figure 2(b), the reference frame *O-xyz* fixed to the base is assumed to be located at the median point of the side b_1b_2 of the fixed base with the *z* axis normal to the base plate and the *y* axis directed along the side b_1b_2. The body-fixed frame *O'-x'y'z'* is attached to the platform with the origin located at the center point of side P_1P_2, the *z'* axis perpendicular to the platform, and the *y'* axis directed along with the line P_1P_2. The movable platform has the mobility of moving along *y* axis, *z* axis, and rotating about *y* axis with respect to the reference frame.

Assuming that the fixed-base and the platform are isosceles triangles described by the parameters R and r, respectively, where $Ob_i=R$, $O'P_i=r$ (i=1,2,3), and that the length of the link for each leg is denoted as L, where $P_ib_i=L$ (i=1,2,3), the configuration of the mechanism can be determined once the input of the actuated prismatic joints is given or the position and orientation of the platform is specified.

According to the proposed method, the position vector and rotation angle of the platform should be selected as the system generalized coordinates,

$$\mathrm{q} = (y, z, \varphi)^T \tag{28}$$

The position vector and direction cosine matrix of the body-fixed frame O'-$x'y'z'$ with respect to the reference frame O-xyz could be obtained as

$$\mathrm{r}_\mathrm{p} = (0, y, z)^T \tag{29}$$

$$\mathrm{A}_\mathrm{rp} = \begin{bmatrix} c_\varphi & 0 & s_\varphi \\ 0 & 1 & 0 \\ -s_\varphi & 0 & c_\varphi \end{bmatrix} \tag{30}$$

The velocity and acceleration of the platform, both translational and rotational, could be derived from the above equations as

$$\dot{\mathrm{r}}_\mathrm{p} = (0, \dot{y}, \dot{z})^T \tag{31}$$

$$\omega_\mathrm{p} = (0, \dot{\varphi}, 0)^T \tag{32}$$

$$\ddot{\mathrm{r}}_\mathrm{p} = (0, \ddot{y}, \ddot{z})^T \tag{33}$$

$$\dot{\omega}_\mathrm{p} = (0, \ddot{\varphi}, 0)^T \tag{34}$$

Then, the position vector of the point P_i with respect to the reference frame could be expressed by

$$\mathrm{r}_{P_i} = \mathrm{r}_\mathrm{p} + \mathrm{A}_\mathrm{rp}\, {}^\mathrm{p}\rho_{P_i} \tag{35}$$

In Eq.(35), ${}^\mathrm{p}\rho_{P_i}$ is the body-joint vector of the corresponding joint P_i in the body-fixed frame, which could obtained from

$${}^\mathrm{p}\rho_{P_1} = \begin{pmatrix} 0 & -r & 0 \end{pmatrix}^T, {}^\mathrm{p}\rho_{P_2} = \begin{pmatrix} 0 & r & 0 \end{pmatrix}^T, {}^\mathrm{p}\rho_{P_3} = \begin{pmatrix} -r & 0 & 0 \end{pmatrix}^T \tag{36}$$

Therefore the velocity and acceleration of the point P_i can be expressed by the time differentiation of Eq.(35) with respect to the reference frame as

$$\dot{\mathrm{r}}_{P_i} = \dot{\mathrm{r}}_{\mathrm{p}} + \omega_{\mathrm{p}} \times \rho_{P_i} \tag{37}$$

$$\ddot{\mathrm{r}}_{P_i} = \ddot{\mathrm{r}}_{\mathrm{p}} + \dot{\omega}_{\mathrm{p}} \times \rho_{P_i} + \omega_{\mathrm{p}} \times (\omega_{\mathrm{p}} \times \rho_{P_i}) \tag{38}$$

where $\rho_{P_i} = \mathrm{A}_{\mathrm{rp}} \, {}^{\mathrm{P}}\rho_{P_i}$ is the body-joint vector with respect to the reference frame.

From the geometric constraint of legs, we could get the following constraint equations of the active prismatic joints:

$$(\mathrm{r}_{P_i} - \mathrm{r}_{B_i})^T (\mathrm{r}_{P_i} - \mathrm{r}_{B_i}) = L^2 \tag{39}$$

where r_{B_i} denotes the position vector of the corresponding joint B_i in the reference frame, which could be obtained from

$$\mathrm{r}_{B_1} = \begin{pmatrix} 0 & -R & z_1 \end{pmatrix}^T, \mathrm{r}_{B_2} = \begin{pmatrix} 0 & R & z_2 \end{pmatrix}^T, \mathrm{r}_{B_3} = \begin{pmatrix} -R & 0 & z_3 \end{pmatrix}^T \tag{40}$$

Thus, the inputs of the active joints could be calculated directly[33] from equation (39), that is

$$z_1 = M_1 \sqrt{L^2 - (-r + y + R)^2} + z \tag{41}$$

$$z_2 = M_2 \sqrt{L^2 - (r + y - R)^2} + z \tag{42}$$

$$z_3 = M_3 \sqrt{L^2 - (-r\cos\varphi + R)^2 - y^2} + r\sin\varphi + z \tag{43}$$

where $M_i = \pm 1 \ (i = 1, 2, 3)$ is the configuration coefficient of the mechanism.

The velocities and the accelerations of joint points B_i could be derived by differentiating Eq.(39) as

$$(\mathrm{r}_{P_i} - \mathrm{r}_{B_i})^T (\dot{\mathrm{r}}_{P_i} - \dot{\mathrm{r}}_{B_i}) = 0 \tag{44}$$

Solving the above equation for the velocity $\dot{z}_i$, we get

$$\dot{z}_i = \frac{(\mathrm{r}_{P_i} - \mathrm{r}_{B_i})^T \dot{\mathrm{r}}_{P_i}}{(\mathrm{r}_{P_i} - \mathrm{r}_{B_i})^T \boldsymbol{e}_3} \tag{45}$$

where $\boldsymbol{e}_3 = (0\ \ 0\ \ 1)^T$

The accelerations of joint points B_i could be obtained by differentiating Eq.(45) as

$$\ddot{z}_i = \frac{(\dot{\mathrm{r}}_{P_i} - \dot{\mathrm{r}}_{B_i})^T (\dot{\mathrm{r}}_{P_i} - \dot{\mathrm{r}}_{B_i}) + (\mathrm{r}_{P_i} - \mathrm{r}_{B_i})^T \ddot{\mathrm{r}}_{P_i}}{(\mathrm{r}_{P_i} - \mathrm{r}_{B_i})^T \boldsymbol{e}_3} \tag{47}$$

Then, the velocity and acceleration of joint points B_i with respect to the reference frame could be obtained as

$$\dot{\mathrm{r}}_{B_i} = \boldsymbol{e}_3 \dot{z}_i = \frac{\boldsymbol{e}_3 (\mathrm{r}_{P_i} - \mathrm{r}_{B_i})^T}{(\mathrm{r}_{P_i} - \mathrm{r}_{B_i})^T \boldsymbol{e}_3} \dot{\mathrm{r}}_{P_i} \tag{48}$$

$$\ddot{\mathrm{r}}_{B_i} = \boldsymbol{e}_3 \ddot{z}_i = \frac{\boldsymbol{e}_3 (\mathrm{r}_{P_i} - \mathrm{r}_{B_i})^T}{(\mathrm{r}_{P_i} - \mathrm{r}_{B_i})^T \boldsymbol{e}_3} \ddot{\mathrm{r}}_{P_i} + \frac{\boldsymbol{e}_3 (\dot{\mathrm{r}}_{P_i} - \dot{\mathrm{r}}_{B_i})^T (\dot{\mathrm{r}}_{P_i} - \dot{\mathrm{r}}_{B_i})}{(\mathrm{r}_{P_i} - \mathrm{r}_{B_i})^T \boldsymbol{e}_3} \tag{49}$$

Assuming that the mass centers for each link and slider are located at the geometric center, then their position vectors could be obtained as

$$\mathrm{r}_{L_i} = \frac{\mathrm{r}_{P_i} + \mathrm{r}_{B_i}}{2} \quad (i = 1, 2, 3) \tag{50}$$

$$\mathrm{r}_{S_i} = \mathrm{r}_{B_i} \quad (i = 1, 2, 3) \tag{51}$$

Then, the velocities and accelerations of the mass center of each link can be expressed as follows:

$$\dot{\mathrm{r}}_{L_i} = \frac{\dot{\mathrm{r}}_{P_i} + \dot{\mathrm{r}}_{B_i}}{2} \quad (i = 1, 2, 3) \tag{52}$$

$$\dot{\mathrm{r}}_{S_i} = \dot{\mathrm{r}}_{B_i} \quad (i = 1, 2, 3) \tag{53}$$

$$\ddot{\mathrm{r}}_{L_i} = \frac{\ddot{\mathrm{r}}_{P_i} + \ddot{\mathrm{r}}_{B_i}}{2} \quad (i = 1, 2, 3) \tag{54}$$

$$\ddot{\mathrm{r}}_{S_i} = \ddot{\mathrm{r}}_{B_i} \quad (i = 1, 2, 3) \tag{55}$$

Let ϕ_i (i=1,2) be the rotation angle of leg i about the joint B_i, ϕ_i could be described as

$$\phi_i = \arcsin\frac{\boldsymbol{e}_2^T(\mathrm{r}_{P_i} - \mathrm{r}_{B_i})}{L} \quad (i = 1, 2) \tag{56}$$

where $\boldsymbol{e}_2 = (0\ 1\ 0)^T$

The angular velocity and acceleration of the legs could be derived by time differentiation of the above equation with respect to the reference frame

$$\dot{\phi}_i = \frac{\boldsymbol{e}_2^T}{L\cos\phi_i}(\dot{\mathrm{r}}_{P_i} - \dot{\mathrm{r}}_{B_i}) \quad (i = 1, 2) \tag{57}$$

$$\ddot{\phi}_i = \frac{\boldsymbol{e}_2^T}{L\cos\phi_i}(\ddot{\mathrm{r}}_{P_i} - \ddot{\mathrm{r}}_{B_i}) + \tan\phi_i\dot{\phi}_i^2 \quad (i = 1, 2) \tag{58}$$

So the rotational velocity and acceleration of the link in leg i could be written as

$$\omega_{L_i} = \boldsymbol{e}_1\dot{\phi}_i = \frac{\boldsymbol{e}_1\boldsymbol{e}_2^T}{L\cos\phi_i}(\dot{\mathrm{r}}_{P_i} - \dot{\mathrm{r}}_{B_i}) \quad (i = 1, 2) \tag{59}$$

$$\dot{\omega}_{L_i} = \boldsymbol{e}_1\ddot{\phi}_i = \frac{\boldsymbol{e}_1\boldsymbol{e}_2^T}{L\cos\phi_i}(\ddot{\mathrm{r}}_{P_i} - \ddot{\mathrm{r}}_{B_i}) + \boldsymbol{e}_1\tan\phi_i\dot{\phi}_i^2 \quad (i = 1, 2) \tag{60}$$

where $\boldsymbol{e}_1 = (1\ 0\ 0)^T$

As for the third leg, because of the geometric constraint of the parallelogram structure, two links along the leg must have the same velocities and accelerations at their mass centers. To simplify the process of dynamics modeling, we consider this two links as one body which is connected to the platform and fixed frame both with universal joint. A body-fixed frame O_3-$x_3y_3z_3$ of the third leg is defined as shown in figure 2(c), the origin of the frame is located at the center of the revolute joint B_3, with x_3 axis points from B_3 to P_3, and the z_3 axis directs vertically to the plane where the parallelogram structure lays on, and y_3 axis is defined by the right-hand rule.The orientation of the body-fixed frame O_3-$x_3y_3z_3$ could be described by rotations about two axis: angle α rotating about y_3 axis followed by another rotation of angle β about the z_3 axis. Then, the direction cosine matrix from frame O_3-$x_3y_3z_3$ to reference frame O-xyz could be written as

$$\mathrm{A}_{r3} = \begin{bmatrix} c_\alpha & 0 & s_\alpha \\ 0 & 1 & 0 \\ -s_\alpha & 0 & c_\alpha \end{bmatrix}\begin{bmatrix} c_\beta & -s_\beta & 0 \\ s_\beta & c_\beta & 0 \\ 0 & 0 & 1 \end{bmatrix} = \begin{bmatrix} c_\alpha c_\beta & -c_\alpha s_\beta & s_\alpha \\ s_\beta & c_\beta & 0 \\ -s_\alpha c_\beta & s_\alpha s_\beta & c_\alpha \end{bmatrix} \tag{61}$$

Transforming the vector ${}^{3}\mathrm{L}_3 = L\boldsymbol{e}_1$ in the body-fixed frame O_3-$x_3y_3z_3$ to the reference frame O-xyz, it yields

$$\mathrm{L}_3 = \mathrm{A}_{\mathrm{r}3}{}^{3}\mathrm{L}_3 = \mathrm{r}_{P_i} - \mathrm{r}_{B_i} \quad (62)$$

By simplifying Eq.(62), following result could be obtained

$$\frac{\mathrm{r}_{P_i} - \mathrm{r}_{B_i}}{L} = \begin{bmatrix} c_\alpha c_\beta \\ s_\beta \\ -s_\alpha c_\beta \end{bmatrix} \quad (63)$$

Solving Eq.(63) for angle α and β, it gives

$$s_\beta = \frac{\boldsymbol{e}_2^T(\mathrm{r}_{P_i} - \mathrm{r}_{B_i})}{L} \quad (64)$$

$$s_\alpha = -\frac{\boldsymbol{e}_3^T(\mathrm{r}_{P_i} - \mathrm{r}_{B_i})}{Lc_\beta} \quad (65)$$

$$c_\alpha = \frac{\boldsymbol{e}_1^T(\mathrm{r}_{P_i} - \mathrm{r}_{B_i})}{Lc_\beta} \quad (66)$$

The relative velocity of P_3 with respect to B_3 can be expressed in two ways in the body-fixed frame O_3-$x_3y_3z_3$ as

$${}^{3}\dot{\mathrm{r}}_{P_3} = \mathrm{A}_{3\mathrm{r}}(\dot{\mathrm{r}}_{P_3} - \dot{\mathrm{r}}_{B_3}) \quad (67)$$

$${}^{3}\dot{\mathrm{r}}_{P_3} = {}^{3}\omega_3 \times {}^{3}\mathrm{L}_3 = {}^{3}\omega_3 \times \boldsymbol{e}_1 L \quad (68)$$

where the direction cosine matrix from frame O-xyz to O_3-$x_3y_3z_3$ is the transposition of the matrix $\mathrm{A}_{\mathrm{r}3}$, that is $\mathrm{A}_{3\mathrm{r}} = \mathrm{A}_{\mathrm{r}3}^T$.

Since the third leg can not rotate about the x_3 axis, namely, ${}^{3}\omega_3 \perp \boldsymbol{e}_1 = 0$. Left cross-multiplying Equation (68) with $\boldsymbol{e}_1$, it turns out

$${}^{3}\omega_3 = \frac{\tilde{\boldsymbol{e}}_1 \mathrm{A}_{3\mathrm{r}}(\dot{\mathrm{r}}_{P_3} - \dot{\mathrm{r}}_{B_3})}{L} \quad (69)$$

where $\tilde{\boldsymbol{e}}_1 = \begin{bmatrix} 0 & 0 & 0 \\ 0 & 0 & -1 \\ 0 & 1 & 0 \end{bmatrix}$ is the coordinate square matrix of the unit vector $\boldsymbol{e}_1$.

Likewise, the relative acceleration of P_3 with respect to B_3 in body-fixed frame could be also obtained in the same way.

$$^3\ddot{\mathrm{r}}_{P_3} = \mathrm{A}_{3\mathrm{r}}(\ddot{\mathrm{r}}_{P_3} - \ddot{\mathrm{r}}_{B_3}) \tag{70}$$

$$^3\ddot{\mathrm{r}}_{P_3} = {}^3\dot{\omega}_3 \times {}^3\mathrm{L}_3 + {}^3\omega_3 \times ({}^3\omega_3 \times {}^3\mathrm{L}_3) = {}^3\dot{\omega}_3 \times \boldsymbol{e}_1 L + {}^3\omega_3 \times ({}^3\omega_3 \times \boldsymbol{e}_1) L \tag{71}$$

Cross-multiplying the both sides of Eq.(71) with $\boldsymbol{e}_1$ yields

$$\boldsymbol{e}_1 \times {}^3\ddot{\mathrm{r}}_{P_3} = \boldsymbol{e}_1 \times {}^3\dot{\omega}_3 \times \boldsymbol{e}_1 L + \boldsymbol{e}_1 \times {}^3\omega_3 \times ({}^3\omega_3 \times \boldsymbol{e}_1) L \tag{72}$$

$$^3\dot{\omega}_3 = \frac{\tilde{\boldsymbol{e}}_1 \mathrm{A}_{3\mathrm{r}}}{L}(\ddot{\mathrm{r}}_{P_3} - \ddot{\mathrm{r}}_{B_3}) \tag{73}$$

Transforming the angular velocity and acceleration of the third leg from the body-fixed frame O_3-$x_3y_3z_3$ to the reference frame, it yields

$$\omega_{L_3} = \mathrm{A}_{\mathrm{r}3}\,{}^3\omega_3 = \frac{\mathrm{A}_{\mathrm{r}3}\tilde{\boldsymbol{e}}_1 \mathrm{A}_{3\mathrm{r}}}{L}(\dot{\mathrm{r}}_{P_3} - \dot{\mathrm{r}}_{B_3}) \tag{74}$$

$$\dot{\omega}_{L_3} = \mathrm{A}_{\mathrm{r}3}\,{}^3\dot{\omega}_3 = \frac{\mathrm{A}_{\mathrm{r}3}\tilde{\boldsymbol{e}}_1 \mathrm{A}_{3\mathrm{r}}}{L}(\ddot{\mathrm{r}}_{P_3} - \ddot{\mathrm{r}}_{B_3}) \tag{75}$$

Thus, the translational and rotational velocities and accelerations for all the bodies in this mechanism have been derived as the functions of the generalized coordinates and their differentiations with respect to the reference frame.

Rewriting the acceleration equations for each body in the mechanism, the formulations are shown as follows:

For the platform:

$$\ddot{\mathrm{r}}_{\mathrm{p}} = \alpha_{\mathrm{p}}\ddot{\mathrm{q}} + \mu_{\mathrm{p}} \tag{76}$$

$$\dot{\omega}_{\mathrm{p}} = \beta_{\mathrm{p}}\ddot{\mathrm{q}} + \sigma_{\mathrm{p}} \tag{77}$$

$$\text{where}\,\alpha_{\rm p} = \begin{bmatrix} 0 & 0 & 0 \\ 1 & 0 & 0 \\ 0 & 1 & 0 \end{bmatrix},\ \beta_{\rm p} = \begin{bmatrix} 0 & 0 & 0 \\ 0 & 0 & 1 \\ 0 & 0 & 0 \end{bmatrix},\ \mu_{\rm p} = \begin{bmatrix} 0 \\ 0 \\ 0 \end{bmatrix},\ \sigma_{\rm p} = \begin{bmatrix} 0 \\ 0 \\ 0 \end{bmatrix}$$

For the bodies of the legs:

The sliders:

$$\ddot{\mathrm{r}}_{S_i} = \alpha_{S_i}\ddot{\mathrm{q}} + \mu_{S_i} \quad (i=1,2,3) \tag{78}$$

$$\dot{\omega}_{S_i} = \beta_{S_i}\ddot{\mathrm{q}} + \sigma_{S_i} \quad (i=1,2,3) \tag{79}$$

where

$$\alpha_{S_i} = \frac{\boldsymbol{e}_3(\mathrm{r}_{P_i} - \mathrm{r}_{B_i})^T(\alpha_{\rm p} - \tilde{\rho}_{P_i}\beta_{\rm p})}{(\mathrm{r}_{P_i} - \mathrm{r}_{B_i})^T\boldsymbol{e}_3},\ \beta_{S_i} = \begin{bmatrix} 0 & 0 & 0 \\ 0 & 0 & 0 \\ 0 & 0 & 0 \end{bmatrix}$$

$$\mu_{S_i} = -\frac{\boldsymbol{e}_3(\mathrm{r}_{P_i} - \mathrm{r}_{B_i})^T\tilde{\omega}_{\rm p}\tilde{\rho}_{P_i}\omega_{\rm p}}{(\mathrm{r}_{P_i} - \mathrm{r}_{B_i})^T\boldsymbol{e}_3} + \frac{\boldsymbol{e}_3(\dot{\mathrm{r}}_{P_i} - \dot{\mathrm{r}}_{B_i})^T(\dot{\mathrm{r}}_{P_i} - \dot{\mathrm{r}}_{B_i})}{(\mathrm{r}_{P_i} - \mathrm{r}_{B_i})^T\boldsymbol{e}_3},\ \sigma_{S_i} = \begin{bmatrix} 0 \\ 0 \\ 0 \end{bmatrix}$$

and $\tilde{\omega}_{\rm p}, \tilde{\rho}_{P_i}$ are the coordinate square matrices of the corresponding vectors.

The links:

$$\ddot{\mathrm{r}}_{L_i} = \alpha_{L_i}\ddot{\mathrm{q}} + \mu_{L_i} \quad (i=1,2,3) \tag{80}$$

where,

$$\alpha_{L_i} = \frac{\alpha_{\rm p} - \tilde{\rho}_{P_i}\beta_{\rm p} + \alpha_{S_i}}{2},\quad \mu_{L_i} = -\frac{\tilde{\omega}_{\rm p}\tilde{\rho}_{P_i}\omega_{\rm p} - \mu_{S_i}}{2}$$

$$\dot{\omega}_{L_i} = \beta_{L_i}\ddot{\mathrm{q}} + \sigma_{L_i} \quad (i=1,2) \tag{81}$$

where

$$\beta_{L_i} = \frac{\boldsymbol{e}_1\boldsymbol{e}_2^T(\alpha_{\rm p} - \tilde{\rho}_{P_i}\beta_{\rm p} - \alpha_{S_i})}{L\cos\phi_i}, \sigma_{L_i} = -\frac{\boldsymbol{e}_1\boldsymbol{e}_2^T(\tilde{\omega}_{\rm p}\tilde{\rho}_{P_i}\omega_{\rm p} + \mu_{S_i})}{L\cos\phi_i} + \boldsymbol{e}_1\tan\phi_i\dot{\phi}_i^2$$

$$\dot{\omega}_{L_3} = \beta_{L_3}\ddot{\mathrm{q}} + \sigma_{L_3} \tag{82}$$

where

$$\beta_{L_3} = \frac{\mathrm{A}_{r3}\,\tilde{e}_1\,\mathrm{A}_{3r}(\alpha_p - \tilde{\rho}_{P_3}\beta_p - \alpha_{S_3})}{L}, \sigma_{L_3} = -\frac{\mathrm{A}_{r3}\,\tilde{e}_1\,\mathrm{A}_{3r}(\tilde{\omega}_p\tilde{\rho}_{P_3}\omega_p + \mu_{S_3})}{L}$$

Equations (76)~(82) of acceleration could be assembled in a matrix form as

$$\ddot{\mathrm{r}} = \alpha\,\ddot{\mathrm{q}} + \mu \tag{83}$$

$$\dot{\omega} = \beta\,\ddot{\mathrm{q}} + \sigma \tag{84}$$

where

$$\ddot{\mathrm{r}} = \begin{bmatrix} \ddot{\mathrm{r}}_P \\ \ddot{\mathrm{r}}_{S_1} \\ \vdots \\ \ddot{\mathrm{r}}_{L_3} \end{bmatrix}_{21\times1}, \dot{\omega} = \begin{bmatrix} \dot{\omega}_P \\ \dot{\omega}_{S_1} \\ \vdots \\ \dot{\omega}_{L_3} \end{bmatrix}_{21\times1}, \alpha = \begin{bmatrix} \alpha_P \\ \alpha_{S_1} \\ \vdots \\ \alpha_{L_3} \end{bmatrix}_{21\times3}, \beta = \begin{bmatrix} \beta_P \\ \beta_{S_1} \\ \vdots \\ \beta_{L_3} \end{bmatrix}_{21\times3}, \mu = \begin{bmatrix} \mu_P \\ \mu_{S_1} \\ \vdots \\ \mu_{L_3} \end{bmatrix}_{21\times1}, \sigma = \begin{bmatrix} \sigma_P \\ \sigma_{S_1} \\ \vdots \\ \sigma_{L_3} \end{bmatrix}_{21\times1}$$

The Newton-Euler equations of all the bodies in this mechanism are

For the platform:

$$m_p\,\ddot{\mathrm{r}}_p = \bar{\mathrm{F}}_p^a + \bar{\mathrm{F}}_p^n \tag{85}$$

$$\mathrm{J}_p\,\dot{\omega}_p + \varepsilon_p = \bar{\mathrm{M}}_P^a + \bar{\mathrm{M}}_P^n \tag{86}$$

where m_p is the mass of the platform and the $\mathrm{J}_p = \mathrm{A}_{rp}\,{}^{p}\mathrm{J}_p\,\mathrm{A}_{pr}$ is the inertial matrix of the platform about the center of mass expressed in the reference frame; $\bar{\mathrm{F}}_p^a$ and $\bar{\mathrm{M}}_p^a$ are the applied forces and moments exerted on the mass center of the platform, including the force of gravity; $\varepsilon_p = \omega_p \times (\mathrm{J}_p\,\omega_p)$ is the item of Coriolis force and centrifugal force; $\bar{\mathrm{F}}_p^n$ and $\bar{\mathrm{M}}_p^n$ are the equivalent constraint forces and moments applied on platform with respect to the mass center, which could be obtained as

$$\bar{\mathrm{F}}_p^n = \sum_{i=1}^{3} S_{pP_i}\,\mathrm{F}_{P_i}^n,\ \bar{\mathrm{M}}_p^n = \sum_{i=1}^{3} S_{pP_i}\,\mathrm{M}_{P_i}^n + \sum_{i=1}^{3} S_{pP_i}\,\mathrm{A}_{rp}\rho_{P_i} \times \mathrm{F}_{P_i}^n \tag{87}$$

where $\mathrm{F}_{P_i}^n$ and $\mathrm{M}_{P_i}^n$ are the constraint forces and moments reacted in the joint P_i, and $S_{pP_i} = \pm 1$ are determined by the definition of the direction of the constraint forces.

Similarly, the Newton-Euler equations of bodies of legs could be obtained by the same method:

For the sliders

$$m_{S_i}\ddot{\mathrm{r}}_{S_i} = \bar{\mathrm{F}}^a_{S_i} + \bar{\mathrm{F}}^n_{S_i} \tag{88}$$

$$\mathrm{J}_{S_i}\dot{\omega}_{S_i} + \varepsilon_{S_i} = \bar{\mathrm{M}}^a_{S_i} + \bar{\mathrm{M}}^n_{S_i} \tag{89}$$

Under the assumption that the axes of the prismatic joint and revolute joint are through the center of mass of the slider, which implys that constraint forces are applied at the mass center and there are no moment on the mass center, the equivalent constraint forces and moments in Eq.(86) and (87) are given by

$$\bar{\mathrm{F}}^n_{S_i} = S_{S_iZ_i}\mathrm{F}^n_{Z_i} + S_{S_iB_i}\mathrm{F}^n_{B_i},\ \bar{\mathrm{M}}^n_{S_i} = S_{S_iZ_i}\mathrm{M}^n_{Z_i} + S_{S_iB_i}\mathrm{M}^n_{B_i} \tag{90}$$

For the links of legs

$$m_{L_i}\ddot{\mathrm{r}}_{L_i} = \bar{\mathrm{F}}^a_{L_i} + \bar{\mathrm{F}}^n_{L_i} \tag{91}$$

$$\mathrm{J}_{L_i}\dot{\omega}_{L_i} + \varepsilon_{L_i} = \bar{\mathrm{M}}^a_{L_i} + \bar{\mathrm{M}}^n_{L_i} \tag{92}$$

where

$$\bar{\mathrm{F}}^n_{L_i} = S_{L_iB_i}\mathrm{F}^n_{B_i} + S_{L_iP_i}\mathrm{F}^n_{P_i},\ \bar{\mathrm{M}}^n_{L_i} = S_{L_iP_i}\mathrm{M}^n_{P_i} + S_{L_iB_i}\mathrm{M}^n_{B_i} + S_{L_iP_i}\mathrm{A}_{ri}\,{}^i\rho_{P_i}\times\mathrm{F}^n_{P_i} + S_{L_iB_i}\mathrm{A}_{ri}\,{}^i\rho_{B_i}\times\mathrm{F}^n_{B_i} \tag{93}$$

The Newton-Euler equations of this 3-DOF parallel manipulator could be generated by assembling the equation of each body as

$$\mathrm{m}\ddot{\mathrm{r}} = \bar{\mathrm{F}}^a + \bar{\mathrm{F}}^n \tag{94}$$

$$\mathrm{J}\dot{\omega} + \varepsilon = \bar{\mathrm{M}}^a + \bar{\mathrm{M}}^n \tag{95}$$

The items in the Eq.(94) and (95) are

$$\mathrm{m} = \begin{bmatrix} m_{\mathrm{P}}\mathrm{I}_3 & 0 & 0 & 0 \\ 0 & m_{S_1}\mathrm{I}_3 & 0 & 0 \\ 0 & 0 & \ddots & 0 \\ 0 & 0 & 0 & m_{L_3}\mathrm{I}_3 \end{bmatrix}_{21\times 21},\ \mathrm{J} = \begin{bmatrix} \mathrm{J}_{\mathrm{P}} & 0 & 0 & 0 \\ 0 & \mathrm{J}_{S_1} & 0 & 0 \\ 0 & 0 & \ddots & 0 \\ 0 & 0 & 0 & \mathrm{J}_{L_3} \end{bmatrix}_{21\times 21}, \tag{96}$$

$$\bar{M}^a = \begin{bmatrix} \bar{M}^a_P \\ \bar{M}^a_{S_1} \\ \vdots \\ \bar{M}^a_{L_3} \end{bmatrix}_{21\times1}, \bar{F}^n = \begin{bmatrix} \bar{F}^n_P \\ \bar{F}^n_{S_1} \\ \vdots \\ \bar{F}^n_{L_3} \end{bmatrix}_{21\times1}, \bar{F}^a = \begin{bmatrix} \bar{F}^a_P \\ \bar{F}^a_{S_1} \\ \vdots \\ \bar{F}^a_{L_3} \end{bmatrix}_{21\times1}, \varepsilon = \begin{bmatrix} \varepsilon_P \\ \varepsilon_{S_1} \\ \vdots \\ \varepsilon_{L_3} \end{bmatrix}_{21\times1}, \bar{M}^n = \begin{bmatrix} \bar{M}^n_P \\ \bar{M}^n_{S_1} \\ \vdots \\ \bar{M}^n_{L_3} \end{bmatrix}_{21\times1} \quad (97)$$

Substituting the acceleration equation into the system Newton-Euler equations, the system dynamic equation can be obtained with a formulation in generalized coordinates as,

$$\Phi H\ddot{q} + K(\dot{q}, q) = \hat{F}^a + \hat{F}^n \quad (98)$$

where

$$\Phi = \begin{bmatrix} m & 0 \\ 0 & J \end{bmatrix}_{42\times42}, H = \begin{bmatrix} \alpha \\ \beta \end{bmatrix}_{42\times3}, K = \begin{bmatrix} m\mu \\ J\sigma + \varepsilon \end{bmatrix}_{42\times1}, \hat{F}^a = \begin{bmatrix} \bar{F}^a \\ \bar{M}^a \end{bmatrix}_{42\times1}, \hat{F}^n = \begin{bmatrix} \bar{F}^n \\ \bar{M}^n \end{bmatrix}_{42\times1}$$

Premultiplying the above equation with the transposition of matrix H, we have

$$H^T \Phi H\ddot{q} + H^T K(\dot{q}, q) = H^T \hat{F}^a + H^T \hat{F}^n \quad (99)$$

Assuming that all joints are ideal, items including constraint forces could be eliminated by $H^T \hat{F}^n = 0$. The equation of motion of the mechanism could be obtained by rearranging Eq.(99) as

$$H^T \Phi H\ddot{q} = H^T \hat{F}^a - H^T K(\dot{q}, q) \quad (100)$$

For the forward dynamics problem, Eq.(100) is a set of 3 second-order ordinary differential equations corresponding to the generalized coordinates on the movable platform. The trajectory of the platform can be predicted by solving this set of equations with numerical integration algorithm when the position and velocity of the platform are given as the initial value. In the case that the position and velocity of the slider are given as the initial value, one step forward kinematic analysis is needed to calculate the position and velocity of the movable platform before the integration of Eq.(100).

While for the inverse dynamics problem, the trajectory of the platform is pre-defined, then the displacement, velocity and acceleration of the generalized coordinates are known as the inputs of the equations of motion. Equation (100) can be transformed as

$$H^T \hat{F}^a = H^T \Phi H\ddot{q} + H^T K(\dot{q}, q) \quad (101)$$

Equation(101) turns into a set of 3 purely algebraic equations. The actuating forces exerted on the active sliders can be solved out directly provided that the motion of movable platform has been planned.

5. Numerical Solutions

A numerical example is given under the following assumptions:

1、 The applied forcees are the actuating forces in the active prismatic joints and Gravity of each body.
2、 The mass center of the platform locates at the origin of the body-fixed frame *O'-x'y'z'*in figure.2. For the other links, the mass centers locate at their geometrical centers.

The geometric parameters and mass properties used in the numerical example are listed in Tab.1, in which all the parameters are assigned in international units. The ideal assignment of some parameters in Tab.1, which makes the computation easily, will certainly not hamper the illustration of the feasibility of the proposed method. The coefficients in Eq.(41), (42), and (43) are assigned to $M_i=1$ (i=1,2,3).

Table 1. Geometric parameters and mass properties of the mechanism

Length of the links	$r = 1.0\,, R = 2.5\,, L = 4.0$
mass and moment of inertia of the platform	$m_p = 1.0\,, \mathrm{J_p} = diag(1.0\,,1.0\,,1.0)$
mass and moment of inertia of the sliders	$m_{S_i} = 1.0\,, \mathrm{J}_{S_i} = diag(1.0\,,1.0\,,1.0)\ (i = 1,2,3)$
mass and moment of inertia of the links	$m_{L_i} = 1.0\,, \mathrm{J}_{L_i} = diag(1.0\,,1.0\,,1.0)\ (i = 1,2,3)$

Based on the method presented above, a computer program is developed by the MATLAB software. For the simulation of forward dynamic problem, the actuating forces applied on the active prismatic joints are assumed as constants, and their values are assigned as

$$\tau_1 = 30\,,\ \tau_2 = 20\,,\ \tau_3 = 15$$

The initial position and velocity of the platform are given as

$$\mathrm{q}_0 = (0\ \ -3\ \ 0)^T,\ \dot{\mathrm{q}}_0 = (0\ \ 0\ \ 0)^T$$

The simulation results of forward dynamics are illustrated in figure 3, including time histories of the displacement of three active prismatic joints, trajectory and angular displacement of the platform and angular displacement of the third leg.

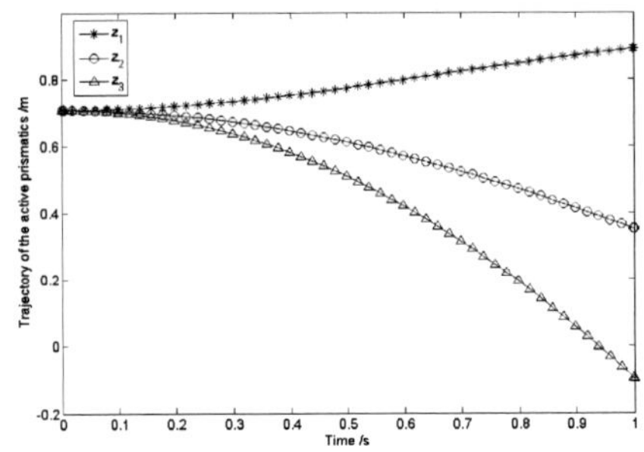

a) Displacements of the active prismatic joints

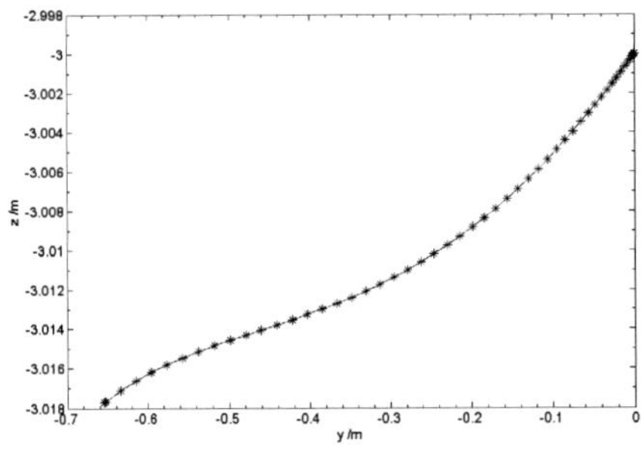

b) Trajectory of mass center of the platform

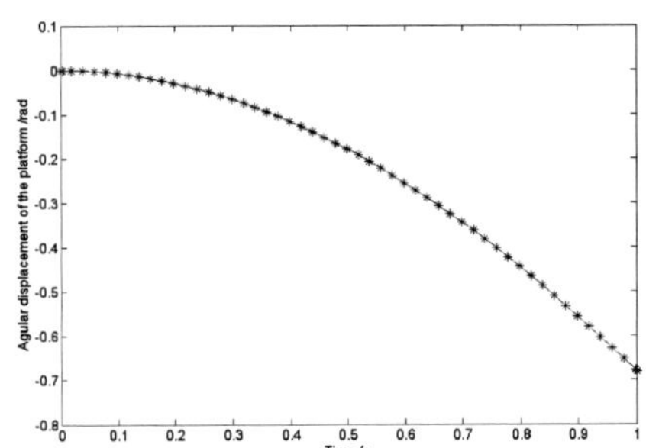

c) Angular displacement of the platform

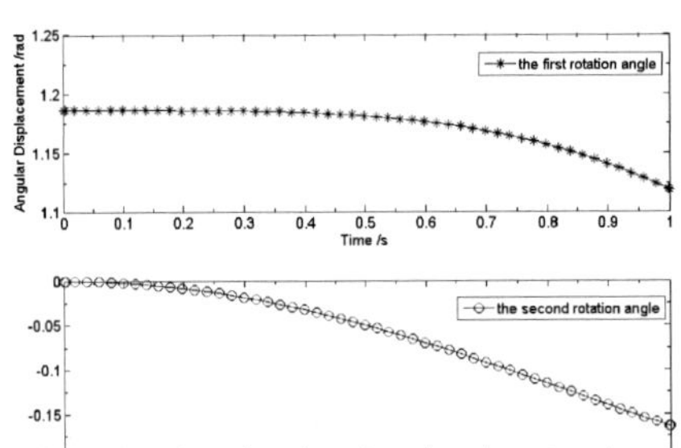

d) Angular displacement of the third leg

Figure 3. Outcome of the forward dynamic analysis of the 3-DOF Parallel manipulator.

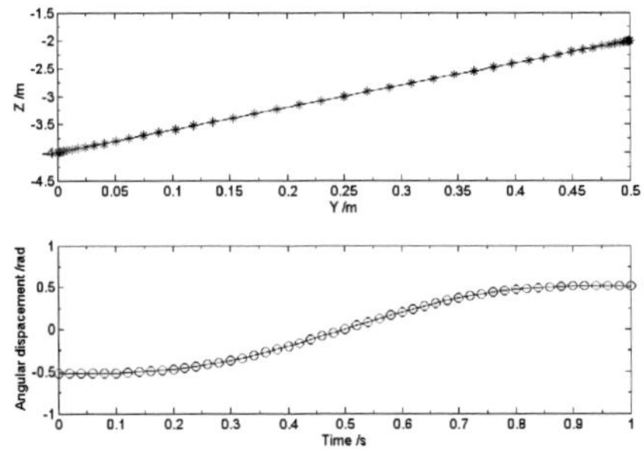

a) Prescribed trajectory of the center of mass of the platform

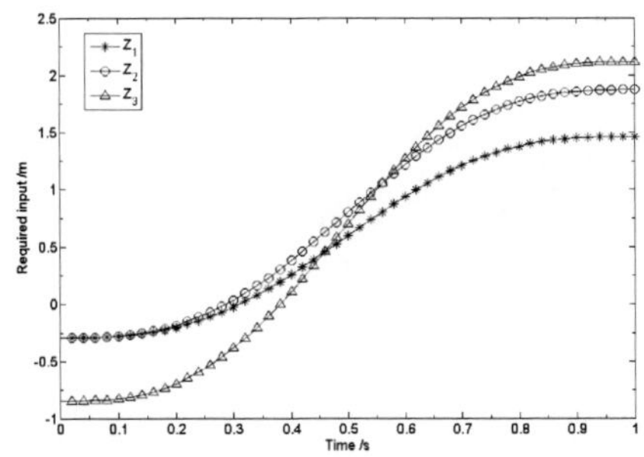

b) Displacements of the active prismatic joints

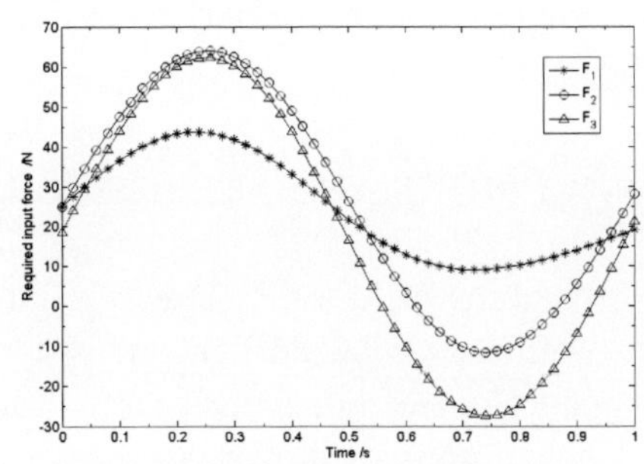

c) Required input forces

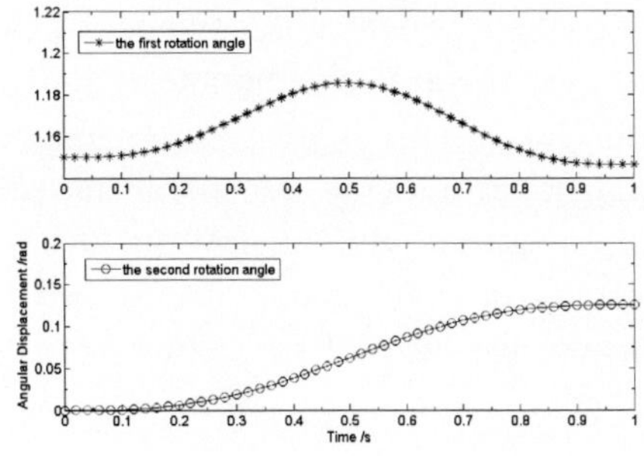

d) Angular displacement of the third leg

Figure 4. Outcome of the inverse dynamic analysis of the 3-DOF Parallel manipulator.

For the simulation of inverse dynamic problem, the prescribed translational and rotational trajectories of the platform are given as a trigonometric function with the property of acceleration continuous.

$$y = -\frac{a_y}{\omega_y^2}\sin\omega_y t + \frac{a_y}{\omega_y}t + y_0$$

$$z = -\frac{a_z}{\omega_z^2}\sin\omega_z t + \frac{a_z}{\omega_z}t + z_0$$

$$\varphi = -\frac{a_\varphi}{\omega_\varphi^2}\sin\omega_\varphi t + \frac{a_\varphi}{\omega_\varphi}t + \varphi_0$$

where the parameters are assigned as

$$a_y = \pi, \omega_y = 2\pi, y_0 = 0, a_z = 4\pi, \omega_z = 2\pi, z_0 = -4, a_\varphi = \frac{2}{3}\pi^2, \omega_\varphi = 2\pi, \varphi_0 = -\frac{\pi}{6}$$

The simulation results of inverse dynamics are illustrated in figure 4, including time histories of the displacement of three active prismatic joints, angular displacement of the third leg and required actuating forces to perform the prescribed trajectory of the platform.

CONCLUSIONS AND FUTURE WORK

This chapter presents a general method for the modeling and simulation on both forward and inverse dynamics of the Parallel Manipulators. Forward dynamics of a 3-DOF Parallel manipulator is studied thoroughly. Computational results are illustrated to show the feasibility of this method. The following conclusions are drawn as a result:

1) By Schiehlen's method, the equation of motion of a Parallel manipulator could be reduced to a set of equations in which the number of ODEs equals to the degree of freedom. In forward dynamic analysis, the integration variables are the minimal set of generalized coordinates.
2) It is recommended that the generalized coordinates be selected on the movable platform rather than on active joints in forward dynamic analysis. Based on this selection, the efficiency of computation in forward dynamic analysis could be improved vastly because the forward kinematics is only needed in the first step to determine the initial value on movable platform from the state of active joints.

The studies of the dynamics of parallel manipulator, especially the forward dynamics, are far from complete. Future work of this topic is suggested on the following aspects: 1) to get more precise model, especially when the manipulator is working on the condition of high speed and undertaking heavy load, the elasticity of the joints (e.g. ball joints) as well as links must be taken into account; 2) optimization of the parallel manipulator considering the dynamic performance is the subsequent work, whereas new performance index of the dynamics of parallel manipulator should be setup beforehand; 3) stable and high efficiency simulation algorithms are needed in forward dynamics to solve the assembly problem caused by numerical errors and continuous calculation in the neighborhood of singularity. Some of these suggestions are under study now and the results will be presented in the future.

ACKNOWLEDGMENTS

The authors are very grateful to Professor Yan-Zhu LIU in Institute of Mechanics of Shanghai Jiaotong University for his directions and discussion on multi-body system dynamics.

The research is jointly supported by National Science Foundation of China (NSFC) under research grand 50605042 and National Basic Research Program of China (973 Program) under research grand 2006CB705400.

REFERENCES

[1] Wang J., Gosselin C.M.; A New Approach for the Dynamic Analysis of Parallel Manipulators. *Multibody System Dynamics*, 1998, 2: 317-334

[2] Tsai L.W.; Solving the Inverse Dynamics of a Stewart-Gough Manipulator by the Principle of Virtual Work. Trans. of ASME, *Journal of Mechanical Design*, 2000, 122: 3-9

[3] Staicu St., Carp-Ciocardia D. C.; Dynamic Analysis of Clavel's Delta Parallel Robot. Proeecdings of the IEEE International Conference on Robotics and Automation, 2003: 4116-4121

[4] Sokolov A., Xirouchakis P.; Dynamics Analysis of a 3-DOF Parallel Manipulator with R-P-S Joint Structure. *Mechanism and Machine Theory*, 2007, 42: 541-557

[5] Yang Z. Y., Huang T., Ni Y. B.; Dynamic Modeling and Trail Tracking Robust Control of 3-HSS Parallel Kinematic Machine. *Chinese Journal of Mechanical Engineering*, 2004, 40(11) : 75-81

[6] Geike T., McPhee J.; Inverse Dynamic Analysis of Parallel Manipulators with Full Mobility. *Mechanism and Machine Theory*, 2003, 38(6): 549-562

[7] Gallardo J., Rico J.M., Frisoli A., et al; Dynamics of Parallel Manipulators by Means of Screw Theory. *Mechanism and Machine Theory*, 2003, 38(11): 1113-1131

[8] Bhattacharya S., Hatwal H., Ghosh A.: An on-line estimation scheme for generalized Stewart platform type parallel manipulators. *Mechanism and Machine Theory*, 1997, 32(1): 79-89

[9] Lee M.K., Shah D.K.; Dynamic Analysis of a Three-Degrees-Of-Freedom In-Parallel Actuated Manipulator. IEEE Journal of Robotics and Automation, 1988, 4(3): 361-367

[10] Xi F.F., Sinatra R., Han W.Z.; Effect of Leg Inertia on Dynamics of Sliding-Leg Hexapods. Trans. of ASME, *Journal of Dynamic Systems, Measurement, and Control*, 2001, 123: 265-271

[11] Fichter E F.; A Stewart Platform Based Manipulator: General Theory and Practical Construction. *International Journal of Robotics Research*, 1986, 5 (2): 157-182

[12] Sugimito K.; Kinematics and Dynamic Analysis of Parallel Manipulator by Means of Motor Algebra. Transactions of the ASME: *Journal of Mechanisms, Transmissions, and Automation in Design,* 1987, 109: 3-7

[13] Sugimoto K.; Computational scheme for dynamic analysis of parallel manipulators. Transactions of the ASME: *Journal of Mechanisms, Transmissions, and Automation in Design,* 1989,111: 29-33

[14] Dasgupta B., Mruthyunjaya T.S.; Closed-Form Dynamic Equations of the General Stewart Platform Through the Newton-Euler Approach. *Mechanism and Machine Theory,* 1998, 33 (7), pp: 993-1012

[15] Dasgupta B., Mruthyunjaya T.S.; A Newton-Euler Formulation for the Inverse Dynamics of the Stewart Platform Manipulator. *Mechanism and Machine Theory*, 1998, 33 (8), pp: 1135-1152

[16] Dasgupta B., Choudhury P.; A General Strategy Based on the Newton-Euler Approach for the Dynamic Formulation of Parallel Manipulators. *Mechanism and Machine Theory,* 1999, 34: 801-824

[17] Khalil W., Ibrahim O.; General Solution for the Dynamic Modeling of Parallel Robots. Proceedings of the 2004 IEEE International Conference on Robotics and Automation, 2004: 3665-3670

[18] Khalil W., Ibrahim O.; General Solution for the Dynamic Modeling of Parallel Robots. *Journal of Intelligent Robot System*, 2007, 49:19-37

[19] Khalil W., Guegan S.; Inverse and Direct Dynamic Modeling of Gough-Stewart Robots. *IEEE Transactions On Robotics*, 2004, 20(4): 754-762

[20] Guo Z. H., Chen W. Y., Chen D. C.; Dynamics Model on the 6-UPS Parallel manipulator. *Chinese Journal of Mechanical Engineering*, 2002, 38(11):53-57

[21] Wang Q., Wang L. P., Wang J. S.; Investigation of the Optimal Control Method Based on the Rigid Dynamic Model of PKM. *China Mechanical Engineering*, 2003, 14(21): 1801-1806

[22] Zhang L. X., Wang J. S., Wang L. P.; Simplified rigid body dynamic model for a parallel kinematics machine under uniform motion. *Journal of Tsinghua University (Science and Technology),* 2003, 43(8): 1041-1044

[23] Zhang L. X., Wang J. S., Wang L. P.; Simplification of the Rigid Body Dynamic Model for a 6-UPS Parallel Kinematic Machine Under the Acceleration Motion and the Decelerated Motion. *Chinese Journal of Mechanical Engineering*, 2003, 39(11): 117-122

[24] Carvalho J., Ceccarelli M.; A Closed-Form Formulation for the Inverse Dynamics of a Cassino Parallel Manipulator. *Multibody System Dynamics*, 2001, 5: 185-210

[25] Liu M.J., Li C.X., Li C.N; Dynamics Analysis of the Gough-Stewart Platform Manipulator. *IEEE Trans. on Robotics and Automation*, 2000, 16(1): 94-98

[26] Miller, K., Optimal Design and Modeling of Spatial Parallel Manipulators. *International Journal of Robotics Resesrch,* 2004, 23(2): 127-140

[27] Liu G.F., Li Z.X., A Unified Geometric Approach to Modeling and Control of Constrained Mechanical Systems. *IEEE Transactions on Robotics and Automation*, 2002, 18(4): 574-587

[28] Wittenburg, J., *Dynamics of System of Rigid Bodies*, Teubner, Stuttgart, 1977

[29] Schiehlen, W., *Technische Mechanik*, Teubner, Stuttgart, 1986

[30] Schiehlen W. Multibody System Dynamics: Roots and Perspectives. *Multibody System Dynamics*, 1997, 1: 149-188

[31] Huang Z., Advanced Spatial Mechanism, *Higher Education Press*, Beijing, 2006 (In Chinese)

[32] Liu Y. Z., Dynamics of Multi-Rigidbody Systems, *High Education Press*, Beijing, 1989 (In Chinese)

[33] Xin-Jun Liu, Jinsong Wang, Feng Gao, and Li-Ping Wang, On the Analysis of a New Spatial Three-Degrees-of-Freedom Parallel Manipulator, *IEEE Transactions on Robotics and Automation*, 2001, 17(6): 959-968

[34] Li Y.W., Wang J.S., Wang L.P.; Inverse Dynamics and Simulation of a 3-DOF Spatial Parallel Manipulator. Proceedings of the 2003 IEEE International Conference on Robotics and Automation, 2003: 4092-4097

In: Parallel Robotics: Recent Advances in Research... ISBN: 978-1-60456-859-2
Editors: Jinsong Wang and Xin-Jun Liu, pp. 123-161

Chapter 6

ANALYSIS AND KINEMATIC OPTIMIZATION OF PLANAR 2-DOF 5R PARALLEL MECHANISMS CONSIDERING THE FORCE TRANSMISSIBILITY

Xin-Jun Liu[1,*], Jinsong Wang[1], Fugui Xie[1], Markus Wabner[2] and Reimund Neugebauer[2]

[1] Institute of Manufacturing Engineering, Department of Precision Instruments, Tsinghua University, Beijing, 100084, P.R. of China

[2] Fraunhofer-Institut fuer Werkzeugmachinen und Umformtechnik (IWU), Reichenhainer Str. 88, D-09126, Chemnitz, Germany

ABSTRACT

This paper treats the problems of kinematic analysis and optimization of the 5R symmetrical parallel mechanism taking into account the force transmissibility. In the design process, the theoretical workspace cannot be used practically due to the inside singularity. The usable workspace without singularity inside cannot still be useful in an optimization problem since a mechanism at and near the boundary of the workspace will be out of control. The local conditioning index (LCI), which is widely used in the field of serial robots and has been applied in many optimizations, has the problem of physical significance and difficulty of measuring the distance to a singularity of a parallel mechanism. For such reasons, the concept of transmission angle is introduced in this paper to the optimization of the planar 5R parallel mechanisms. Based on this concept, three new indices are proposed. After presenting the performance charts, the optimization process of the mechanisms is finally presented by considering the workspace and force transmissibility.

Keywords: parallel mechanism, optimization, performance chart, workspace, singularity

* Email: xinjunliu@mail.tsinghua.edu.cn

1. INTRODUCTION

Analysis and kinematic design are two important issues in the development of a parallel mechanism. Without exhaustive analysis, a design will not be perfect or even will be lost. Then, analysis is primary. In the design process, kinematics and workspace are two typical problems since they are the basic model and reference to define and evaluate the performances of a mechanism. For such a reason, the kinematics and workspace are the most studied issues in the field [1-6]. The efforts were motivated by the fact that they would be finally applied to the design and application of the devices.

Kinematic optimization is one of the most important and challenging problems in the design of parallel kinematics, and is increasingly attracting the attention of researchers [7]-[9]. There are two issues involved: performance evaluation and dimension synthesis. Having designed a mechanism, it is necessary to evaluate its performances. A second problem is to determine the dimensions (link lengths) of the mechanism, which is suitable for the task at hand. The latter one is one of the most difficult issues in this field.

Several well-defined performance indices, such as workspace [8], singularity [7], dexterity [10,11], accuracy [12], stiffness [13], and conditioning index [7], have been extensively developed and applied to both serial and parallel mechanisms. A recent study [14] reviewed the most common of these indices that have been applied in the optimum design of parallel mechanisms: the condition number [10] of the Jacobian matrix and the global conditioning index (GCI) [15]. The study found serious inconsistencies when these indices are applied to parallel mechanisms with combined translational and rotational degrees of freedom, and concluded that these indices should not be used in parallel mechanisms with mixed types of degrees of freedom (translational and rotational). What is more, to eliminate the singularity and its near configurations, most researchers used the local conditioning index (LCI) [15]. Usually, a good-condition workspace or effective workspace was defined with respect to a specified minimum LCI [9,16]. However, the minimum is still arbitrary or comparative since we cannot give it a definite value. Generally, it is not possible to define a mathematical distance to a singularity for a parallel mechanism [14]. Instead of LCI, another index will be defined in this paper.

As is well known, the *transmission angle* is an important index for the design of a planar four-bar mechanism as was pointed out by Alt [17], who defined the concept, using the forces tending to move the driven link and tending to apply pressure to the driven link bearings as a simple index, to judge the force-transmission characteristics of a mechanism. By means of the index of *transmission angle*, the quality of motion/force transmission in a mechanism can be judged in the design stage. Therefore, the *transmission angle* is an index evaluating the quality of motion/force transmission. It helps to decide the "best" from among a family of possible mechanisms for the most effective force transmission [18]. Actually, a good *transmission angle* is the solution to most of the problems in planar mechanisms. For example, Alt [17] used the transmission angle to isolate better chains for various linkage applications. As is pointed out in [19], though a good *transmission angle* is not a cure-all for every design problem, for many mechanical applications it can guarantee the performance of a linkage at higher speed without unfavorable vibrations.

The study in [20] shows that when the *transmission angle* is equal to 90°, the most effective force transmission takes place and the output motion becomes less sensitive to the manufacturing tolerances on the link lengths, clearance between joints, and change of dimensions due to thermal expansion. Mechanisms having a *transmission angle* too far from 90° exhibit poor operational characteristics such as noise and jerk at high speeds [21,22]; and if it is 0, self-locking takes place. Thus the *transmission angle* of a mechanism provides a very good indication of the quality of its motion, the accuracy of its performance, its expected noise output, and its costs in general [23]. A large *transmission angle* usually leads to reasonable mechanical advantages and a high quality of motion transmission. The study of link mechanisms shows that *transmission angle* is significant not only as an indicator of good force and motion transmission but also as a prime factor in the linkage sensitivity to small design parameter errors. The smaller the *transmission angle*, the more sensitive the linkage will be [24]. In [18], the authors gave a summary description of the effect of transmission angle on velocity, input crank angle, friction, mechanical advantage, velocity, tolerance, clearance, and performance sensitivity. Many studies have reached the conclusion that if the *transmission angle* becomes too small, the mechanical advantage becomes small, and even a very small amount of friction will cause the mechanism to jam. For the purpose of high speed, high accuracy, and high quality of motion transmission, the most widely accepted design limits for the transmission angle are (45°, 135°) [21] or (40°, 140°) [17].

A planar four-bar mechanism is a single-closed-loop system. A parallel mechanism belongs to the multi-closed-loop mechanism group. Usually, a fully parallel mechanism has more or less the characteristic of a planar four-bar mechanism. We suggest that the design concept of the four-bar mechanisms could be used in the design of a parallel mechanism.

Many methods have been proposed for the dimension optimization of specified mechanisms. The most common one is the objective-function-based optimal design [8,13,16]. According to this methodology, an objective function with specified constraints must be established, and then a search conducted to find the result utilizing an optimization algorithm. Not only is this method time-consuming, but it is difficult to reach the globally optimum target because of the non-finiteness of the individual parameters, the antagonism of multiple criteria and the assignment of its initial value. The most serious drawback is that it provides only one solution for a design problem. This is actually a fatal one for practical design purposes, since it is impossible to predict any application condition in advance and very difficult to be sure that a particular design is the only possible solution for a given problem. The ideal dimension optimization method, or optimal method, would be the performance chart (atlas), which is widely used in classical design and in most design manuals. A performance atlas can show, globally and visually, the relationship between a performance index and the associated design parameters in a limited space [25]. Moreover, it can show how antagonistic the involved indices actually are. Compared with the result achieved by the objective-function-based method, the optimal method result is comparative and fuzzy. However, it is more flexible, due to the fact that it provides not only a single solution, but all possible solutions to a design problem. This means that the designer can adjust the optimum result appropriately according to the particular design conditions he is dealing with [25].

The five-bar mechanism is a typical parallel mechanism with the minimal degrees of freedom (DoFs) in the field, which can be used for positioning a point on a region of a plane that is known as the workspace. The 5R (R-revolute joint) parallel mechanism consists of five bars that are connected end to end by five revolute joints, two of which that are connected to the base are actuated. Such a mechanism with a symmetric structure attracted many researchers, who have investigated its position analysis [26], workspace [27, 28], assembly modes [28], singularity [27, 29], performance atlases [28, 30] and kinematic design [30]-[32]. What is more, the symmetrical mechanism has been applied in the development of the MELFA RP Series robots in the Mitsubishi Electric Corporation. In this paper, based on the concept of *transmission angle*, three indices will be introduced to analyze and design the planar 2-DOF 5R parallel mechanisms. In order to study all possible 5R parallel mechanisms, a design space is established according to the concept in [24]. The design space is used to investigate the indices of each parallel mechanism systematically.

The remainder of this paper is organized as follows. The next section describes the kinematic problems of the planar 2-DOF 5R parallel mechanism. Section 3 introduces the parameter design space of the mechanisms. Section 4 recalls the traditional analysis, including theoretical workspace, Jacobian matrix, singularity and usable workspace. The concept of transmission angle is presented in section 5. Section 6 gives the definition of some indices, i.e. local transmission index (LTI), good-transmission workspace (GTW), and global transmission index (GTI). In section 7, the performance charts of the mechanisms have been represented. Section 8 presents the optimization process of the mechanisms using the obtained performance charts. Conclusions are given in the last section.

2. Kinematic Problems

The planar 5R parallel mechanism, as shown in figure 1(a), is such a mechanism that the output point is connected to the base by two legs, each of which consists of three revolute joints and two links. The two legs are connected to a common point with the common revolute joint at the end of each leg. In each of the two legs, the revolute joint connected to the base is actuated. Such a mechanism can position a point freely in a plane.

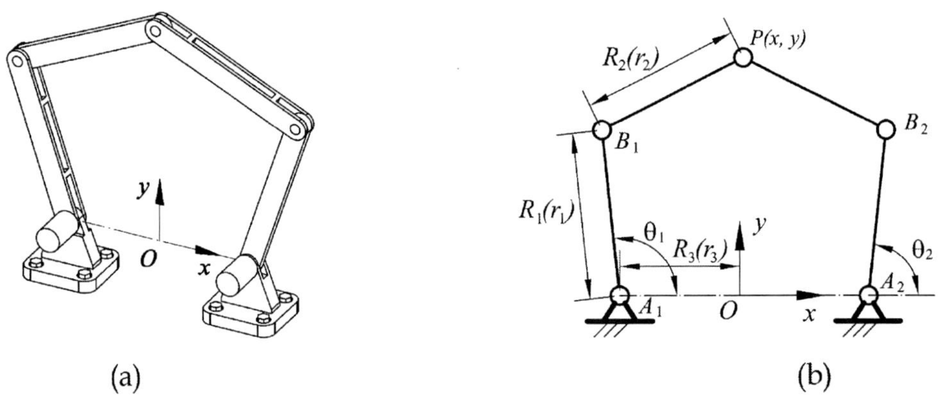

Figure 1. The planar 5R parallel mechanism

A kinematics model of the mechanism is developed as shown in figure 1(b). Each actuated joint is denoted as A_i ($i = 1, 2$), the other end of each actuated link is denoted as B_i and the common joint of the two legs is denoted as P, which is also the output point. A fixed global reference system $O-xy$ is located at the center of A_1A_2 with the y-axis normal to A_1A_2 and the x-axis directed along A_1A_2. For the structure symmetry, there are

$OA_1 = OA_2$, $OA_1 = OA_2$ and $B_1P = B_2P$. The length of the actuated link for each leg is denoted as $A_iB_i = R_1(r_1)$. Additionally, $B_iP = R_2(r_2)$ and $OA_i = R_3(r_3)$. Here, R_j ($j = 1,2,3$) are link lengths with dimension and r_j ($j = 1,2,3$) are those with non-dimension.

2.1. Inverse Kinematics

The position of the output point P in the reference system $O-xy$ can be described by the position vector $\boldsymbol{p}$, and there is

$$\boldsymbol{p} = (x \quad y)^{\mathrm{T}} \tag{1}$$

In the reference frame $O-xy$, the position vectors $\boldsymbol{b}_i$ of points B_i $(i = 1,\ 2)$ can be written as

$$\boldsymbol{b}_1 = (r_1 \cos\theta_1 - r_3 \quad r_1 \sin\theta_1)^{\mathrm{T}} \text{ and } \boldsymbol{b}_2 = (r_1 \cos\theta_2 + r_3 \quad r_1 \sin\theta_2)^{\mathrm{T}} \tag{2}$$

where θ_1 and θ_2 are the input angles of the two legs. Then, the inverse kinematic problem can be solved by writing following constraint equation

$$|\boldsymbol{pb}_i| = r_2,\ i = 1,\ 2 \tag{3}$$

in another form

$$(x - r_1 \cos\theta_1 + r_3)^2 + (y - r_1 \sin\theta_1)^2 = r_2^2 \tag{4}$$

$$(x - r_1 \cos\theta_2 - r_3)^2 + (y - r_1 \sin\theta_2)^2 = r_2^2 \tag{5}$$

from which, if the position of output point P is known, the inputs to reach the position can be obtained as

$$\theta_i = 2\tan^{-1}(z_i),\ i = 1,\ 2 \tag{6}$$

Where

$$z_i = \frac{-b_i + \sigma_i\sqrt{b_i^2 - 4a_ic_i}}{2a_i},\ i = 1,\ 2 \tag{7}$$

in which

$$
\begin{aligned}
&\sigma_i = 1 \text{ or } -1 \\
&a_1 = r_1^2 + y^2 + (x + r_3)^2 - r_2^2 + 2(x + r_3)r_1 \\
&b_1 = -4yr_1 \\
&c_1 = r_1^2 + y^2 + (x + r_3)^2 - r_2^2 - 2(x + r_3)r_1 \\
&a_2 = r_1^2 + y^2 + (x - r_3)^2 - r_2^2 + 2(x - r_3)r_1 \\
&b_2 = b_1 = -4yr_1 \\
&c_2 = r_1^2 + y^2 + (x - r_3)^2 - r_2^2 - 2(x - r_3)r_1
\end{aligned}
$$

From Eq.(7), one can see that there are four solutions for the inverse kinematic problem of the 5R mechanism. The configuration shown in figure 1 can be obtained if $\sigma_1 = 1$ and $\sigma_2 = -1$. Such a configuration is denoted as the "+ – " model. Then there are three others, which are "– +", "– –", and "++" models, respectively. These four inverse kinematics models correspond to four types of working modes.

2.2. Forward Kinematics

The Forward kinematic problem is to obtain the output with respect to a set of given inputs. From Eqs.(4) and (5), one obtains

$$x^2 + y^2 - 2(r_1 \cos\theta_1 - r_3)x - 2r_1 \sin\theta_1 y - 2r_1 r_3 \cos\theta_1 + r_3^2 + r_1^2 - r_2^2 = 0 \quad (8)$$

$$x^2 + y^2 - 2(r_1 \cos\theta_2 + r_3)x - 2r_1 \sin\theta_2 y + 2r_1 r_3 \cos\theta_2 + r_3^2 + r_1^2 - r_2^2 = 0 \quad (9)$$

Eq.(8)-Eq.(9) yields

$$x = ey + f \quad (10)$$

in which $e = \dfrac{r_1(\sin\theta_1 - \sin\theta_2)}{2r_3 + r_1\cos\theta_2 - r_1\cos\theta_1}$ and $f = \dfrac{r_1 r_3(\cos\theta_2 + \cos\theta_1)}{2r_3 + r_1\cos\theta_2 - r_1\cos\theta_1}$. Substituting Eq.(10) to Eq.(8) yields

$$dy^2 + gy + h = 0 \quad (11)$$

where

$$d = 1 + e^2$$

$$g = 2\left(ef - er_1 \cos\theta_1 + er_3 - r_1 \sin\theta_1\right)$$
$$h = f^2 - 2f\left(r_1 \cos\theta_1 - r_3\right) - 2r_1 r_3 \cos\theta_1 + r_3^2 + r_1^2 - r_2^2$$

Then, y can be obtained from Eq.(11) as

$$y = \frac{-g + \sigma\sqrt{g^2 - 4dh}}{2d} \tag{12}$$

in which $\sigma = 1$ or -1. From Eqs. (10) and (12), one sees that there are two solutions for the forward kinematic problem of the mechanism. They correspond to types of assembly modes. The corresponding configurations are called the *up-* and *down-configurations*. The *up-configuration* can be achieved when $\sigma = 1$.

For example, if $r_1 = 1.2$, $r_2 = 1.0$ and $r_3 = 0.8$, the four inverse kinematic models of the mechanism are shown in figure 2, where the specified position of the output point is $(x = -0.37,\ y = 1.44)$. And the two forward kinematic models are shown in figure 3, where the inputs are given as $\theta_1 = 4\pi/9$ and $\theta_2 = 7\pi/18$.

In this paper, what we are concerned about is the mechanism with the "+ – " model and, at the same time, the *up-configuration*.

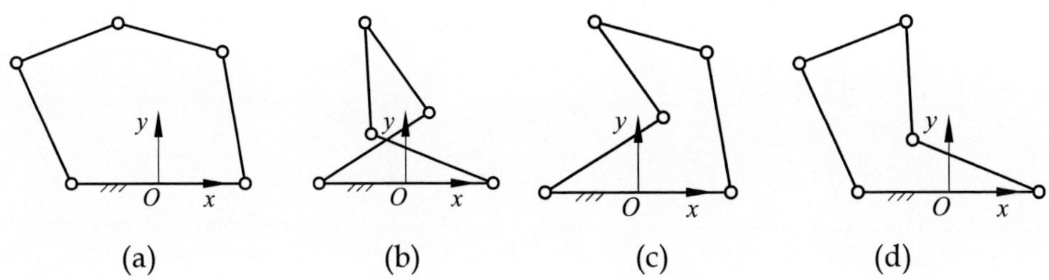

Figure 2. The four inverse kinematic models: (a) "+ – " model; (b) "– +" model; (c) "– –" model; (d) "++" model.

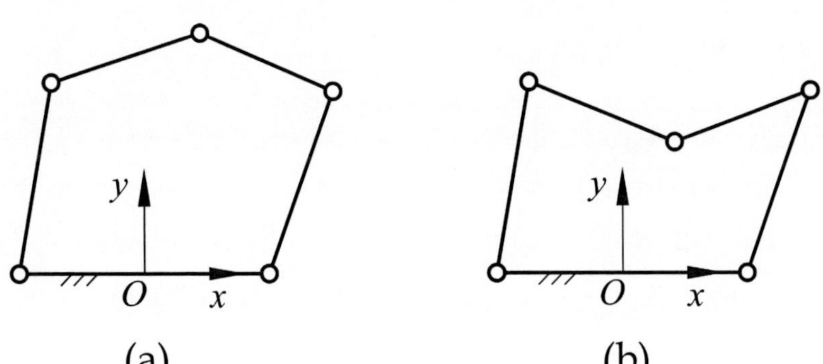

Figure 3. The two forward kinematic models: (a) the up-configuration; (b) the down-configuration

3. Design Space of the 5R Parallel Mechanisms

As it is well known, the performance of a parallel mechanism depends on not only the pose of the moving platform but also the link lengths (dimensions). Disregarding the pose, each of the links can be the length between zero and infinite. And there are always several links in a parallel mechanism. Then the combination of the links with different lengths will be infinite. They undoubtedly have different performance characteristics. To apply a specified mechanism in practice, we usually should determine the link lengths with respect to a desired application. This is actually the so-called optimum design (parameter synthesis) of the mechanism. In such a process, one of the most classical tools that has been using is the chart, which can show the relationship between performances and link lengths. To make it work, we should first develop a space that contains all involved links. Next is to plot the chart considering a desired performance. In this paper, the space is referred to as *the design space*. The chart that can show the relationship between performances and link lengths is referred to as *atlas*. The *index* is used to evaluate a performance. Normally, several indices will be considered in the design process.

For the parallel mechanism considered here, due to the symmetric structure, there leaves three parameters, which are R_1, R_2 and R_3 as shown in figure 1. Any one of the parameters R_1, R_2 and R_3 can have any value between zero and infinite. This is the biggest difficulty to develop a design space that can embody all mechanisms (with different link lengths) within a finite space. For this reason, we must eliminate the physical link size of the mechanisms [25]. Let

$$D = (R_1 + R_2 + R_3)/3 \tag{13}$$

One can obtain 3 non-dimensional parameters r_i by means of

$$r_1 = R_1/D,\ r_2 = R_2/D,\ r_3 = R_3/D \tag{14}$$

This would then yield

$$r_1 + r_2 + r_3 = 3 \tag{15}$$

Theoretically, from Eq.(15), the three non-dimensional parameters r_1, r_2 and r_3 can have any value between 0 and 3. For the 5R parallel mechanism studied here, the analysis on the workspace and singularity shows that r_1 and r_2 cannot be 0 and $r_1 + r_2$ cannot be less than r_3, otherwise they will result in the failure of mechanism assembly. Therefore, the three parameters should be

$$0 < r_1,\ r_2 < 3 \text{ and } 0 \le r_3 \le 1.5 \tag{16}$$

Based on Eqs. (15) and (16), one can establish a design space as shown in figure 4(a), in which the trapezoid $ABCD$ is actually the design space. In figure 4(a), the trapezoid $ABCD$ is restricted by r_1, r_2 and r_3. Therefore it can be figured in another form as shown in figure 4(b), which is referred to as the planar-closed configuration of the design space.

For convenience, two orthogonal coordinates s and t are utilized to express r_1, r_2 and r_3. Thus, by using

$$\begin{cases} s = 2r_1/\sqrt{3} + r_3/\sqrt{3} \\ t = r_3 \end{cases} \tag{17}$$

coordinates r_1, r_2 and r_3 can be transformed into s and t. Eq. (17) is useful for constructing a performance atlas.

In order to study the performance of the parallel mechanism in detail, five lines $r_1 = r_3$, $r_2 = r_3$, $r_2 = r_1 + r_3$, $r_1 = r_3 + r_2$ and $r_1 = r_2$ can be used to divide the design space into 10 sub-regions, i.e., Ia, Ib, IIa, IIb, IIIa, IIIb, IVa, IVb, Va and Vb, as shown in figure 4(b). The former four lines indicate four types of change point mechanisms, respectively. The reason why $r_1 = r_2$ is also used is that if $r_1 = r_2$ the boundaries of the theoretical workspace are only two circles C_{1o} and C_{2o}. Such a design space will help us to investigate detailedly the performance characteristics of all 5R parallel mechanisms with possible combinations of $R_1(r_1)$, $R_2(r_2)$ and $R_3(r_3)$.

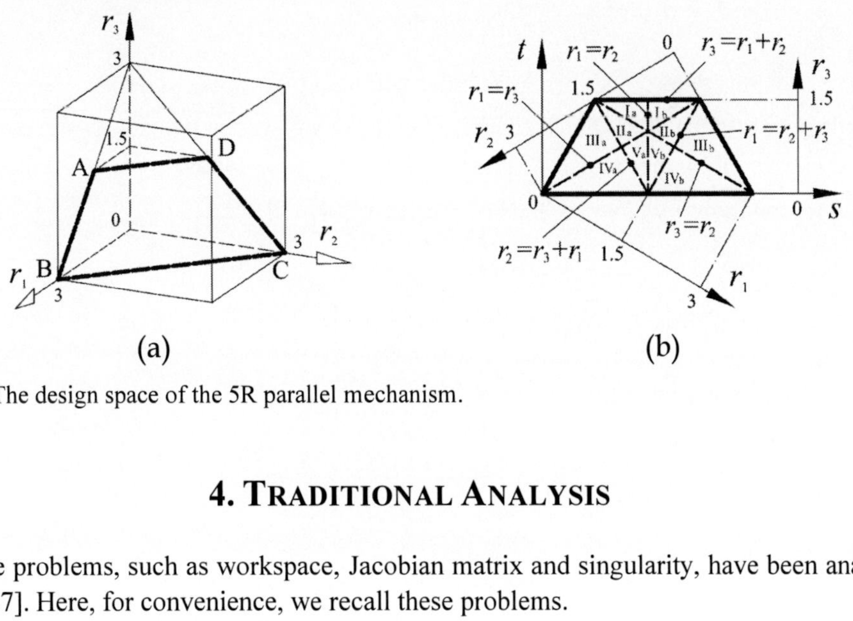

Figure 4. The design space of the 5R parallel mechanism.

4. Traditional Analysis

Some problems, such as workspace, Jacobian matrix and singularity, have been analyzed in Ref. [27]. Here, for convenience, we recall these problems.

4.1. Theoretical Workspace

Theoretical workspace is defined as the region that the output point can reach if θ_i changes from 0 to 2π without the consideration of interference between links and the singularity.

From Eq. (4), one can see that if θ_1 is specified, the workspace of the first leg is a circle centered at the point $(r_1 \cos\theta_1 - r_3, \quad r_1 \sin\theta_1)$ with a radius of r_2. The circle is denoted as C_{11}. If θ_i changes from 0 to 2π, the center point is located at a circle centered at point $A_1(-r_3 \quad 0)$ with a radius of r_1. The circle is denoted as C_{12}. Then, the workspace of the leg is the enveloping region of the circle C_{11} when its center rolls at the circle C_{12}. Actually, the enveloping region is an annulus bounded by two following circles

$$C_{1o}: (x + r_3)^2 + y^2 = (r_1 + r_2)^2 \tag{18}$$

$$C_{1i}: (x + r_3)^2 + y^2 = (r_1 - r_2)^2 \tag{19}$$

For the second leg, the workspace is an annulus bounded by circles

$$C_{2o}: (x - r_3)^2 + y^2 = (r_1 + r_2)^2 \tag{20}$$

$$C_{2i}: (x - r_3)^2 + y^2 = (r_1 - r_2)^2 \tag{21}$$

When $r_1 = r_2$ the workspaces are two circles defined by Eqs. (18) and (20), respectively. The theoretical workspace of the mechanism is the intersection of the two annuluses. For example, the workspace of a mechanism with $r_1 = 1.2$, $r_2 = 1.0$ and $r_3 = 0.8$ is shown as the hatched region in figure 5, where the boundaries are also illustrated.

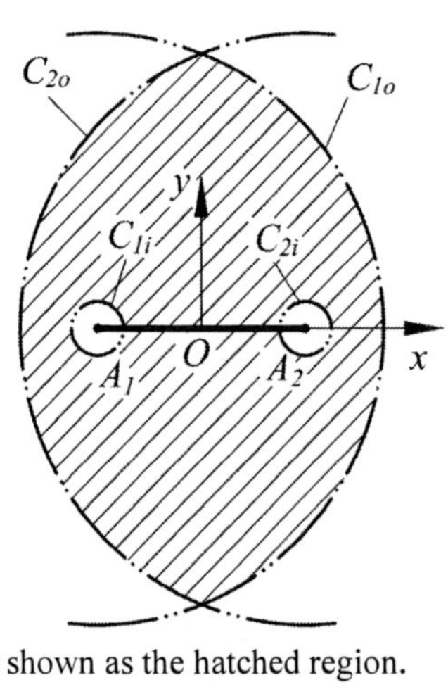

Figure 5. Theoretical workspace is shown as the hatched region.

From the above analysis, one can see that if $r_3 > r_1 + r_2$, there is no intersection, i.e., no workspace. Especially, if $r_3 = r_1 + r_2$, the workspace is only one point, which is the center of the line segment A_1A_2.

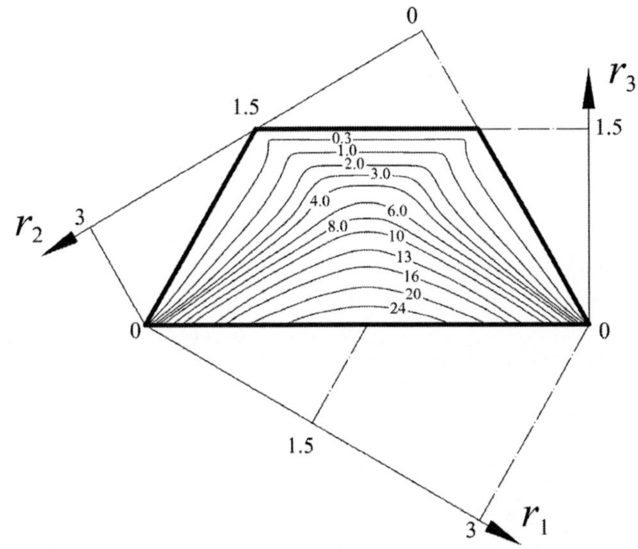

Figure 6. Performance chart of the theoretical workspace area.

4.1.1. Performance Chart of the Theoretical Workspace Area

The theoretical workspace can be determined based on Eqs. (18)-(21). The chart for the theoretical workspace area is shown in figure 6, from which one can see that

- The theoretical workspace area is inverse proportional to parameter r_3;
- The area atlas is symmetric with respect to $r_1 = r_2$, which means that the area of a mechanism with $r_1 = a$, $r_2 = b$ $(a, b < 3)$ and $r_3 = 3 - a - b$ is identical to that of a mechanism with $r_1 = b$, $r_2 = a$ $(a, b < 3)$ and $r_3 = 3 - a - b$;
- When $r_3 = 1.5$ (or $r_1 + r_2 = r_3$), the area is 0;
- The area reaches its maximum value when $r_1 = r_2 = 1.5$ and $r_3 = 0$. The maximum value is 9π.

4.1.2. Distribution Characteristics of the Theoretical Workspace Shape in the Design Space

Meanwhile, we can also obtain the workspace of a typical mechanism in each of the 10 sub-regions. The distribution characteristics of the workspace shape can be summarized based on these workspaces. The distribution in the design space is shown in figure 7, from which one can see that the distribution is symmetric with respect to the line $r_1 = r_2$. This means that the theoretical workspace shape of a mechanism with $r_1 = a$, $r_2 = b$ $(a, b < 3)$ and $r_3 = 3 - a - b$ is the same as that of a mechanism with $r_1 = b$, $r_2 = a$ $(a, b < 3)$ and

$r_3 = 3 - a - b$. This characteristic is what it should be, as the radii of boundary circles C_{1o}, C_{2o}, C_{1i} and C_{2i} of the theoretical workspace are just related to $r_1 + r_2$ and $|r_1 - r_2|$. Then there are only five types of theoretical workspace shapes for the 5R parallel mechanism. What is more, each theoretical workspace is symmetric about the x and y axes.

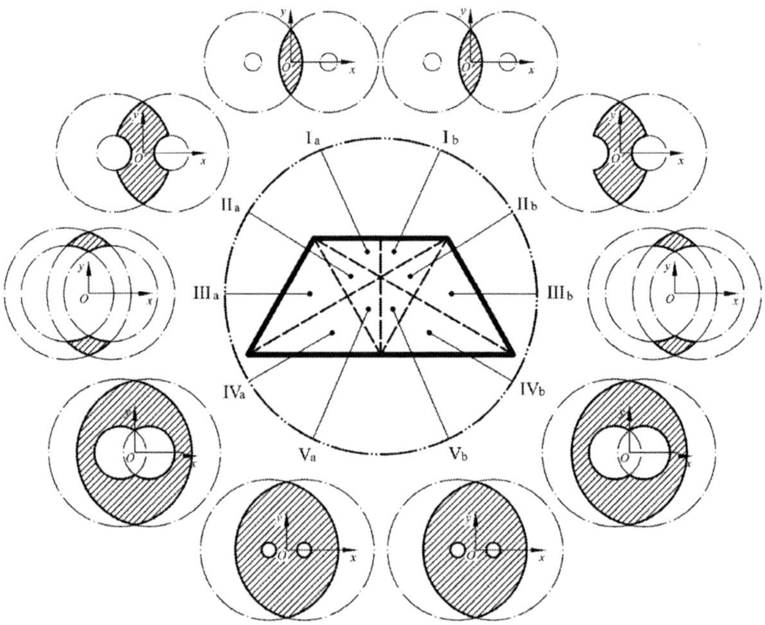

Figure 7. Distribution of the theoretical workspace shape in the design space.

4.2. Jacobian Matrix

Equations (4) and (5) can be differentiated with respect to time to obtain the velocity equations, which yields

$$r_1[y\cos\theta_1 - (x + r_3)\sin\theta_1]\dot{\theta}_1 = (x + r_3 - r_1\cos\theta_1)\dot{x} + (y - r_1\sin\theta_1)\dot{y} \quad (22)$$

$$r_1[y\cos\theta_2 + (r_3 - x)\sin\theta_2]\dot{\theta}_2 = (x - r_3 - r_1\cos\theta_2)\dot{x} + (y - r_1\sin\theta_2)\dot{y} \quad (23)$$

Rearranging Eqs.(22) and (23) leads to an equation of the form

$$\boldsymbol{A}\dot{\boldsymbol{\theta}} = \boldsymbol{B}\dot{\boldsymbol{p}} \quad (24)$$

where $\dot{\boldsymbol{p}}$ is the vector of output velocities defined as

$$\dot{\boldsymbol{p}} = (\dot{x} \quad \dot{y})^{\mathrm{T}} \quad (25)$$

and $\dot{\boldsymbol{\theta}}$ is the vector of input velocities defined as

$$\dot{\boldsymbol{\theta}} = \begin{pmatrix} \dot{\theta}_1 & \dot{\theta}_2 \end{pmatrix}^{\mathrm{T}} \tag{26}$$

Matrices $\boldsymbol{A}$ and $\boldsymbol{B}$ are, respectively, the 2×2 matrices of the mechanism and can be expressed as

$$\boldsymbol{A} = \begin{bmatrix} y\cos\theta_1 - (x + r_3)\sin\theta_1 & 0 \\ 0 & y\cos\theta_2 + (r_3 - x)\sin\theta_2 \end{bmatrix} r_1 \tag{27}$$

$$\boldsymbol{B} = \begin{bmatrix} x + r_3 - r_1\cos\theta_1 & y - r_1\sin\theta_1 \\ x - r_3 - r_1\cos\theta_2 & y - r_1\sin\theta_2 \end{bmatrix} \tag{28}$$

The Jacobian matrix of the mechanism can be written as

$$\boldsymbol{J} = \boldsymbol{A}^{-1}\boldsymbol{B} \tag{29}$$

4.3. Singularity Analysis

4.3.1. Three Kinds of Singularities

In the parallel mechanism, singularities occur whenever $\boldsymbol{A}$, $\boldsymbol{B}$ or both, become singular. As a singularity leads to an instantaneous change of the mechanism's DoF, the analysis of parallel mechanisms has drawn considerable attention.

The singularity of the 5R symmetrical parallel mechanism has been studied by many researchers. Here, we summarize this issue as following.

1) The stationary singularity occurs when $\boldsymbol{A}$ becomes singular but $\boldsymbol{B}$ remains invertible. Physically, this corresponds to the configuration whenever one of the legs A_1B_1P and A_2B_2P is completely extended or folded. From the analysis of the workspace, this singularity occurs when the output point P reaches its limit or is at the boundary of the workspace. This singularity is also referred to as the *serial singularity*. For example, for the mechanism with the parameters $r_1 = 1.2$, $r_2 = 1.0$ and $r_3 = 0.8$, some configurations of this kind of singularity are shown in figure 8. The loci of this kind of singularity are actually given by circles C_{1o}, C_{1i}, C_{2o} and C_{2i}, as shown in figure 5. Note that, $r_1 = 0$ leads to $\det(\boldsymbol{A}) = 0$ as well. But this also results the mechanism in an unmovable status.

2) The uncertainty singularity, occurring only in closed kinematics chains, arises when $\boldsymbol{B}$ becomes singular but $\boldsymbol{A}$ remains invertible. Usually, there are two cases for this singularity. The first case is that B_1PB_2 is completely folded, i.e., the points B_1 and B_2 are

coincident. Such a configuration is shown in figure 9(a), from which one can see that when this condition occurs the locus of the point P is a circle centered at point $\left(0,\ \sqrt{r_1^2 - r_3^2}\right)$ or $\left(0,\ -\sqrt{r_1^2 - r_3^2}\right)$ with a radius of r_2, namely,

$$x^2 + \left(y - \sqrt{r_1^2 - r_3^2}\right)^2 = r_2^2 \text{ or } x^2 + \left(y + \sqrt{r_1^2 - r_3^2}\right)^2 = r_2^2 \tag{30}$$

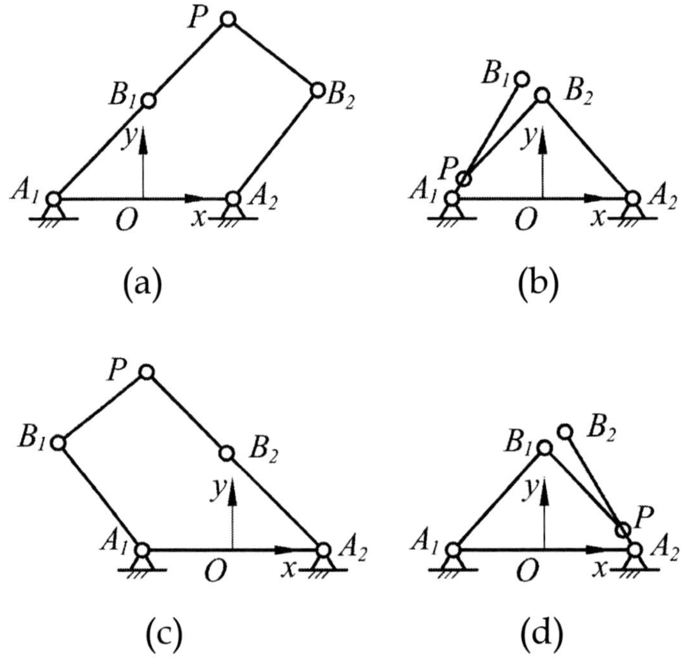

Figure 8. Some configurations of the stationary singularity: (a) the first leg is completely extended; (b) the first leg is completely folded; (c) the second leg is completely extended; (d) the second leg is completely folded.

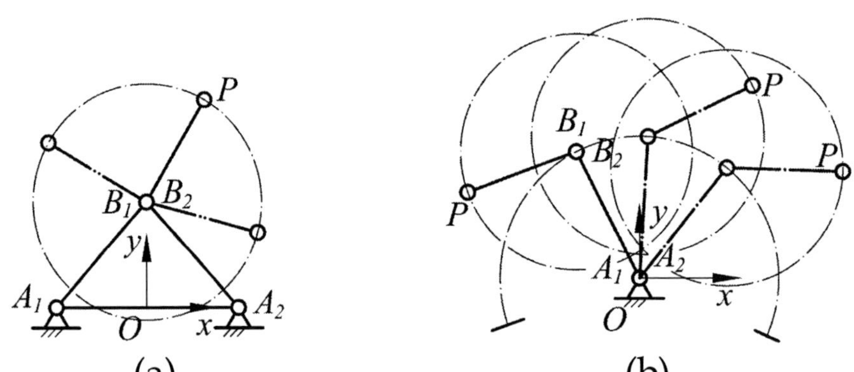

Figure 9. The configurations when B_1 and B_2 are coincident.

Please note that, if $r_3 = 0$, this type of singularity can occur easily when links A_1B_1 and A_2B_2 are coincident as shown in figure 9(b). For this case, the loci of point P is an annulus, which is bounded by two circles

$$x^2 + y^2 = (r_1 + r_2)^2 \text{ and } x^2 + y^2 = (r_1 - r_2)^2 \tag{31}$$

The second case is that when B_1PB_2 is completely extended. Such a singular configuration is shown in figure 10. The locus of point P for this type of singularity can be described by

$$\begin{cases} x = r_1(\cos\theta_2 + \cos\theta_1)/2 \\ y = r_1(\sin\theta_2 + \sin\theta_1)/2 \end{cases} \tag{32}$$

Where

$$\begin{aligned}
&\theta_1 \in [0,\ 2\pi] \\
&\theta_2 = 2\tan^{-1}(z) \\
&z = \left(-b \pm \sqrt{b^2 - 4ac}\right)/(2a) \\
&a = 4r_2^2 + 4r_1r_3\cos\theta_1 + 4r_1r_3 - 4r_3^2 - 2r_1^2 - 2r_1^2\cos\theta_1 \\
&b = 4r_1^2\sin\theta_1 \\
&c = 4r_2^2 + 4r_1r_3\cos\theta_1 - 4r_1r_3 - 4r_3^2 - 2r_1^2 + 2r_1^2\cos\theta_1
\end{aligned}$$

Based on the analysis in [33], Eq. (32) can be also rewritten as the following sextic:

$$(x+2r_3)^2\left(x^2+4r_3x+3r_3^2+r_2^2-r_1^2+y^2\right)^2+y^2\left(x^2+4r_3x+5r_3^2+r_2^2-r_1^2+y^2\right)^2-4r_3^2r_2^2y^2=0 \tag{33}$$

For the uncertainty singularity, if $r_1 < r_3$, the singularity that B_1 and B_2 are coincident will not occur. If $r_2 > r_1 + r_3$, there is no the singularity that B_1PB_2 is completely extended.

3) The third kind of singularity occurs when both $\boldsymbol{A}$ and $\boldsymbol{B}$ become simultaneously singular. This singularity is of a slightly different nature than the first two since it is not only configuration but also architecture dependent.

From the analysis on the stationary and uncertainty singularities, one can see that the third kind of singularity occurs when the five points A_1, B_1, P, B_2 and A_2 are collinear. There are six cases for this kind of singularity, as shown in figure 11. The parameter conditions to those singularities are as following

(a) $r_1 = r_3$, one singularity is shown in figure 11(a);

(b) $r_2 = r_3$, a corresponding singular configuration is shown in figure 11(b);

(c) $r_3 = r_1 + r_2$, see figure 11(c) for the singular configuration;

(d) $r_2 = r_1 + r_3$, figure 11(d) shows such a singularity;

(e) $r_1 = r_3 + r_2$, see figure 11(e);

(f) $r_3 = 0$, one singular configuration is shown in figure 11(f).

The mechanism satisfies any one of the five parameter conditions (a)~(e) is referred to as *a change point mechanism*.

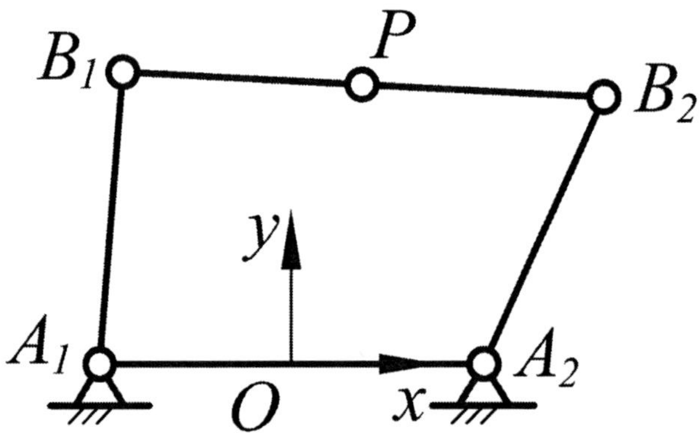

Figure 10.The configuration when B_1PB_2 is completely extended.

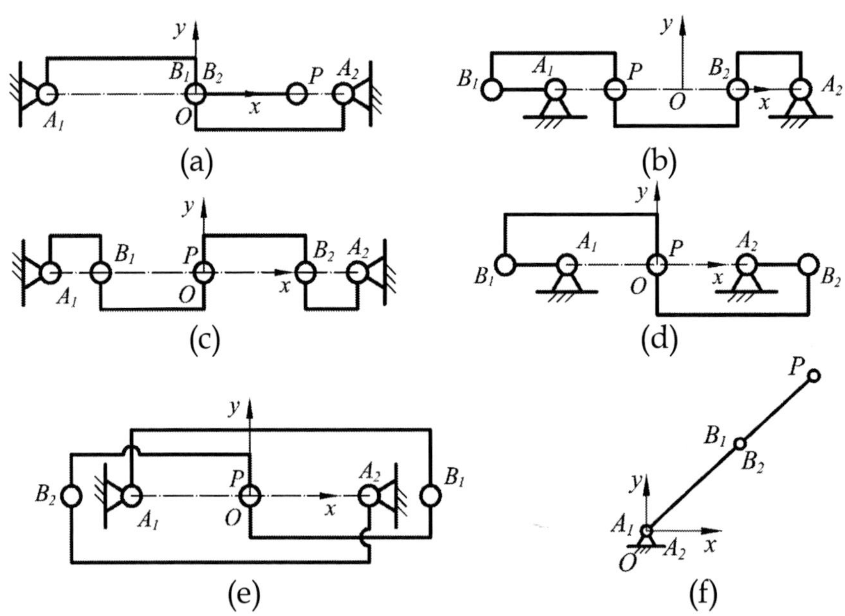

Figure 11. Six configurations of the stationary singularity.

The analysis on the kinematics of the mechanism shows that there are four solutions for the inverse kinematics and two solutions for the forward kinematics. Any one of the singularities will result in the change of solution number of the kinematics. For example, the stationary singularity leads to the loss of solution number of the inverse kinematics. While in the uncertainty singular configuration, the solution number of the forward kinematics can be less or more than two. In the third kind of singularity, both the inverse and forward kinematics will be different. Then the stationary singularity can be called the inverse kinematic singularity, the uncertainty singularity the forward kinematic singularity, and the third kind of singularity the inverse and forward kinemtic singularity. Notably, the loci of the uncertainty singularity must be within the workspace.

4.3.2. Distribution of the Singular Loci in the Design Space

Based on the analysis on the singularity, there exist two kinds of singular loci. The first kind of locus is the inverse singular locus. The loci are C_{1o}, C_{1i}, C_{2o} and C_{2i} given by Eqs. (18)-(21). They are actually the boundaries of the theoretical workspace. The second one is the forward singular locus. There are two cases for this kind of singular locus. One is that when the points B_1 and B_2 are coincident. The loci are actually two circles given by Eq. (30), which are denoted as C_{Coin-U} and C_{Coin-D}. The second case occurs when B_1PB_2 is completely extended. The locus is presented by Eq. (32) or Eq. (33), which is denoted as C_{Col}. For example, the singular loci of the mechanism with $r_1 = 1.2$, $r_2 = 1.0$ and $r_3 = 0.8$ are shown in figure 12.

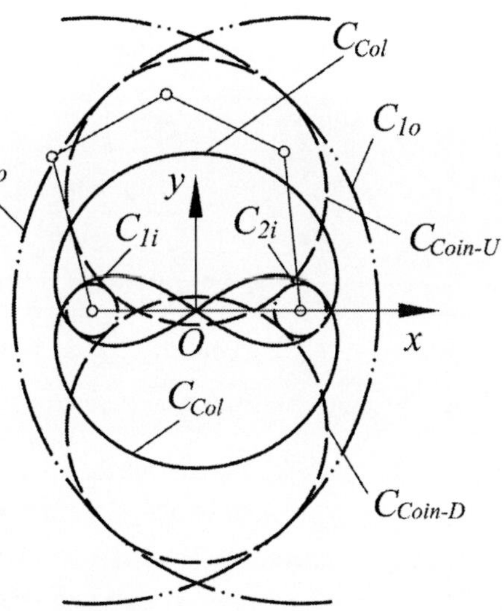

Figure 12. The singular loci of a mechanism.

However, with specified working and assembly modes, a mechanism cannot have all the singularities. According to the singularity analysis, the stationary singularity is actually the boundary of a theoretical workspace. Then, a mechanism with every working mode can have such singular loci. However, as the uncertainty singularity occurs inside the workspace, not every working mode has all such singularities. Normally, for most 5R parallel mechanisms, there are four tangent points between the first and second kinds of singular loci. At these points, the mechanism is in the change point's mechanism. The points can be used to identify which singular loci a specified working mode can have. For example, figure 13 shows some singular configurations and singular loci of the mechanism with $r_1 = 1.2$, $r_2 = 1.0$ and $r_3 = 0.8$. As shown in figure 13 (a), there are four tangent points m, n, q and k between the singular curve C_{Coin-U} and the first kind of singular loci. At these four points, both of the stationary and uncertainty singularities occur. The four points divide the singular curve C_{Coin-U} into four parts. At the arc $m1n$, the mechanism is in singular only when it is with the "– –" mode. At the arcs $n2q$, $q3k$ and $k4m$, the working modes "– +", "++" and "+ –" are in singular, respectively. Similarly, at C_{Coin-D}, arcs $m'1'n'$, $n'2'q'$, $q'3'k'$ and $k'4'm'$ are the singular curves for the working modes "++", "+ –", "– –" and "– +", respectively. For the singular locus C_{Col}, as shown in figure 13 (b) and (c), curves $t'5v$, $t76v$, $t5'v'$ and $t'7'6'v'$ are the singular loci for the working modes "+ –", "++", "– +" and "– –", respectively. For the mechanisms in sub-region IVb where $r_1 > r_2 + r_3$ and $r_2 > r_3$, the parameter condition does not allow the case that the singular locus C_{Col} and the first kind of singular locus are tangent to occur. These mechanisms with the working modes "++" and "– –" don't have the singularity that B_1PB_2 is completely extended.

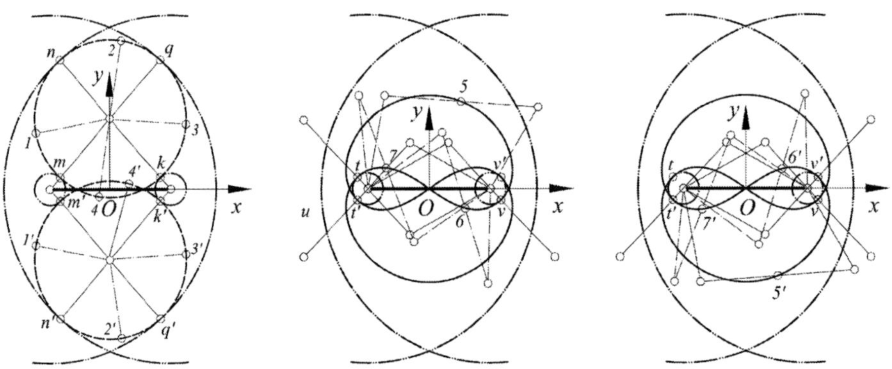

(a) (b) (c)

Figure 13. The second kind of loci of a mechanism with different working modes.

In this paper, we are just concerned about the mechanism with the working mode "+ –". Figure 14 shows the singular loci for all mechanisms with the working mode in the design space.

4.4. Usable Workspace

As we know, the practical workspace where a real mechanism works is different from the theoretical workspace. The region should contain no singular loci inside. From the distribution of theoretical workspace area and shape, one can see that the workspace area and shape of the mechanism with parameters $r_1 = 0.7$, $r_2 = 1.3$ and $r_3 = 1.0$ are identical to those of the mechanism with parameters $r_1 = 1.3$, $r_2 = 0.7$ and $r_3 = 1.0$. But from the analysis on singularity, the singularity loci for these two mechanisms are different. This will directly result in different usable workspaces. Therefore, based on figures 6 and 7, one cannot give any design result for a practical mechanism.

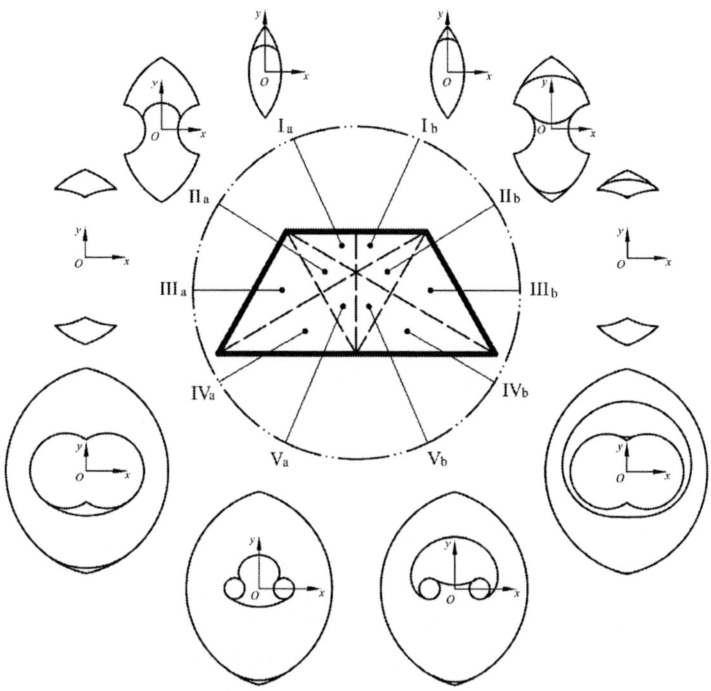

Figure 14. Distribution of singular loci in the design space for the mechanism with the working mode "+ –"

4.4.1. Definition

Usually, there exist singular loci inside the theoretical workspace (see figure 12). A mechanism wants to move from one point to another it maybe should pass a singular configuration. That means it should change from one working mode to another. In practice, changing working mode during the working process is definitely impossible. Therefore, we should find out a working space without singularity.

The *usable workspace* is defined as the maximum continuous workspace that contains no singular loci inside but bounded by singular loci outside. According to this definition, not every point within the usable workspace can be available for a practical mechanism. The mechanism will be out of control at the points on the boundaries and their neighborhoods. But within this region, the mechanism with a specified working mode can move freely.

The singular loci shown in figure 14 can be used to determine the usable workspace of a 5R parallel mechanisms with both the working mode "+ –" and the up-configuration. The workspace of a mechanism should be continuous. As shown in figure 14, the singular loci divide the theoretical workspace into several regions. The regions for the mechanisms in all sub-regions are separated. A practical mechanism cannot access the singular loci from one region to another. Even though, from figure 14, one can see that the two actuated links of the mechanism in sub-regions IVa, IVb, Va and Vb are both crank links. If one wants to find a mechanism that has two actuated crank links and has a surrounding workspace, he can select such a mechanism in these four sub-regions. However, in this paper we define the region above the x-axis as our usable workspace. For example, the usable workspace of the mechanism with $r_1 = 1.2$, $r_2 = 1.0$ and $r_3 = 0.8$ is shown as the hatched region in figure 15. Within this workspace, the mechanism has the configuration with both the inverse working mode "+ –" and the forward up-configuration. For the mechanism in one of the sub-regions Ia, Ib, IIa, IIb, IIIa and IIIb, the theoretical workspace above the x-axis is divided by the singular loci into several non-continuous regions, in which the region on the most top, where the mechanism is with the up-configuration, is considered as its usable workspace.

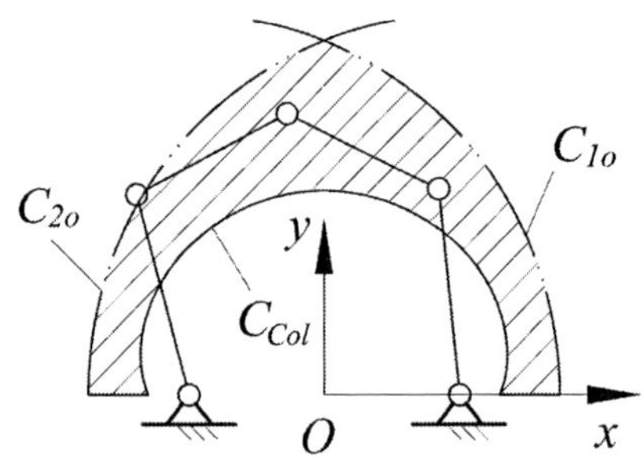

Figure 15. The usable workspace of a mechanism.

4.4.2. Distribution of the Usable Workspace Shape in the Design Space

The usable workspace shape for the 5R parallel mechanism is classified in the design space as shown in figure 16. From the distribution one sees that

- In figure 16, each of the hatched regions is the usable workspace of the mechanism with the "+ –" working mode and the up-configuration, simultaneously.
- Each usable workspace is symmetric about the y-axis.
- The distribution in the design space is not symmetric about $r_1 = r_2$. The usable workspace shape of a mechanism with $r_1 = a$, $r_2 = b$ ($a, b < 3$) and $r_3 = 3 - a - b$ is no longer the same as that of a mechanism with $r_1 = b$, $r_2 = a$ ($a, b < 3$) and $r_3 = 3 - a - b$.
- The usable workspace of a mechanism in sub-region IVa is half of its theoretical workspace.
- The flattest part in the workspace is always located at the region where $y = 0$.

According to the definition, the usable workspace of a mechanism shown in figure 16 still cannot be used in practice since it is surrounded by singularities. At and near the boundary, the mechanism will be out of control. We should find a workspace in which every point or every pose is under control. Most importantly, we should find a solution to bound a workspace. In Refs. [9,16], a good-condition workspace or effective workspace was defined with respect to a specified minimum LCI. However, the minimum is still arbitrary or comparative since we cannot give it a definite value. Generally, it is not possible to define a mathematical distance to a singularity for a parallel mechanism [14]. Instead of LCI, here we will define an index by considering the force transmission, which is evaluated by the *transmission angle* in four-bar mechanisms.

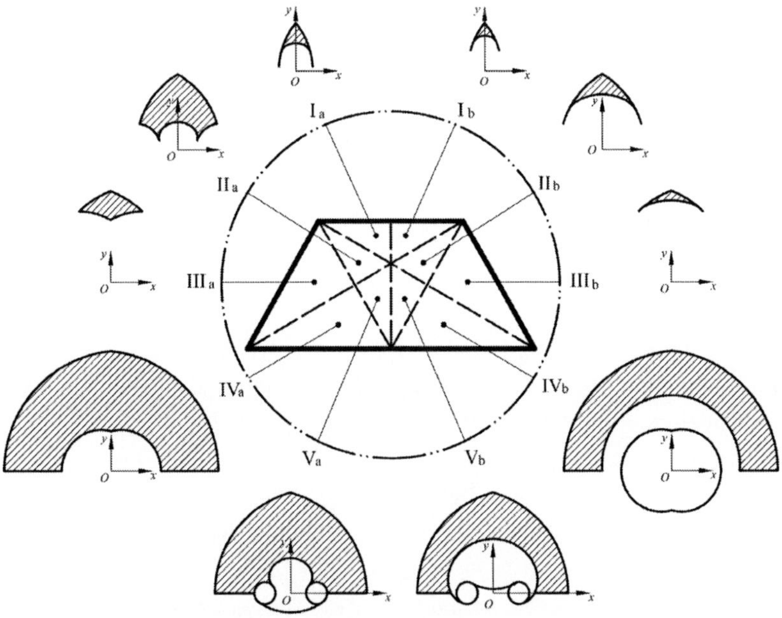

Figure 16. The usable workspace shape in the design space.

5. Transmission Angle

5.1. Definition

The *transmission angle* is something we are very familiar with without realizing it. In everyday life, we frequently try to move an object that is somehow constrained, which cannot move freely but is attached to something: the handle of a crank, a curtain on a rail, a sliding door. In all of these cases, the object may not be able to move even when we exert pressure against it. Let's take the case of the handle of a crank as an example. As shown in figure 17, the crank is attached to the base with a constant counterclockwise moment ***M***. To move it, we must apply a right-hand force F at the end of the crank. When the direction of the force is constant, depending on the position of the end point it will be more or less easy to start rotating. This is the reason why it sometimes feels comfortable and sometimes laborious when we ride a bicycle. This is actually a matter of the transmission angle. Since the direction of motion of a crank is always perpendicular to the crank, the smaller angle between the force and crank is defined as the *transmission angle*, denoted as μ. When the force is normal to the crank, i.e. identical to the direction of motion, force transmission is most effective; when the force is perpendicular to the direction of motion, force transmission is very inefficient.

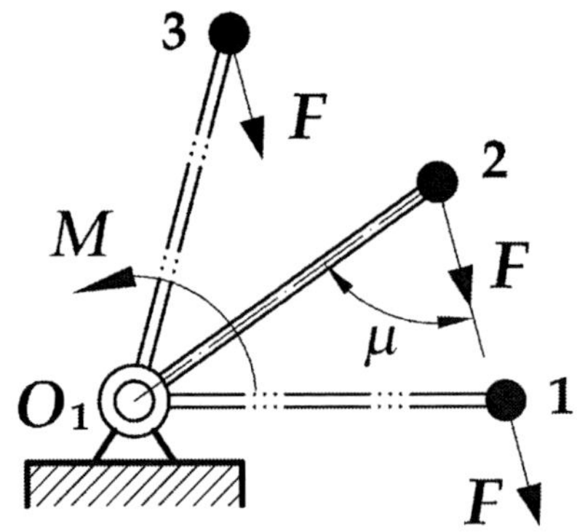

Figure 17. The handle of a crank.

For the planar four-bar mechanism shown in figure 18, if O_1A is the input link, the force applied to the output link BO_2 is transmitted through the coupler link AB. For sufficiently slow motions (negligible inertia forces), the force in the coupler link is pure tension or compression (negligible bending action) and is directed along AB. For a given force in the coupler link, the torque transmitted to the output bar (about point O_2) is at a maximum when the angle μ between the coupler bar AB and the output bar BO_2 is $90°$. Therefore, angle ABO_2 is called the *transmission angle*. In [21], the transmission angle is defined as the smaller angle between the direction of the velocity difference vector of the driving link and the direction of the absolute velocity vector of the output link, both taken at the point of connection. Although there are other definitions (see [17,34], for instance), all these definitions are somehow related to a joint variable(s) of the mechanism.

When the *transmission angle* deviates significantly from $90°$, the torque on the output bar decreases and may not be sufficient to overcome the friction in the system. For this reason, the deviation angle $\alpha = |90° - \mu|$ should not be too great.

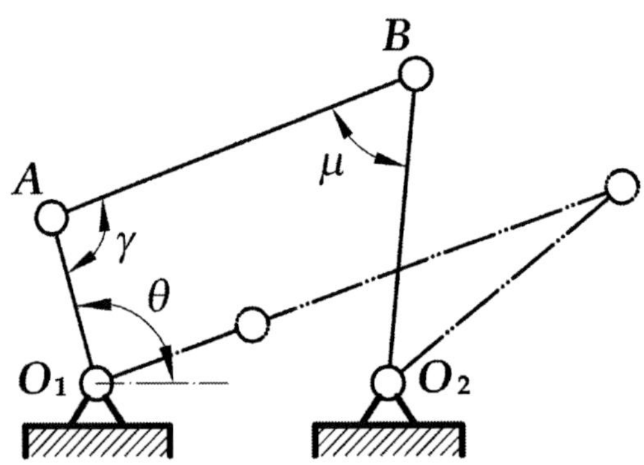

Figure 18. Transmission angle.

Meanwhile, at the moment that the angle γ between the input link O_1A and the coupler link AB is 0 or $180°$, as shown in figure 18, the output point B will not move whatever the input is. Then, the motion of the input cannot be transmitted to the output effectively. This means the output point will lose a degree of freedom. Therefore, the deviation angle

$\beta = |90° - \gamma|$ should not be too great either. In this paper, the angle μ is defined as the *forward transmission angle*, and the angle γ is referred to as the *inverse transmission angle*. If BO_2 is an active link, the angle O_1AB is the *forward transmission angle* and angle ABO_2 the *inverse transmission angle*.

Take the slider-crank mechanism shown in figure 19 as an example, where angle $O'P_3B_3$ is the *forward transmission angle* and the angle between the coupler link and the normal to the straight-line path of the slider is the *inverse transmission angle* if the slider is actuated. For the 2-DOF 5R parallel mechanism shown in figure 1, the *forward transmission angle* is defined as the angle between the two couplers *AP* and *PC* [35]; and the *inverse transmission angle* is the angle between the input link and the coupler, as illustrated in figure 20. For the mechanism, there are two *inverse transmission angles*. By using the concept of *transmission angle*, as of 1961, the design of a 2-DOF five-bar mechanism has been investigated [36]. Recently, a synthesis method for the 5R variable topology mechanism has been proposed with transmission angle control [18].

5.2 Relationship between Transmission Angle and Singularity

The *transmission angle* is an important index that can evaluate the quality of motion/force transmission. For this reason, the concept has been used in the design of almost all types of four-bar mechanisms, some other planar five-, six- and seven-bar mechanisms [37], the spatial four-link mechanisms [38] and even the 2-DOF spatial RSSRP mechanism [39].

Singularity is one of the most important problems in mechanisms, especially in parallel mechanisms. Since singularity leads to a loss of controllability and degradation of the natural stiffness, the analysis of parallel mechanisms has attracted considerable attention [4, 29, 40]. In a singularity, the condition number of the Jacobian matrix is normally null.

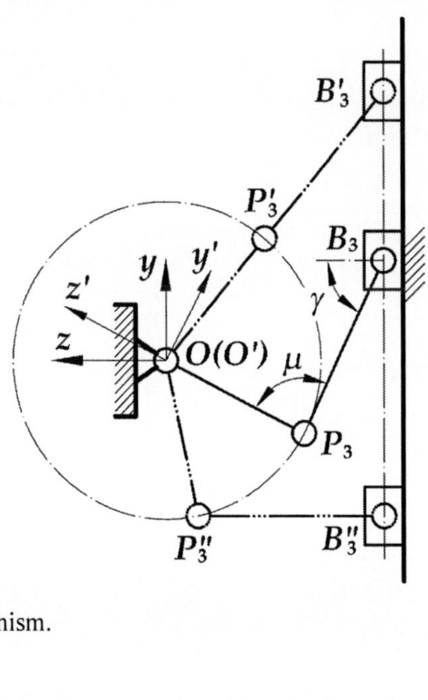

Figure 19. Slider-crank mechanism.

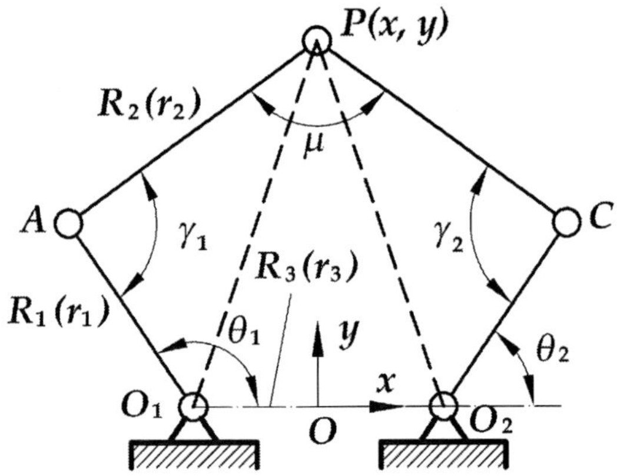

Figure 20. The 5R parallel mechanism.

There is a close relationship between the *transmission angle* and singularity. Taking the four-bar mechanism shown in figure 18 as an example, when $\mu = 0$ or $180°$ the mechanism is in the "dead point" which is the second kind of singularity according to the classification in [4], while $\gamma = 0$ or $180°$ corresponds to the first kind of singularity where the mechanism has the self-jamming characteristic. For the slider-crank mechanism shown in figure 19, the configuration $O'P'_3B'_3$ where $\mu = 180°$ is the second kind of singularity, whereas the configuration $O'P'_3B'_3$ where $\gamma = 0$ is the first kind. The singularity of the 5R parallel mechanism shown in figure 20 has been listed systematically in section 4.3. One can see that $\gamma = 0$ or $180°$ and $\mu = 0$ or $180°$ represent the first and second kinds of singularities of the mechanism, respectively. Therefore, the *forward* and *inverse transmission angles*, μ and γ, are close relatives of the two kinds of singularities of the parallel mechanism. It is easy to conclude that if $\gamma = 0$ or $180°$ and $\mu = 0$ or $180°$ occur at the same time, the mechanism will be in the third kind of singularity. This case is not only architecture- but dimension-dependent as well.

In [41], the authors also investigated the relationship between pressure angle and singularity, and reached the conclusion that in the singularity the pressure angle is at $90°$. However, according to their definition of pressure angles, their discussion was actually only about the second kind of singularity.

6. Definition of Some New Indices

6.1. Local Transmission Index (LTI)

The condition number of the Jacobian matrix is an index that has been used successfully in the design of serial robots. The condition number has the main advantage of describing the kinematic behavior of a robot by means of a number. The index has also been applied in the analysis and design of parallel mechanisms. It was used as an index to evaluate the accuracy/dexterity [10,11], to describe the closeness of a pose to a singularity of a parallel mechanism [42]. In an optimal design, the condition number (or its reciprocal) was used to define a useful, *good-condition* or *effective workspace* with a specified minimum [9,16]. However, the minimum is still arbitrary or comparative since we cannot give it a definite value. Generally, it is not possible to define a mathematical distance to a singularity for parallel mechanisms, whose DOF is a mix between translation and rotation [14].

Following the definition of transmission index proposed in [43], an index with respect to the *transmission angle* is defined as

$$\chi = \sin(TA) \tag{34}$$

where $TA = \mu$ or $TA = \gamma$. Then, there is

$$0 \le \chi \le 1 \tag{35}$$

A larger χ indicates better motion/force transmission. Since at a different pose the *transmission angles* will be different, the index χ is referred to as the *local transmission index* (LTI) in this paper. The angle is defined as the figure formed by two lines diverging from a common point, or as that formed by two planes diverging from a common line. Thus, the angle is usually measured by the ratio of two linear parameters. Therefore, the LCI is definitely independent of any coordinate frame. This is one of its advantages and is most important for the optimal design of mechanisms.

For the purpose of high speed and high quality of motion/force transmission, the most widely accepted ranges for the *transmission angle* are $(45°,\ 135°)$ [21] and $(40°,\ 140°)$ [17]. Therefore, the LTI limits will correspondingly be

$$\chi > \sin(\pi/4) \text{ and } \chi > \sin(2\pi/9) \tag{36}$$

Then, unlike the LCI, the LTI has a significative limitation to its application.

6.2. Good-Transmission Workspace (GTW)

With the minimum of LTI, i.e. $\sin(\pi/4)$ or $\sin(2\pi/9)$, we can identify a workspace for any parallel mechanism. The corresponding workspace is referred to as a *good-transmission workspace* (GTW), which is defined as the set of pose where the transmission index for every *transmission angle* is greater than $\sin(\pi/4)$ or $\sin(2\pi/9)$. It is obvious that the minimum LTI, $\sin(2\pi/9)$, will lead to a larger *GTW* than $\sin(\pi/4)$.

As an example, consider the planar 5R symmetrical parallel mechanism shown in figure 20. As analyzed in section 2, for a given position vector $\boldsymbol{p} = (x \quad y)^{\mathrm{T}}$ of the output point *P*, the position vectors $\boldsymbol{a}_{\Re}$, and $\boldsymbol{c}_{\Re}$, in the reference frame $\Re : O - xy$, of points *A* and *C*, respectively, can be obtained. Then, there are

$$\mu = \cos^{-1}\left(\frac{2r_2^2 - |\boldsymbol{a}_{\Re}\boldsymbol{c}_{\Re}|^2}{2r_2^2}\right) \tag{37}$$

$$\gamma_i = \cos^{-1}\left(\frac{r_1^2 + r_2^2 - |\boldsymbol{o}_{i\Re}\boldsymbol{p}|^2}{2r_1 r_2}\right), \; i=1,2 \tag{38}$$

where $\boldsymbol{o}_{i\Re}$ (i=1,2) are the position vectors of points O_i. By letting $\sin\mu > \sin(\pi/4)$ and $\sin\gamma_i > \sin(\pi/4)$, we can obtain the GTW of the mechanism numerically. This means, for a given position $(x,\ y)$ of point *P*, to calculate the *forward* and *inverse transmission angles* μ and γ_i using Eqs. (37) and (38), if they all are subject to the LTI constraint given in Eq. (36), the point belongs to the GTW. For example, the GTW of the 5R parallel mechanism with parameters $r_1 = 0.85$, $r_2 = 1.6$ and $r_3 = 0.55$ is shown as the region bounded by the locus of LTI=sin(π /4) in figure 21(a). This normalized mechanism is an optimized solution when workspace, global conditioning index (GCI) and stiffness performances are the design indices [30]. According to the singular analysis result of the said paper, this mechanism does not have the second kind of singularity where points *A*, *P* and *C* are collinear. The region between the two loci of LTI=sin(π /4) in figure 21(b) is the GTW of the mechanism $r_1 = 1.5$, $r_2 = 1.2$ and $r_3 = 0.3$, which has the second kind of singularity. This non-dimensional mechanism is an optimum solution when workspace, GCI and velocity performances are all considered [30]. For comparison, in figure 21, the *good-condition workspace* specified by the loci of LCI=0.5 (LCI is defined as the reciprocal of the condition number of the Jacobian matrix) is also illustrated. One may see that the two kinds of workspaces defined by LCI and LTI are different from others. There are two points we should notice. Firstly, from a given LCI we don't know how far the pose is from the singularity and we cannot even be sure whether the *good-condition workspace* (or the specified LCI value 0.5) is suitable or not. For example, as shown in figure 21(a), at and near the point *a* the LCI distribution is very dense. In this area, the points where LCI=0.5 are very near to the singularity, while some other points with 0.5 are far from the singularity. However, the *transmission angle* has a definite physical significance. We can be sure that in the GTW defined by the LTI the motion/force transmission of the mechanism is effective. Secondly, for some points with a worse LCI value (for example less than 0.5 in the two examples), the mechanism is still effective in motion/force transmission. For example, figure 22 shows the configuration at the point (1.7310, 0) of the mechanism with $r_1 = 0.85$, $r_2 = 1.6$ and $r_3 = 0.55$, and all the *transmission angle* values. They all fall within the acceptable range $(45°,\ 135°)$, but the LCI in this configuration is lower than 0.5 (see figure 21(a)). This means if we design a workspace, e.g. the *good-condition workspace*, with respect to the LCI, some points where the mechanism is good at motion/force transmission may be lost. Since the *transmission angle* has specific significance, we here prefer the design using the LTI.

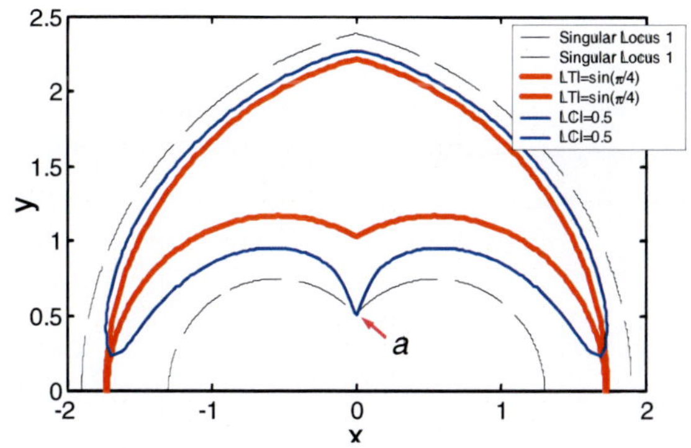

(a)

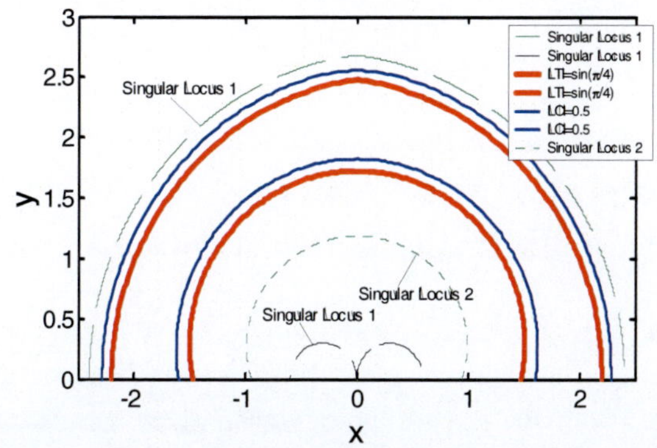

(b)

Figure 21. *Good-transmission workspace* of the 5R parallel mechanisms with the parameters (a) $r_1 = 0.85$, $r_2 = 1.6$ and $r_3 = 0.55$; (b) $r_1 = 1.5$, $r_2 = 1.2$ and $r_3 = 0.3$.

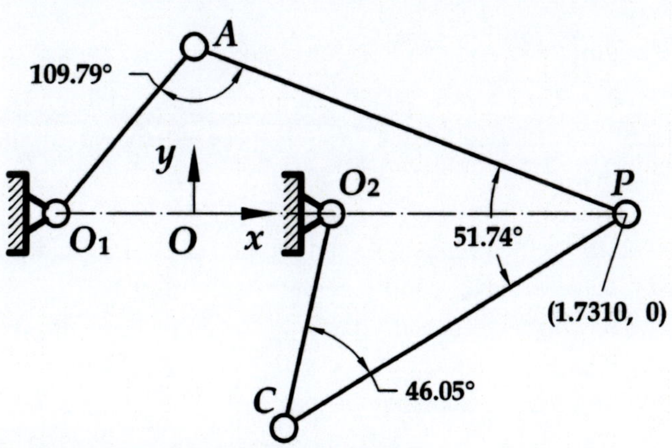

Figure 22. One configuration of the 5R parallel mechanism.

6.3. Global Transmission Index (GTI)

The LTI χ can only judge the effectiveness of motion/force transmission of a mechanism at a pose. In order to measure the global behavior of the motion/force transmission over the whole *good-transmission workspace* (GTW), a *global transmission index* (GTI) is defined as

$$\Gamma = \frac{\int_W \sum_i^n \chi_i \Big/ n \, dW}{\int_W dW} \tag{39}$$

in which W denotes the GTW, n the number of transmission angles and $\Gamma_{\min} < \Gamma < 1$ ($\Gamma_{\min}$ equals $\sin(\pi/4)$ or $\sin(2\pi/9)$). It is obvious that the GTI is also independent of any coordinate frame.

7. Performance Chart of the 2-DOF 5R Parallel Mechanism

As is well known, a performance chart (atlas) can show the relationship between a performance index and associated design parameters in a limited space, globally and visually. The design method by using the performance chart (atlas) has being widely used in the classical mechanical design and most design manuals. Here, performance charts of the indices for the 2-DOF 5R parallel mechanism will be presented in this section. Some of them will be used in the optimization of the mechanism.

7.1. Distribution of the GTW Shape in the Design Space

Letting $\sin\mu > \sin(\pi/4)$ and $\sin\gamma_i > \sin(\pi/4)$, one can numerically obtain the *good-transmission workspace* (GTW) of a 2-DOF 5R parallel mechanism. The GTW for the 5R parallel mechanisms is classified in the design space as shown in figure 23. From the distribution one sees that

- In figure 23, each of the hatched regions is the GTW of the mechanism with the "+ – " working mode and the up-configuration, simultaneously.
- For some mechanisms, especially those in the sub-regions Ia, Ib, IIb and IIIb, their GTWs are zero.
- Each GTW is symmetric about the y-axis.
- The distribution in the design space is not symmetric about $r_1 = r_2$.
- The fattest part in the workspace is always located at the y-axis.

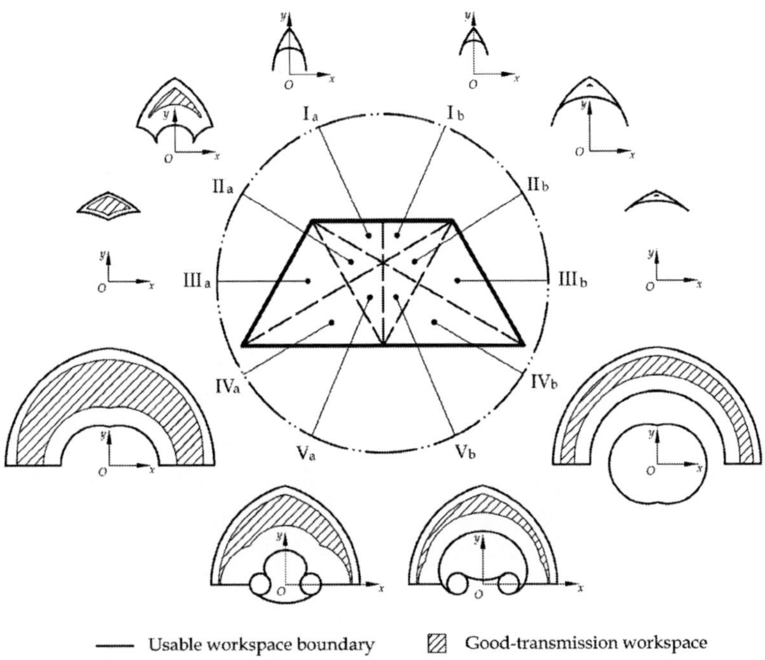

Figure 23. The GTW shape in the design space.

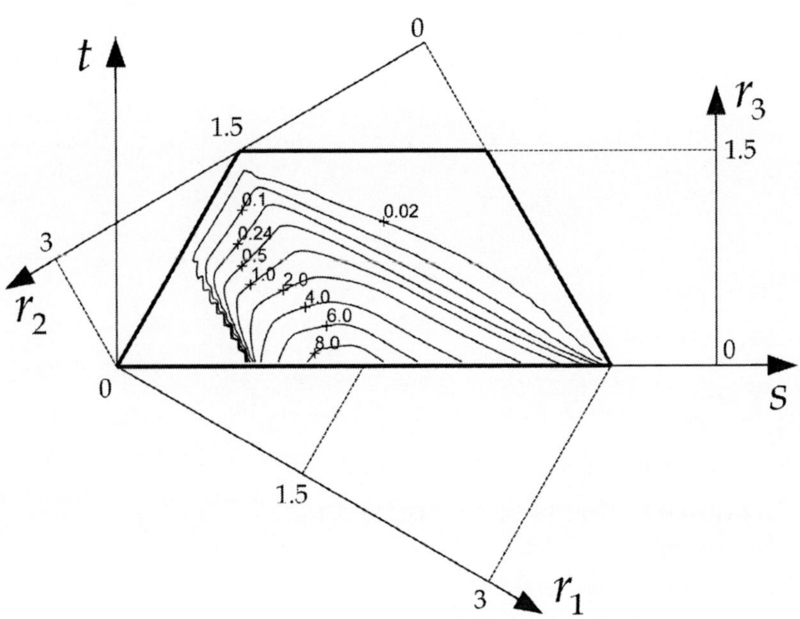

Figure 24. Performance chart of the *good-transmission workspace* (GTW).

7.2. Performance Chart of the GTW

According to the definition of GTW in Sec. 6.2, it is possible to calculate the GTW of any one non-dimensional mechanism in the parameter design space shown in figure 4(b). The performance chart illustrating the relationship between the normalized parameters r_i (i=1, 2, 3) and the GTW, letting $\sin\mu > \sin(\pi/4)$ and $\sin\gamma_i > \sin(\pi/4)$, is illustrated in figure 24, from which we may see that (a) the index is inversely proportional to the parameter r_3 when r_2 is specified, and (b) the GTW of some mechanisms in the design space is zero.

7.3. Performance Chart of the GTI

By using the Eq. (39) and the GTW result, we can also represent the relationship between the GTI and the non-dimensional mechanisms in the parameter design space shown in figure 24(b). The corresponding performance chart is illustrated in figure 25, which shows that, in the design space, the mechanisms near the area $r_2 = 1.69$ usually have better GTI performance.

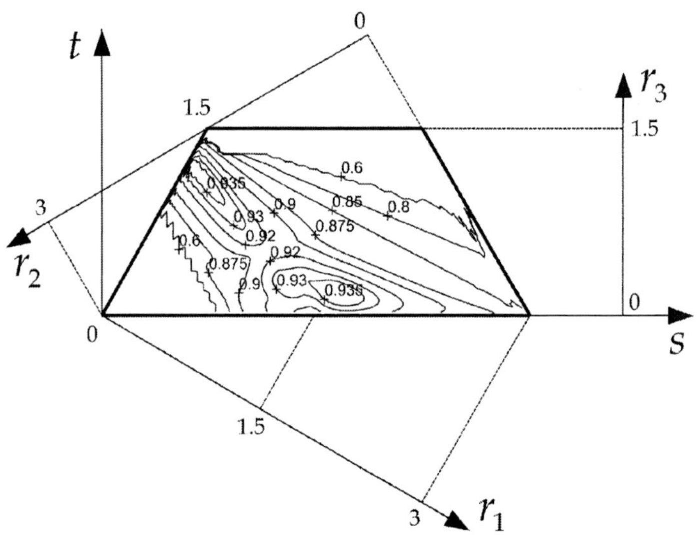

Figure 25. Performance chart of the *global transmission index* (GTI).

7.4. Performance Chart of the Fatness Index (FI) of the GTW

7.4.1. Definition of the Fatness Index (FI)

From figure 23, one may see that with different link lengths the GTW shape may be different. The GTW of some mechanisms is in a narrow shape, while for some mechanisms it is broad. Comparing figures 23 and 24, we may observe that a GTW with larger volume does not mean broad, whereas a GTW with smaller volume may be in a broad shape.

Undoubtedly, in some applications a narrow or circular workspace will be required. However, in some industrial applications, a regular workspace such as a rectangular or foursquare workspace is needed. For this case, a broader workspace is welcome. For this purpose, we here should introduce an index to evaluate the fatness of the GTW. From figure

23, one may see that the fattest part in the GTW is always located at the y-axis. In this paper, the GTW length along the y-axis, denoted as d_y, is defined as the *fatness index* (FI) of the GTW. The larger the FI value, the broader the GTW.

It is undoubted that two mechanisms with same GTW volumes may have different FIs. For example, figure 26 (a) and (b) show the GTW of a mechanism with $r_1 = 1.57$, $r_2 = 1.161$ and $r_3 = 0.269$, and that of a mechanism with $r_1 = 0.795$, $r_2 = 1.93$ and $r_3 = 0.275$, respectively. The volume of both of them is 4.0, however, their FIs are different, d_{y1} =0.7 and d_{y2} =0.95.

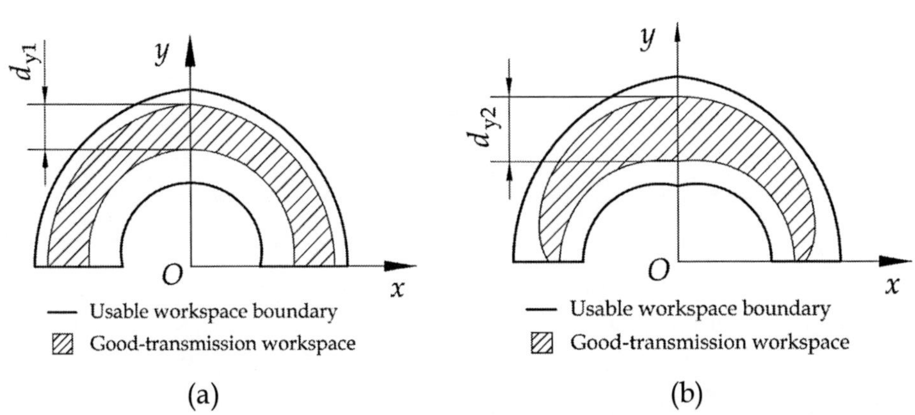

Figure 26. GTW and fatness index of two mechanisms with: (a) $r_1 = 1.57$, $r_2 = 1.161$ and $r_3 = 0.269$; and (b) $r_1 = 0.795$, $r_2 = 1.93$ and $r_3 = 0.275$.

7.4.2. Performance Chart of the Fatness Index

Figure 27 illustrates the FI performance chart of the 2-DOF 5R parallel mechanisms. It shows that the index is inversely proportional to the parameter r_3 when r_2 is specified. The performance chart can be used in the optimization of the mechanism if a regular workspace is required in a practical design.

8. Optimization Process

8.1. Relationship between a Non-Dimensional Mechanism and Its Corresponding Dimensional Mechanism

In the design space, the parameters r_1, r_2 and r_3 have no dimension. They are the ratios of the dimensional parameters R_i to a dimensional scale D, which is the average of R_1, R_2 and R_3 as shown in Eq. (13). As given in Eq. (16), the parameters r_1, r_2 and r_3 are limited. But each of the parameters R_1, R_2 and R_3 can be infinite. Practically, we are concerned with the dimensional mechanism but not the non-dimensional one. But it is not difficult to find that the ratio $R_1 : R_2 : R_3$ is always equal to $r_1 : r_2 : r_3$. For any possible dimension combination of R_1, R_2 and R_3, one can always find its corresponding group of r_1, r_2 and r_3 in the design space. The design space defined by r_1, r_2 and r_3 makes it possible for us to investigate the performance of all possible 5R parallel mechanisms. For example, the performance chart of figure 27 shows the global result about the fatness of GTW, which can characterize the GTW with the help of figures 23 and 24. From figures 23, 24 and 27, we can know the GTW result of a non-dimensional 5R parallel mechanism. Then, how about the dimensional mechanisms? We know that the key coefficient between R_i and r_i is the dimensional scale D. Once we select a suitable group of r_1, r_2 and r_3, the parameter D will lead it to a dimensional mechanism by Eq. (14). The problem is how the GTW of the dimension mechanism will be if one knows that of a non-dimensional mechanism from figures. 23, 24 and 27.

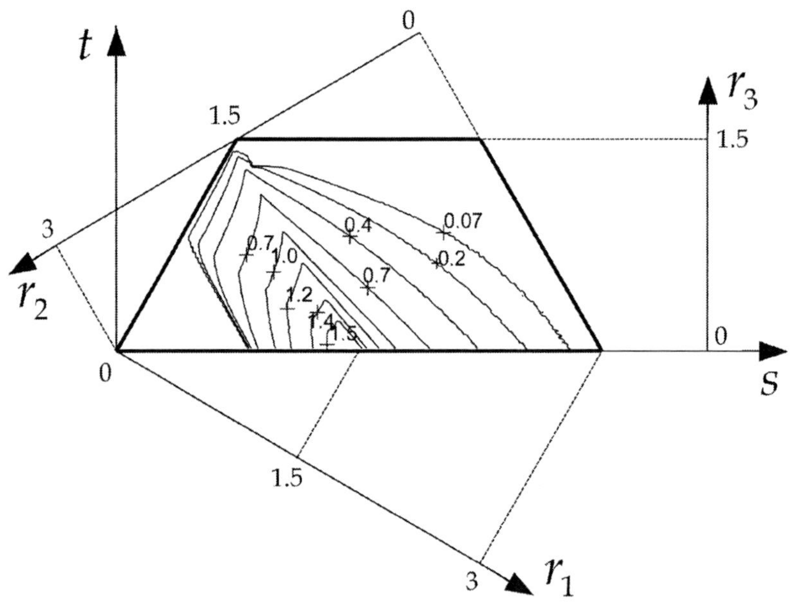

Figure 27. Performance chart of the *fatness index* (FI) for GTW.

Angle is defined as the figure formed by two lines diverging from a common point, or as that formed by two planes diverging from a common line. Thus, the angle is usually measured by the ratio of two linear parameters. In this paper, the transmission angle is used to define the index LTI. And we use the LTI limit, i.e. $\sin(\pi/4)$, to identify the GTW. Therefore, the GTW shape of a mechanism with R_i will be similar to that of its corresponding non-dimensional mechanism with r_i, and their GTW volumes will have the relationship as following

$$GTW_D = D^2\, GTW_{non\text{-}D} \quad (40)$$

in which GTW_D and GTW_{non-D} denote the GTW volumes of the dimensional mechanism with parameters R_i and its corresponding non-dimensional mechanism with parameters r_i, respectively. However, from Eq. (39), we may see that their GTI values will be the same.

Therefore, all mechanisms with parameters $R_i = Dr_i$ have similar performances to that with the non-dimensional parameters r_i. For example, the GTW of the mechanism with R_i is D^2-time that of the mechanism with r_i. Within the similar workspaces, the GTI values are the same as that of the other. The mechanisms, which have such a performance similarity, are referred to as the *similarity mechanisms* (SMs). Then, the ratios of geometric parameters R_i of all SMs are constant. Any one of the non-dimensional mechanisms in the parameter design space stands for all of its possible SMs. Therefore, the mechanisms with parameters r_i are defined as *basic similarity mechanisms* (BSMs). After normalization, a mechanism with any kind of link lengths can find its position or its BSM in the design space. For such a reason, the design space is a useful tool to present a global comparative result in terms of performances for all mechanisms. This also guarantees a global optimum result in the design issue, and the performance charts can be used in the optimization. Since the BSMs are in a limited space, i.e. the parameter design space, in the design process we should first be concerned about the BSMs.

8.2. Optimization by Using the Performance Charts

The objective of optimization is to determine the geometrical parameters of a mechanism for a desired workspace. In this process, such other performances may be considered. For the parallel mechanism studied here, the optimization taking into account the desired workspace and force transmissibility based on the performance charts in Sec. 7 can be summarized as follows:

Step 1: Identify an optimum region in the *parameter design space*. The identification of optimum region is depended on the design requirement. For different practical design, the optimum region may be different. Anyway, we usually would like the mechanism candidate to have a workspace as large as possible. For this case, we can identify an optimum region with large GTW and better GTI. For example, letting GTW$\geqslant$5.0 and GTI$\geqslant$0.925, an optimum region denoted as $\Omega_{GTW-GTI}$ can be identified (shown as the hatched region in figure 28(a)) by using the performance charts of figures 24 and 25. The optimum region $\Omega_{GTW-GTI}$ contains all possible optimum solutions with the non-dimensional parameters r_i (i=1, 2, 3). However, the optimum region is not unique. It can be changed when the constraints are different. For example, if we would like to reduce the optimum region, we can use stricter constraints, such as GTW$\geqslant$6.0 and GTI$\geqslant$0.93. The optimum region denoted as $\Omega'_{GTW-GTI}$ is shown in figure 28(b).

For a practical design that a regular workspace is required, the fatness of GTW should be considered. The corresponding optimum region can be identified by using the performance charts shown in figures (24), (25) and (27). For example, letting GTW≥6.0, GTI≥0.93 and FI≥1.2, an optimum region denoted as $\Omega_{GTW-GTI-FI}$ can be obtained as shown in figure 29.

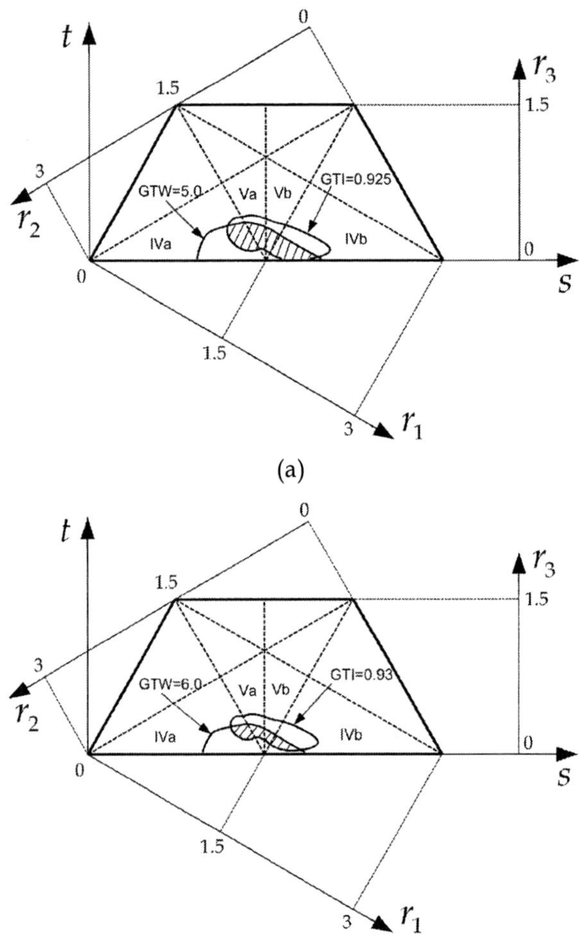

Figure 28. Optimum regions for the 2-DOF 5R parallel mechanism when (a) GTW≥5.0 and GTI≥ 0.925, and (b) GTW≥6.0 and GTI≥0.93.

Step 2: Select a solution candidate, i.e. BSM, from the optimum region. The optimum region obtained contains all possible solutions with better performances. Since there is no best but only a comparatively better solution for a design problem, we can pick any non-dimensional mechanism from the region. For example, the BSM with parameters $r_1 = 1.25$, $r_2 = 1.49$ and $r_3 = 0.26$ is selected here from the optimum region $\Omega_{GTW-GTI-FI}$. The GTW, GTI and FI of the mechanism are 6.4125, 0.9337 and 1.3911, respectively. Figure 30 shows the mechanism, its GTW and the distribution of LTI within the GTW.

Step 3: Determine the dimensional parameters R_i. According to Eq. (14), we first need to determine the normalization factor *D*. As described in [25], this factor can be obtained by comparing the desired workspace of a design problem to the workspace of the non-dimensional mechanism selected from the optimum region. Therefore, how much the normalization factor D will be depends on the desired workspace. Once the desired workspace is given, the factor can be calculated by using Eq. (40). For example, if the desired workspace is similar to the GTW shown in figure 30 and its volume GTW_D=200mm^2, since $GTW_{non\text{-}D}$=6.4125, then $D=\sqrt{GTW_D/GTW_{non\text{-}D}}\approx$ 5.5847mm. Rewriting Eq. (14) as $R_i = Dr_i$, then R_1=5.5847mm × 1.25≈6.98mm, R_2=5.5847mm × 1.49≈8.32mm, and R_3=5.5847mm×0.26≈1.45mm.

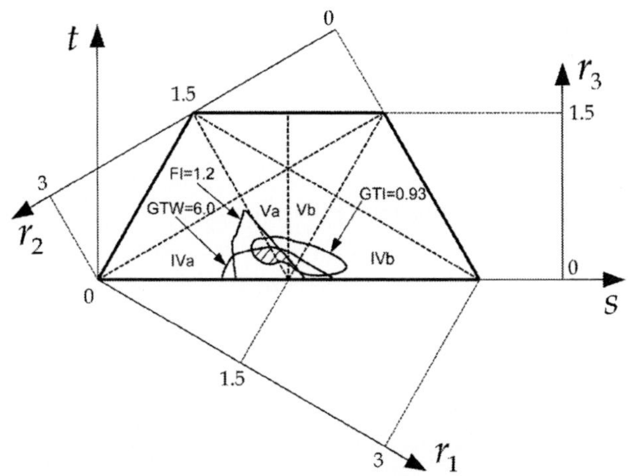

Figure 29. An optimum region for the 2-DOF 5R parallel mechanism when GTW⩾6.0, GTI⩾0.93 and FI⩾1.2.

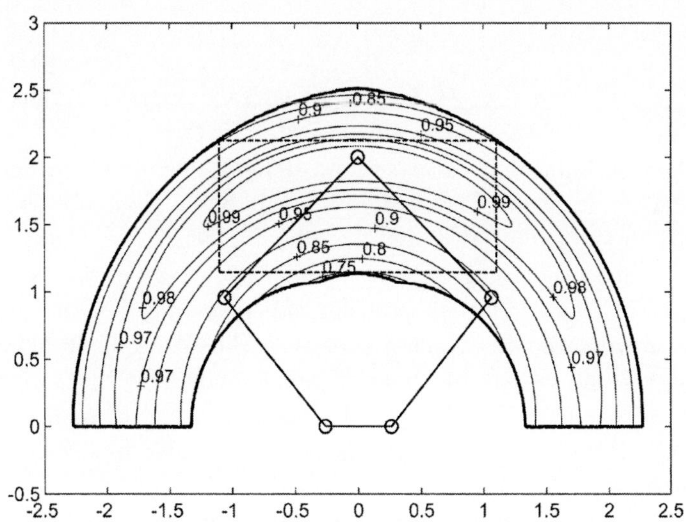

Figure 30. Mechanism, GTW, LTI distribution, and included rectangular workspace of one optimum BSM.

For the above example, as shown in figure 30, an included rectangular workspace with the volume 2.2072(length)×0.9815(width)≈2.16 can be identified in the GTW. Then, there must exist a similar workspace in the whole GTW of the dimensional mechanism with R_1=6.98mm, R_2=8.32mm, and R_3=1.45mm. Its volume should be $D^2 \times 2.17=(D\times 2.2072)\times(D\times 0.9815)\approx 67.56\text{mm}^2$. This means that, if a rectangular workspace is desired in a practical design, one should first find an included rectangle in the GTW of the selected BSM by taking into account the desired workspace.

Step 4: Check whether the dimensions obtained in Step 3 is enough or not for a practical design, and, if necessary, revise the design solution. In the previous step, we determined the dimensional parameters, which are R_1=6.98mm, R_2=8.32mm, and R_3=1.45mm. In this step, one must check the dimensions to make it sure that all components can be placed well and any other item needed in the practical design problem is feasible. If it is okay, proceed to Step 5; otherwise, return to Step 2 and pick up another group of non-dimensional parameters from the optimum region, and repeat Steps 3 and 4.

Step 5: Calculation of the input range. The input range required to reach the desired workspace of the optimized mechanism can be calculated by means of the inverse kinematics of Eq. (6). For example, for the solution obtained in Step 3, the input ranges are $\theta_1 \in [-13.68°,\ 97.64°]$ and $\theta_2 \in [82.36°,\ 193.68°]$ for the desired workspace of $12.3265(\text{length})\times 5.4813(\text{width})\approx 67.56\text{mm}^2$.

Step 6: Finish the optimization problem.

From the above process, one may see that the design method may provide all possible solutions since it gives an optimum region not just one solution, and allows the designer to adjust his design solution. They are actually the advantages of this approach.

9. Conclusions

In this paper, the kinematic analysis and optimization problems for the 2-DOF 5R symmetrical parallel mechanisms are addressed again by taking into account the concept of force transmissibility. Firstly, it shows that the theoretical workspace cannot be used practically due to the inside singularity. The usable workspace without singularity inside cannot be useful in an optimization problem either since a mechanism at and near the boundary of the workspace will be out of control. Additionally, the local conditioning index (LCI), which is widely used in the field of serial robots and has been applied in many optimizations, has the problem of physical significance and difficulty of measuring the distance to a singularity of a parallel mechanism. The usable workspace limited by a specified LCI is then doubtable.

Since the transmission angle that can evaluate the force transmissibility of planar four-bar mechanisms, the concept is applied into the analysis of the 5R parallel mechanisms in this paper. Three new indices, i.e. the local transmission index (LTI), good-transmission workspace (GTW) and global transmission index (GTI), are introduced accordingly. It shows that the LTI, unlike the LCI, has a valuable significance to its application. That is, it can evaluate the force transmissibility of a mechanism. In addition, the fatness index (FI) that is used to measure the fatness of a GTW is also proposed here.

Finally, based on the performance charts of GTW, GTI and FI, the optimization process of determining the link lengths of the 5R parallel mechanism is presented accordingly. The method may not only provide all possible optimum solutions but allow the designer to adjust his solution.

The application of three new indices, i.e. LTI, GTW and GTI, shall be said to be extended to other parallel mechanisms.

ACKNOWLEDGMENTS

This work was supported by the National Natural Science Foundation of China under Grant 50775118, by the High Technology Research and Development Program (863 Program) of China (No. 2006AA04Z227), and partly by the National Basic Research Program (973 Program) of China (No. 2006CB705400 and No. 2007CB714000), and was partly carried out during a one-month visit by the first author to the Fraunhofer-Institut Werkzeugmaschinen und Umformtechnick, Chemnitz, Germany, in the summer of 2007 supported by the Ministry of Education of China and the Deutsche Forschungsgemeinschaft.

REFERENCES

[1] K.M. Lee and D.K. Shah, Part 1: kinematic analysis of a 3-DOF in-parallel actuated manipulator, IEEE *Journal of Robotics and Auotmation* 4 (1988) 354-360.

[2] J.-P. Merlet, Direct kinematic and assembly motion parallel manipulators, *Int. J. Rbotics Research 11* (2) (1992) 150-162.

[3] J.P. Merlet, C.M. Gosselin and N. Mouly, Workspace of planar parallel manipulators, *Mech. Mach. Theory* 33 (1/2) (1998) 7-20.

[4] C.M. Gosselin and J. Angeles, Singularity analysis of closed loop kinematic chains, IEEE *Trans. on Robotics and Automation* 6 (3) (1990) 281-290.

[5] X.-J. Liu, J. Wang, F. Gao and L.-P. Wang, On the analysis of a new spatial three degrees of freedom parallel manipulator, IEEE *Transactions on Robotics and Automation* 17 (6) (2001) 959-968.

[6] I.A. Bonev, D. Zlatanov, C. M. Gosselin, Singularity analysis of 3-DOF planar parallel mechanisms via screw theory, ASME *Journal of Mechanical Design* 125 (2003) 573-581.

[7] C.M. Gosselin and J. Angeles, The optimum kinematic design of a spherical three degree-of-freedom parallel manipulator, *J. Mech. Transm. Autom. Des.* 111 (1989) 202-207.

[8] E. Ottaviano and M. Ceccarelli, Optimal design of CaPaMan (Cassino Parallel Manipulator) with a specified orientation workspace, *Robotica* 20 (2002) 159–166.

[9] X.-J. Liu, Optimal kinematic design of a three translational DOFs parallel manipulator, *Robotica* 24 (2) (2006) 239-250.

[10] C. M. Gosselin, The optimum design of robotic manipulators using dexterity indices, *Robotics and Autonomous Systems* 9 (1992) 213-226.

[11] R.S. Stoughton and T. Arai, A modified Stewart platform manipulator with improved dexterity, IEEE *Transactions on Robotics and Automation* 9 (2) (1993) 166-173.

[12] J. Ryu and J. Cha, Optimal architecture design of parallel manipulators for best accuracy, in: Proceedings of the 2001 IEEE/RSJ International Conference on Intelligent *Robots and Systems*, Maui, Hawwaii, USA, 2001, pp. 1281-1286.

[13] H. S. Kim, and L.-W. Tsai, Design optimization of a Cartesian parallel manipulator, *Journal of Mechanical Design* 125 (1) (2003) 43-51.

[14] J. P. Merlet, Jacobian, manipulability, condition number, and accuracy of parallel robots, *Journal of Mechanical Design* 128 (2006) 199-206.

[15] C. Gosselin, J. Angeles, A global performance index for the kinematic optimization of robotic manipulators, *Journal of Mechanical Design* 113 (1991) 220-226.

[16] Y.J. Lou, G.F. Liu, N. Chen, and Z.X. Li., Optimal design of parallel manipulators for maximum effective regular workspace, in: Proc. of IEEE/RSJ International Conference on Intelligent Robots and Systems, Edmonton, Canada, 2005, pp.1208-1213.

[17] Von H. Alt, Der uberstragungswinkel und seine bedeutung fur dar konstruieren periodischer getriebe, Werksstattstechnik 26 (Heft 4) (1932) 61-65.

[18] S. S. Balli and S. Chand, Transmission angle in mechanisms, *Mechanism and Machine Theory* 37(2002) 175-195.

[19] P.W. Eschenbach and D. Tesar, Link length bounds on the four bar chain, *J. Eng. Ind. Trans. ASME* Ser. B 93(1971) 287-293.

[20] A.S. Hall, Kinematics and linkage design, Prentice-Hall, Englewood Cliffs, NJ, 1961, pp.41.

[21] R.S. Hartenberg, J. Denavit, Kinematic synthesis of linkages, McGraw-Hill, New York, 1964, pp. 46-47.

[22] D.C. Tao, Applied linkage synthesis, Addison-Wesley, Reading, MA, 1964, pp.7-12.

[23] J.T. Kimbrella, Kinematics analysis and synthesis, McGraw-Hill, NY, 1991, pp.14-15.

[24] G. H. Sutherland and J. N. Siddall, Dimensional synthesis of linkage by multifactor optimization, *Mechanism and Machine Theory* 9 (1974) 81-95.

[25] X.-J. Liu and J. Wang, A new methodology for optimal kinematic design of parallel mechanisms, *Mechanism and Machine Theory*, Vol.42, No.9, pp.1210-1224, 2007.

[26] G. Alıcı, An inverse position analysis of five-bar planar parallel manipulators, Robotica 20 (2002) 195-201.

[27] X.-J. Liu, J. Wang and G. Pritschow, Kinematics, singularity and workspace of planar 5R symmetrical parallel mechanisms, *Mechanism and Machine Theory*, Vol.41, No.2, pp.145-169, 2006.

[28] J. J. Cervantes-Sánchez, J. C. Hernández-Rodríguez and J. G. Rendón-Sánchez, On the workspace, assembly configurations and singularity curves of the RRRRR-type planar manipulator, *Mechanism and Machine Theory* 35 (2000) 1117-1139.

[29] F.C. Park and J.W. Kim, Singularity analysis of closed kinematic chains, *Journal of Mechanical Design* 121 (1999) 32-38.

[30] X.-J. Liu, J. Wang and G. Pritschow, Performance atlases and optimum design of planar 5R symmetrical parallel mechanisms, *Mechanism and Machine Theory*, Vol.41, No.2, pp.119-144, 2006.

[31] J. J. Cervantes-Sánchez, J. C. Hernández-Rodríguez and J. Angeles, On the kinematic design of the 5R planar, symmetric manipulator, *Mechanism and Machine Theory* 36 (2001) 1301-1313.

[32] G. Alıcı and B. Shirinzadeh, Optimum synthesis of planar parallel manipulators based on kinematic isotropy and force balancing, Robotica 22 (2004) 97–108.

[33] I.A. Bonev and C.M. Gosselin, Singularity loci of planar parallel manipulators with revolute joints, 2nd Workshop on Computational Kinematics, Seoul, South Korea, May 20-22, 2001.

[34] K. Hain, Applied kinematics, McGraw-Hill, NY, 1967.

[35] K.L. Ting and G.H. Tsai, "Mobility and synthesis of five-bar programmable linkages," in Proceedings of 9th OSU Applied Mechanisms Conference, Kanas City, MD, 1985, pp.III-1-III-8.

[36] R. Stanley, "Five-bar loop synthesis," Mach. Design, pp.189-195, 1961.

[37] M.R.E. Philipp and F. Preudenstein, "Synthensis of two-degree-of-freedom linkages—a feasibility study of numerical methods of synthesis of bivariate function generators," *Journal of Mechanisms*, vol. 1, pp.9-21, 1965.

[38] E. Söylemez and F. Freudenstein, "Transmission optimization of spatial 4-link mechanisms," *Mechanism and Machine Theory*, vol. 17, pp.263-283, 1982.

[39] R.I. Alizade, A.V. Mohan Rao, G.N. Sandor, "Optimum synthesis of two degree of freedom planar and spatial function generating mechanism using the penalty function approach," *J. Eng. Ind. Trans. ASME*, pp.629-634, 1975.

[40] G. Liu, Y. Lou, and Z. Li, "Singularities of parallel manipulators: a geometric treatment," *IEEE Transactions on Robotics and Automation*, vol. 19, pp. 579-594, 2003.

[41] H. Funabashi and Y. Takeda, "Determination of singular points and their vicinity in parallel manipulators based on the transmission index," in Proc. 9th World Congress on the Theory of Machines and Mechanisms, Milan, Italy, 1995, pp. 1977–1981.

[42] P. A. Voglewede and I. Ebert-Uphoff, "Measuring 'closeness' to singularities for parallel manipulators," in Proc. IEEE Int. Conf. on Robotics and Automation, New Orleans, 2004, pp. 4539–4544.

[43] G. Sutherland and B. Roth, "A transmission index for spatial mechanism," *J. Eng. Ind. Trans. ASME,* pp.589-597, 1973.

In: Parallel Robotics: Recent Advances in Research... ISBN: 978-1-60456-859-2
Editors: Jinsong Wang and Xin-Jun Liu, pp. 163-215

Chapter 7

DESIGN, ANALYSIS AND APPLICATIONS OF FLEXURE-BASED COMPLIANT PARALLEL KINEMATIC MACHINES

Shusheng Bi, Jingjun Yu* and Guanghua Zong
Robotics Institute, School of Mechanical Engineering and Automation, Beihang University, Beijing, 100083, P.R. of China.

ABSTRACT

As a new promising mechanism type, flexure-based compliant parallel kinematic machines (FCPKMs) gather the advantages of both compliant mechanism and parallel architecture. It is surely a primary choice in designing an ultra high-accuracy micromanipulator or micropositioner/nanopositioner. Accordingly design theory and technology on the FCPKMs have become one of the most important issues to be discussed in the past few years. This paper will demonstrate a comprehensive design process for a general FCPKM, which covers topology and size synthesis of the FCPKMs, design of compliant joints employed in the FCPKMs, kinetostatic analysis for the FCPKMs, and design methodology for the general FCPKMs. At last, in terms of the proposed theoretical analysis and design method, some macro-scale compliant in-parallel stages are designed and used in various applications as the high-precision micromanipulators.

Keywords: Parallel kinematic machine, Compliant mechanism, Compliant joint, Stiffness, Optimization

1. INTRODUCTION

In many applications, such as such as manufacturing equipment and MEMS/MOEMS devices, there is an urgent need for three-dimensional positioning and manipulation

* jjyu@buaa.edu.cn

mechanisms (often called micromanipulators or micro-andnano- positioners) to perform an ultra high-precision motion. One of the most essential parts in the micromanipulator is the mechanism that transfers the actuator motions to the end-effector motion. At the meantime, it is also the main error source that may be detrimental to the high accuracy of the micromanipulator. Generally, the errors caused by inappropriate mechanical structure such as due to backlash, hysteresis and manufacturing errors in the conventional joints are hardly traded off by controllers and sensors. There is therefore a need for new design concept of structure free from backlash and friction in the design of micromanipulators. Of all alternatives that may improve the accuracy of the mechanical structure, a compliant mechanism [1] using parallel architecture and compliant joints is by far the best choice.

As defined by Her and Midha, compliant mechanisms (CMs) represent a class of mechanical systems which makes use of flexible elements in design, as opposed to the use of only rigid-body members. Current applicable CMs can be classified into two categories in gross: flexure-based CMs [2] and distributed-compliance fully CMs [3], depending on distribution characteristics of compliance among the mechanism. The motion of the flexure-based compliant mechanism, which characterizes that all kinematic pairs are designed as compliant joints to play the same role as conventional joints, is achieved via the deflection of these flexible elements in the mechanism. This kind of mechanism can be designed as a complex but monolithic structure, which makes it possible to achieve a multi-DOF, a very high positioning accuracy as well as to decrease the manufacturing and assembly costs. While parallel architecture is preferably employed in that it just can reinforce the advantages and make up for the deficiencies of general compliant mechanisms due to its high structural stiffness, the possibility of compensating errors for symmetric structure design, the improvement of its loading status as a consequence of the existence of passive joints in the mechanism. In this paper, we called this kind of mechanism as flexure-based compliant parallel kinematic machine (FCPKM) as a special class of PKM.

There are a majority of proposed studies and design examples on FCPKMs in the literature in recent years. Similar to rigid-body counterparts, FCPKMs have several familiar structures such as 6-DOF, 3-DOF (XYZ, $XY\theta_Z$, and $\theta_X\theta_YZ$), and 2-DOF (XY). Hudgens and Tesar [4], Bi and Zong [5], McInroy and Hamann[6], Hesselbach *et al.*[7], Culpepper *et al.*[8], etc., have extensively studied 6-DOF motion stages. Zhang *et al.* [9], Chang *et al.* [10], Ryu *et al.*[11], etc., focused the study on $XY\theta_Z$ stages. Lee [12], Canfield [13], Clavel *et al.* [14] etc. have paid more attention to $\theta_X\theta_YZ$ stage. Nevertheless, more research focuses on XYZ or XY stages. For example, Arai *et al.* [15] designed two type of XYZ stage with RPPR structure or RPRP structure, respectively. Others include the work of Clavel *et al.* [16], Yu *et al.* [17], Chen *et al.* [18], Awtar *et al.* [19], etc. Notice that a motion-decoupled FCPKM seems advantageous over other types because of satisfactory stiffness and decoupled characteristics.

Of all research on FCPKMs, the modeling of FCPKMs is definitely a critical issue. For leaf-type flexure mechanisms, the deformed leaf beams under different loads and constraints have been modeled by Smith [20] and Awtar *et al.* [21]. For notch hinge-type mechanisms, the first study was by Paros and Weisbord [22]. Lobontiu [23] derived the exact compliances for bending, compression, tension, and shear of various types of notch hinges. For simplicity, the pseudo-rigid-body model (PRBM) method was proposed by Howell [24] to approximately model flexure mechanisms. However, the maximum difference between PRBM prediction and the actual value reaches 9%. Up to date, the PRBM method is by far the most common approach compared with other modeling methods of FCPKMs include nonlinear method [18], matrix method [25]. Finite Element Analysis (FEA) method [26], etc.

Much of the work on FCPKMs has also focused on tools for synthesis and optimal design (see for example Ananthasuresh [3], Liu [27], Yu [28], etc., and for kinetostatic and dynamics analysis (see for example Arai *et al.* [15], Lobontiu [23], Zhang *et al.* [9], Clavel *et al.* [16], McInroy *et al.* [6], Yu *et al.* [17], etc.). For instance, Arai *et al.* developed a serial of 3~6 DOF platform-type compliant mechanism, including Stewart platform, translational 3 DOF mechanism with RPPR structure or RPRP structure etc. Their kinematics and calibrations, which are based on the PRBM and constant Jacobian matrix, are also discussed accordingly.

In addition to the above areas, researchers have also been working on other issues involved in the development of FCPKMs, such as calibration, control algorithms, and fabrication, etc. [29-31].

As early in 1995, Precision-Micro Mechatronic and System Group at ARMS lab, directed by G. H. Zong, Professor of Mechanical Engineering at BUAA and Prof. S.S. Bi, Director of Robotics Institute of BUAA, initiated the studies on FCPKM. Analysis and design methodology for the FCPKM are investigated systematically, a theoretical framework, whose core is based on the compliance matrix of the compliant joint and the mechanism, is built up correspondingly. In terms of the proposed the theoretical analysis and design methodology for the FCPKM, Some compliant in-parallel stages are designed and used for some micro/nano applications as the high-precision micromanipulators.

Therefore, this research work concentrates on a comprehensive developing procedure using a flexure-based compliant parallel kinematic scheme for a macro-scale micro/nano positioning stage, also including the development and state-of-art on FCPKMs at ARAM lab. It is expected that such an approach and process would steer an effective design of the FCPKM with a relatively large workspace, high accuracy, high dynamic stiffness of the structure and compact size. Section 2 of this paper reports on type synthesis of FCPKMs and presents some effective methods for new topology of the FCPKM design. Section 3 develops novel compliant joint design based on some perfect performances such as a large work range, high stiffness, and high accuracy. Section 4 describes three prototype stages characteristics developed at ARMS lab, which are used as typical instances both in Section 5 for kinetostatic analysis and in Section 6 for the optimal design. Section 7 summarizes the general design methodology for FCPKMs based on intuition, PRBM, and compliant building block, respectively, and conclusions are drawn in the last section.

2. Type Synthesis of FCPKMs

The first step when design a FCPKM is the choice or construction of its specified kinematics configuration. High precision mechanisms always merge the advantages of parallel kinematics into those of compliant mechanism.

2.1. Classifications

Configuration of any a flexure-based compliant mechanism can be formed by connecting rigid links and compliant joints. Depending on the design and the overall motion of the mechanism, the flexure-based compliant mechanisms can formally be divided into two main categories: planar and spatial one. Single-DOF compliant joints which mainly exhibit a two-dimensional motion are included in planar mechanisms, whereas two- and multi-DOF compliant joints are practical solutions to spatial compliant mechanisms in that these flexure configurations are designed to perform three-dimensional motion by their geometry. Each category can be further subdivided into serial, parallel, and hybrid (serial-parallel) type, thus the classification of FCPKMs can be illustrated in table 1.

Table 1. Classifications of FCPKM

	Planar	Spatial
Serial		
Parallel		
Hybrid		

2.2. Parallel Architectures and Compliant Mechanisms

There are two well-known configurations employed in the current design of multi-DOF flexure-based mechanisms, i.e. serial kinematics and parallel kinematics. Each configuration may stem from its rigid-body counterpart and has its own set of cons and pros. Compared to serial kinematics machines (SKMs), PKMs have the advantages of (i) low inertia, (ii) high structural stiffness, (iii) balanced mechanical structures without large overhanging masses, and (iv) compactness. On the other hand, these advantages come at a price as PKMs have (i) small workspaces, (ii) poor dexterity, (iii) non-linear kinematics and dynamics, and (iv) difficult-to-solve forward kinematics. One of the major challenges in the realization of a PKM is the synthesis of a mechanism with the appropriate DOF and characteristics. While a SKM is readily realized by stacking appropriate single-DOF joints-link pairs together, a PKM requires the intersection of the degrees of freedom of its parallel kinematic chains to produce the desired degree-of-freedom at its end-effector.

In serial design, multiple DOF are achieved by stacking single DOF systems, one on other. Serial kinematic mechanisms are relatively simple to design, and have substantially decoupled degrees of freedom, but at the same time incorporate moving actuators and cables that deteriorate their dynamic performance. Moving cables are sources of disturbance, which is detrimental for nanometric positioning. Moving actuators, especially when large range of motion is required, are bulky and reduce the bandwidth of the axes that carry them. Furthermore, moving actuators and connections are also undesirable for flexure mechanisms used in MEMS applications because of fabrication reasons. Parallel kinematic designs are free of these problems due to ground mounted actuators, and are also usually more compact, but on the other hand, provide smaller ranges of motion and exhibit significant cross-axis coupling. Moreover, the stiffness of one axis varies with motion or force along the other axes.

This surely affects the static as well as dynamic performance of the mechanism. In spite of the difficulties and challenge associated with parallel kinematic schemes, they are well-suited for micro/nano-positioning applications by perfectly merging it with compliant joints and compliant mechanisms. In this paper, we only focus on flexure-based compliant PKMs.

2.3. Topology Synthesis

As mentioned above, the design of a compliant PKM may stem from a rigid-body linkage topology in most cases. Therefore, an effective type synthesis for the conventional PKMs is a primary task. Notice that the controlled joints of the kinematic chains must span the desired DOF of the PKM. This makes the synthesis of a PKM considerably more complex than that of a SKM. In the context of micropositioning where monolithic designs are often adopted to improve the positioning resolution, these difficulties in the synthesis and realization of a useful PKM will be further aggravated by challenging design and fabrication problems.

Fortunately, there are increasing studies on type synthesis of rigid PKMs in recent years, and several systematic approaches have been proposed correspondingly. As a result, a majority of novel PKM configurations have been found. Generally, according to the different mathematic tools adopted, these type synthesis approaches can be mainly divided into four categories: the enumeration approach based on the general Grübler-Kutzbach mobility criterion, the motion-based synthesis approach by means of displacement subgroup and displacement submanifold theory or single-open-chain theory, the constraint-based synthesis approach aided with the reciprocal screw theory.

Hunt [32] is perhaps the first to use an enumeration method to find some new 3-DOF PKMs such as 3-RPS type. In 1999, Tsai [33] proposed an enumeration of 3-DOF PKMs based on the general Grübler-Kutzbach mobility criterion. In fact, the enumeration method belongs to number synthesis of mechanism in essence and has its limitation. For instance, the use of the enumeration method may often lead to numerous incorrect results since it fails to describe the geometrical arrangement of kinematic pairs both in a limb and in a whole mechanism.

As of 1978, Hervé [34] proposed a comprehensive classification of kinematic subchains based on the algebraic structure of the displacement group, which laid down the foundation of using group theory in type synthesis of PKMs. Meng et al. [35] presented a geometric synthesis theory of lower-mobility PKMs by extending the displacement subgroup to the regular submanifolds field, and substantially broadened the spectrum of manipulators available to design applications. In addition, Jin and Yang [36] put forward a structural synthesis method of lower-mobility PKMs based on the units of single-opened-chain (SOC) limbs, and Gao et al [37] proposed a "GF" method to synthesize some new 2~5 DOF PKMs. All of these may resort to the motion-based synthesis category.

The key idea behind the constraint-based synthesis approach lies in the fact that the common effect of constraints contributed by all limbs in a lower-mobility PKM determines mobility characteristics of the moving platform. Screw system theory just reflects the reciprocal relationship between twists and constraining wrenches [38] applied whether in a single limb or in a whole mechanism. This method has shown its advantage in the type synthesis of lower-mobility PKMs. In 2002, Huang and Li [39] proposed a general methodology for type synthesis of lower-mobility PKMs based on reciprocal screws and implemented type synthesis of nine types of lower-mobility PKMs. Fang and Tsai [40] derived geometric conditions that all joints of each limb should be satisfied and developed a systematic approach for the structural synthesis of a class of lower-mobility overconstrained PKMs. Similar research can also be found in the work of Kong and Gosselin [41], Zhao, Dai and Huang [42], and Yu *et al.* [43]. Nevertheless, it should be noted that other than the motion-based synthesis method based on displacement subgroups and submanifolds, which can describe the finite motion of kinematic pairs or kinematic chains, the constraint-based synthesis method based on reciprocal screws falls into the category of instantaneous motion.

Notice that when designing a FCPKM, choice and determination of artful configuration layout is of particularly a necessity compared with its rigid counterpart, since even if the topology corresponding to two FCPKMs is identical, the trivial change of configuration layout may lead to vast difference in performance. For instance, as for a 6-PSS compliant micromanipulator, though the same structure (6-PSS) is used, the characteristics between a Stewart platform-type mechanism (figure 1a) and a mechanism with an orthogonal architecture (figure 1b) have a clear difference in kinematics, e.g. the latter is kinematic decouple, which means an easy control.

(a) Stewartplatform-type (b) Orthogonal configuratior

Figure 1. Two 6-PSS type compliant micromanipulators.

Although the above-mentioned approaches provide a universal framework for structural design of PKMs, type synthesis is still a comparably challenge for the PKMs with particular geometry or with some prominent characteristics, such as PKMs with decouple motion, or those with isotropic feature, or those with high rotation capability.

In this section, we take a simple but representative example, i.e. a synthesis of 2-DOF PKM with orthogonal arrangement, and the applied mathematical tool is the displacement Lie group theory.

Based on the basic properties of the displacement group and displacement manifold, we propose the concept of *Cartesian DOF-characteristic Matrix* (CDM), corresponding to the canonical class in all subgroups and submanifold of Euclidean group and reflecting the motion output of a rigid body. A new approach based on combination of the atlas of the CDM and displacement group theory is addressed for both exhaustive numerations and type synthesis of the Orthogonal PKMs (OPKMs). The detailed description about this method can be found from reference [44].

The group $\mathcal{T}_2(z)$ has one type of proper Lie subgroup and two types of improper Lie subgroups. They are proper Lie subgroup $\mathcal{T}(x)$ (or $\mathcal{T}(y)$) representing any set of rectilinear translations parallel to any given direction $\boldsymbol{x}$ (or $\boldsymbol{y}$), $\mathcal{T}_2(z)$ set of planar translations normal to given direction z and $\mathcal{E}$ are two improper subgroup. Another, the group $\mathcal{T}_2(z)$ has two types of submanifolds. They are $\mathcal{T}_a(xy)$ (the mechanical generator is P_a) set of 1-dimensional linear motion along an open subset of the unit circle, and $\mathcal{T}_{2a}(z)$ (the mechanical generator is U^* [37]) set of 2-dimensional linear motion along an open subset of the unit sphere.

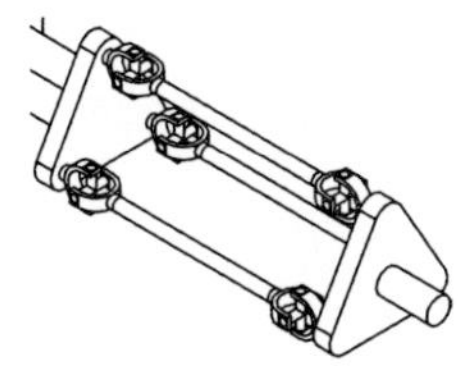

Figure 2. U^*

Due to the product closure in the subgroup $\mathcal{T}_2(z)$, we can deduce a series of set equalities.

$$\mathcal{T}_2(z)=[\mathcal{T}(x)\cdot\mathcal{T}(y)]=[\mathcal{T}(x)\cdot\mathcal{T}_a(xy)]=[\mathcal{T}(y)\cdot\mathcal{T}_a(xy)]=[\mathcal{T}_a(xy)\cdot\mathcal{T}_a(xy)]=\mathcal{T}_{2a}(z) \tag{1}$$

$$\begin{aligned}\mathcal{T}_2(z)&=[\mathcal{T}(x)\cdot\mathcal{T}(y)]\cap[\mathcal{T}(x)\cdot\mathcal{T}(y)]\\&=[\mathcal{T}(x)\cdot\mathcal{T}_a(y)]\cap[\mathcal{T}(x)\cdot\mathcal{T}_a(y)]=[\mathcal{T}_a(x)\cdot\mathcal{T}_a(y)]\cap[\mathcal{T}_a(x)\cdot\mathcal{T}_a(y)]=\mathcal{T}_{2a}(z)\cap\mathcal{T}_{2a}(z)\end{aligned} \tag{2}$$

According to Eq. (1), the mechanical generators of $\mathcal{T}_2(z)$ can be obtained by placing in series of kinematic pairs generating subsets of $\mathcal{T}_2(z)$, as shown in table 2.

Table 2. Mechanical generators of $\mathcal{T}_2(z)$

Kinematic bond	Mechanical generator	Kinematic bond	Mechanical generator
$[\mathcal{T}(x)\cdot\mathcal{T}(y)]$	$[{}^x\mathrm{P}\,{}^y\mathrm{P}]$	$[\mathcal{T}_a(x)\cdot\mathcal{T}_a(y)]$	$[{}^x\mathrm{P}_a\,{}^y\mathrm{P}_a]$
$[\mathcal{T}(x)\cdot\mathcal{T}_a(y)]$	$[{}^x\mathrm{P}\,{}^y\mathrm{P}_a]$	$\mathcal{T}_{2a}(z)$	${}^z\mathrm{U}^*$

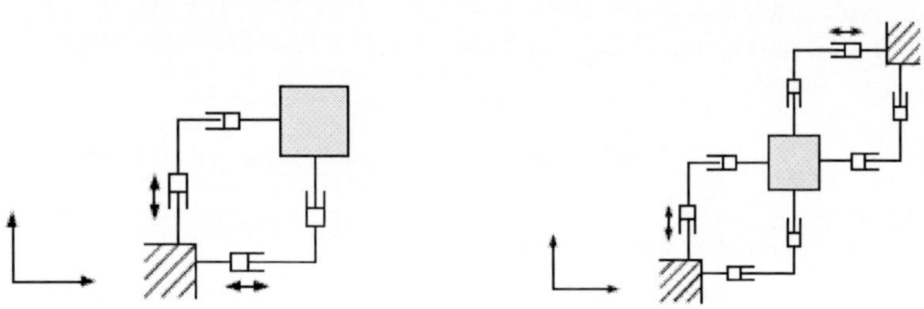

(a) Non-overconstraint (b) Overconstraint

Figure 3. Mechanical generators of $\mathcal{T}(x)\cdot\mathcal{T}(y)$.

3. DESIGN OF COMPLIANT JOINTS

Compliant joints, also called flexure hinges, or flexible joints, are a class of joints other than conventional mechanical joints. They utilize the inherent compliance of a material rather than restrain such deformation. These joints eliminate the presence of friction, backlash, and wear. Further benefits include up to sub-micron accuracy due to their continuous monolithic construction. Such accuracy is important in many micro-, or nano-applications. The monolithic construction also simplifies production, enabling low-cost fabrication.

However, the benefits gained from using compliant joints come at the cost of several disadvantages including small displacements, relatively low out-of-plane stiffness, unavoidable axis drift, all kinds of problems (buckling, fatigue, etc.) relative to strain and stress. To overcome these drawbacks and develop better-performance flexures, a set of criterion must be established for benchmarking. The four most important criteria [45] are (1) the range of motion, (2) the amount of axis drift, (3) the ratio of off-axis stiffness to axial stiffness, and (4) stress concentration effects.

3.1. Classifications

A. Flexural Pivots

A flexural pivot, also called flexible bearing, or flexure, is a rotary mechanical device characterized by a high compliance with respect to the “in-plane” rotational degree of freedom (DOF) and high stiffness in all the other DOFs. The motion of a flexural pivot is achieved by the elastic deformation of a solid material, other than a rigid joint used in rigid-body kinematics.

Flexure pivots are increasingly popular with designs requiring one-piece (monolithic) manufacturing, reduced weight, motion smoothness, virtually infinite resolution, zero backlash, friction and lubrication. However, four main drawbacks, including limited range of motion, imprecision of rotation referred to as axis-shift, the finite ratio of off-axis stiffness to axial stiffness, and visible stress concentration effects, commonly exist in flexural pivots.

In the last decades, a large majority of flexural pivots have been studied and developed, most of which fall into two categories: primitive flexures and complex flexures (usually the combination of two or more flexure primitives).

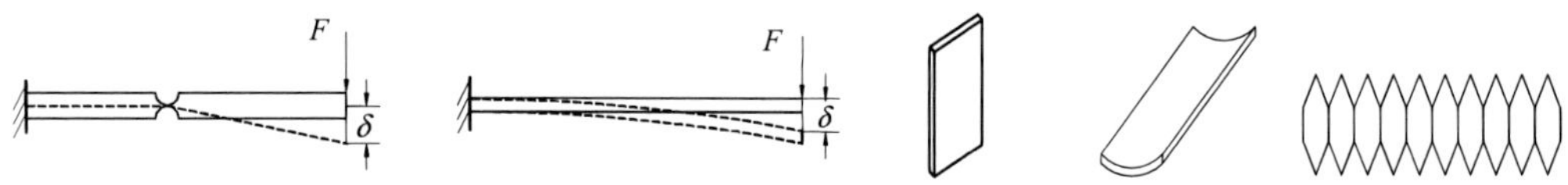

(a) Notch-type (b) Cantilever beam (c) Leaf-type spring (d) Tape spring (e) Bellow

Figure 4. Primitive flexures.

The primitive flexure category includes small-length flexural pivots such as notch-type joints, long flexible segments such as cantilever beam, leaf-type springs, tape spring, or angle spring, and a compliant bellow. The complex flexure category includes cross-axis flexural pivots [46] characterized by connection of two rigid segments with two long flexible segments arranged in a cross-shape configuration, split-tube flexural pivots [47] incorporated by torsion as the primary mode of deformation, and butterfly flexures [48], etc.

The profiles of conventional notch-type flexures experience a long-time evolvement (figure 5). Circular flexures were first analyzed by Paros and Weisbord in 1965, Ragulskis *et al.* [49] used numerical methods to optimize the circular joint geometry by defining optimal flexure behavior. A similar but more detailed work was done by Lobontiu [23] for other profiles, such as corner-filleted, elliptical, parabolic, hyperbolic, secant designs with a conic-section. In addition, some flexures with hybrid profile were also proposed recently. Today these notch-type flexures are widely used for high-precision, and small-displacement mechanisms.

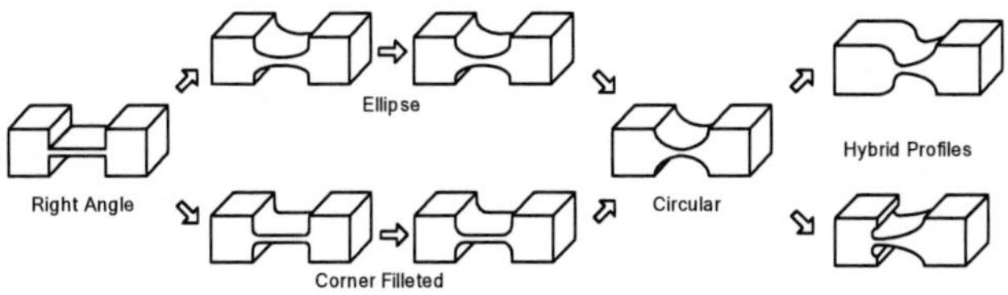

Figure 5. Evolvement of a notch-type flexural pivot.

In a general case, as two most important criteria of flexures, a large motion range and a high rotational precision are always antagonistic to each other. For instance, a conventional notch flexural pivot has a high precision of rotation but a very limit angular displacement. On the contrary, a leaf-type flexure has a large-deflection stroke but a serious axis drift.

With the advent of CNC milling machines and particularly CNC wire electro-discharge machining (WEDM), the monolithic machining of complex flexures makes possible, resulting of a large number of complex flexible joints including cross-axis flexural pivot, cartwheel hinge, split-tube revolute joint, C-Flex, Q-joint, and butterfly flexural pivot, et al. (figure 6).

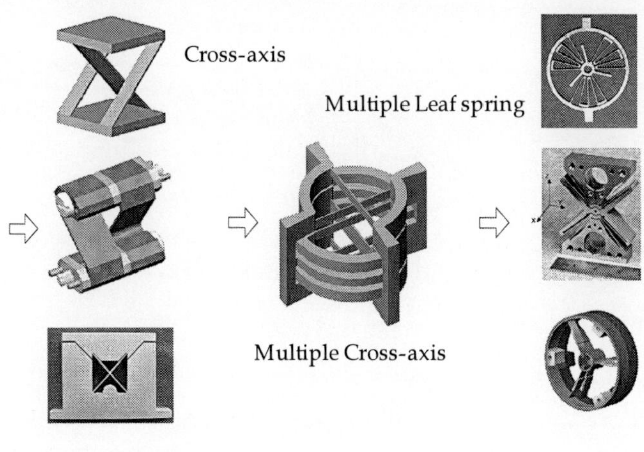

Figure 6. Evolvement of large-deformation flexural pivots.

However, these complex flexures also remain the advantages including a large stroke, a high precision of rotation, or a high ratio of off-axis stiffness to axial stiffness to make up for part of drawbacks existing in the flexure primitives. The main disadvantage of this flexure classes lies in their complexity and prone to voluminous in structure.

B. Compliant Translational Joint

Compliant translational joint (CTJ) is also one of the most common joint types and widely used in FCPKM. The simplest CTJ is a cantilever beam; however, most of the existing CTJs come from a parallel four-bar building block (figure 7a-b). Their flexibility is derived from notch joints or leaf springs. Composition of parallel four-bar building block can make up some compound CTJs (figure 7c-g). Y. Moon and S. Kota from the UM Compliant Systems Design Laboratory [45] designed a novel CTJ (figure 7h) as an improvement to above translational joints.

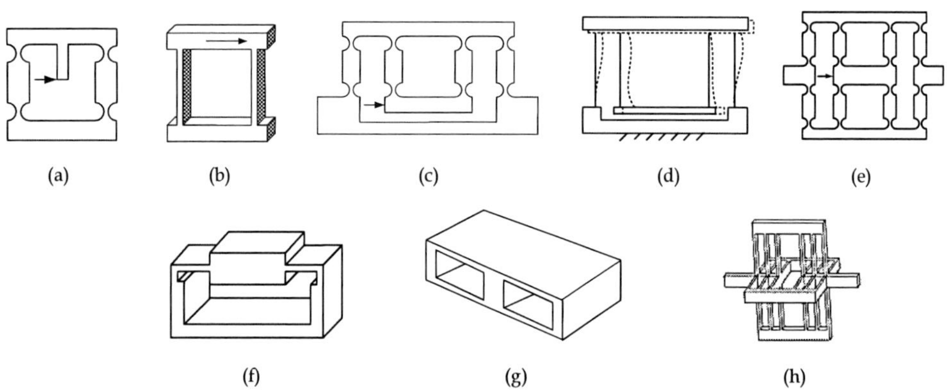

Figure 7. Different types of CTJs.

C. Compliant Spherical Joint

As illustrated in figure 8, compliant spherical joint (CSJ) is a special compliant joint type with three rotational DOFs whose axes are converged into a common point theoretically.

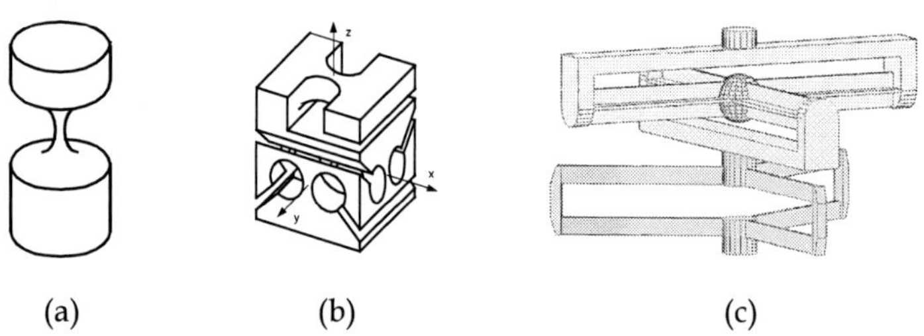

Figure 8. Different types of CSJs.

D. Compliant Universal Joint (CUJ)

A compliant universal joint (CUJ) allows only two rotational degrees of freedom, as does its traditional mechanical counterpart. Therefore, two CRJs whose axes are converged into a point are concatenated to create a CUJ, as shown in figure 9a, but more common useful type is a compact structure as illustrated in figure 9b-f.

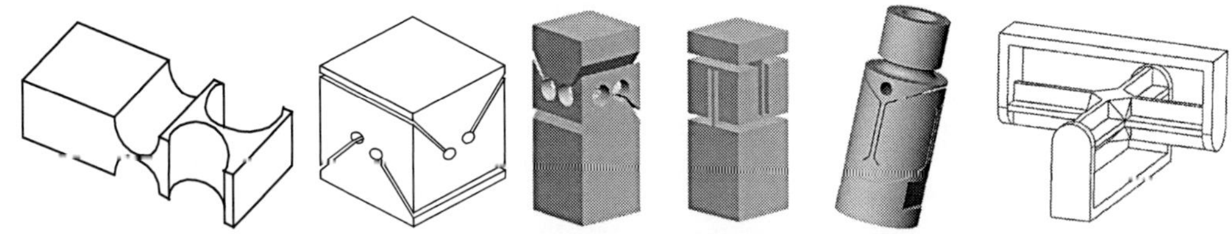

(a) (b) (c) (d) (e) (f)

Figure 9. Different types of CUJs.

3.2. Performance Measures

The compliant joints can be designed and analyzed to verify their effectiveness during working according to performance measures. While detailed performance measures can be laid out depending on specific applications. As pointed out in reference [19], these measures include DOF and DOC (degree of constraint), accuracy of motion, stiffness ratio of DOC/DOF, range of motion, sensitivity, etc. as the meantime, these measures also characterize the performance of a FCPKM.

A. Compliance/Stiffness Model

The loading-displacement characteristics of any a compliant joint can be reflected from its compliance or stiffness model in a matrix form. As for a single-axis revolute CJ, however, the axial stiffness or compliance is usually viewed as its characterized stiffness/compliance, In this case, a method based on pseudo-rigid-body (PRB) model is used to simplify the model of the CJ. Nevertheless, a more general but rigorous method is established to calculate the spatial compliance/stiffness matrix by means of Castigliano's displacement theorem in solid mechanics, just as done by Paros and Lobontiu. In this paper, we use screw theory to achieve the same purpose as Lobontiu.

1. Pseudo-Rigid-Body Model

As said by L. Howell, the pseudo-rigid-body (PRB) concept is used to model the deflection of flexible members using rigid-body components that have equivalent force-deflection characteristics. For example, a single-axis small-length circular notched flexure may be modeled as a pin joint with a characteristic torsional spring constant $K_\alpha = 2Ebt^{\frac{5}{2}} \Big/ 9\pi R^{\frac{1}{2}}$ according to the PRB model theory. But for a CTJ composed of four circular notched flexures, its equivalent linear spring constant may be modeled as $K_t = \dfrac{F}{\delta} = \dfrac{4K_\alpha}{l^2}$.

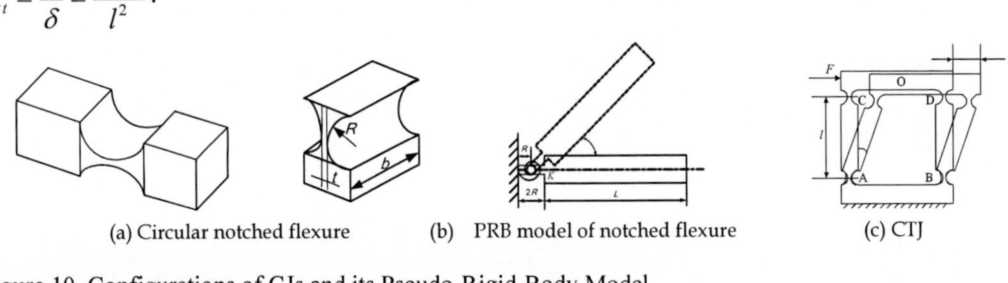

(a) Circular notched flexure (b) PRB model of notched flexure (c) CTJ

Figure 10. Configurations of CJs and its Pseudo-Rigid-Body Model.

Therefore, the PRB model of a CJ provides a simple but accurate method to analyze the deflection of the CJ. It provides the designer with a means of visualizing the deflection. The PRB method can also provide a simple parameterized model to make structure's optimization more easily. A detailed description about the PRB model can be found in reference [24].

2. Stiffness Matrix Model Based on Screw Theory

Consider the case that a rigid body 1 to be connected to a rigid body 0 by a general but initially unloaded spring (indeed a CJ) that restricts some relative spatial motions between bodies via its compliance or stiffness. Then the resultant mapping of compliance or stiffness is a one-to-one correspondence that associates a *twist* describing the relative displacement between the bodies with the corresponding *wrench* which interacts between them [50, 51].

The transformation between the twist and the wrench requires the use of a 6×6 compliance (or stiffness) matrix to perform the mapping of compliance (or stiffness). Here $\boldsymbol{C}$ is used to establish a potential energy-based inner product for twists, and the analytical representation $\boldsymbol{C}$ of the mapping of compliance is general correlation of screws. As well known, in screw theory, a wrench and twist may be used to describe the motion of a rigid body supported by an elastic structure. An applied wrench $\boldsymbol{F}$ produces a twist deformation $\boldsymbol{\varsigma}$ (a small change in home position), a compliance matrix $\boldsymbol{C}$ relates the wrench with the twist.

$$\varsigma = \boldsymbol{CF} \tag{3}$$

where ray coordinates $\boldsymbol{F}$ are assigned to the wrench and represented by a vector $\boldsymbol{F} = \begin{bmatrix} \boldsymbol{f}^{\mathrm{T}} & \boldsymbol{\tau}^{\mathrm{T}} \end{bmatrix}^{\mathrm{T}}$ or by Plücker coordinates $\boldsymbol{F} = (\boldsymbol{f}; \boldsymbol{\tau})$. The first three elements represent the axis of the wrench in the form of three projections of the force attached with the force intensity, and the last three elements represent the location of the axis in the form of three moment elements. Axial coordinates $\boldsymbol{\varsigma}$ which is an element of the Lie algebra, $\varsigma \in se(3)$, are assigned to the twist and represented by a vector $\varsigma = \begin{bmatrix} \boldsymbol{\delta}^{\mathrm{T}} & \boldsymbol{\theta}^{\mathrm{T}} \end{bmatrix}^{\mathrm{T}}$ or by Plücker coordinates $\varsigma = (\boldsymbol{\delta}\,; \boldsymbol{\theta})$. The primary part $\boldsymbol{\delta} = \begin{bmatrix} \delta_x, \delta_y, \delta_z \end{bmatrix}^{\mathrm{T}}$ gives the three translational deflection elements along the corresponding axes of the reference coordinate frame, and the secondary part $\boldsymbol{\theta} = \begin{bmatrix} \theta_x, \theta_y, \theta_z \end{bmatrix}^{\mathrm{T}}$ gives the three rotational deflection elements about the corresponding axes of the reference coordinate frame. $\boldsymbol{C}$ is the compliance matrix of the whole system, it is generally a symmetric and nonsingular 6×6 matrix.

The potential energy stored in the spatial spring system can be expressed in terms of the axis coordinates of the twist and the ray coordinates of the wrench, i.e.

$$\boldsymbol{f} \cdot \boldsymbol{\delta} + \boldsymbol{\tau} \cdot \boldsymbol{\theta} = \boldsymbol{F}^{\mathrm{T}} \varsigma > 0 \tag{4}$$

Substituting $\boldsymbol{F}$ from (3) yields that a given nonsingular matrix $\boldsymbol{C}$ is not only symmetric but also positive-definite.

$$\boldsymbol{F}^{\mathrm{T}} \boldsymbol{CF} > 0 \tag{5}$$

The expression in Eq. (5) introduces a quadratic form that can be used to establish a potential-energy-based, positive-definite inner product for the axial coordinates of twists.

Since the mapping of compliance and its inner product does not induce a Euclidean metric, it must undergo a change in representation whenever a change of coordinate system is encountered.

For a general Euclidean change of coordinate system, the axis coordinates ς of a twist are transformed by the relation,

$$\varsigma = \begin{bmatrix} \boldsymbol{R} & \hat{\boldsymbol{t}}\boldsymbol{R} \\ \boldsymbol{0} & \boldsymbol{R} \end{bmatrix} \varsigma' = \boldsymbol{A}\varsigma' \tag{6}$$

where ς' is the axial coordinates of the twist expressed in the new coordinate frame, $\boldsymbol{R}$ is a 3×3 rotation matrix, and $\hat{\boldsymbol{t}}$ is a 3×3 skew-symmetric translation matrix.

In similar, the ray coordinates $\boldsymbol{F}$ of a wrench are transformed by the relation,

$$\boldsymbol{F} = \begin{bmatrix} \boldsymbol{R} & \boldsymbol{0} \\ \hat{\boldsymbol{t}}\boldsymbol{R} & \boldsymbol{R} \end{bmatrix} \boldsymbol{F}' = \boldsymbol{B}\boldsymbol{F}' \tag{7}$$

where $\boldsymbol{F}'$ is the ray coordinates of the wrench expressed in the new coordinate frame.

Note that $\boldsymbol{B}^{-1} = \boldsymbol{A}^{\mathrm{T}}$, Substituting (6) and (7) into (3) yields

$$\boldsymbol{C}' = \boldsymbol{B}^{\mathrm{T}}\boldsymbol{C}\boldsymbol{B} \tag{8}$$

The 6×6 matrix $\boldsymbol{A}$ which transforms twists from one coordinate frame to another is referred to as the adjoint transformation written as Ad_g.

$$\mathrm{Ad}_g = \boldsymbol{B} = \begin{bmatrix} \boldsymbol{R} & \boldsymbol{0} \\ \hat{\boldsymbol{t}}\boldsymbol{R} & \boldsymbol{R} \end{bmatrix} \tag{9}$$

Thus

$$\boldsymbol{C}' = \mathrm{Ad}_g^{\mathrm{T}}\boldsymbol{C}\mathrm{Ad}_g \tag{10}$$

Eq. (10) indicates some important properties of the compliance matrix: (i). It is symmetric, regardless of coordinate frame; (ii). It is likewise positive-definite; (iii). Eq. (8) relates congruent matrices that in general must have the same invariant signature which is defined as the number of positive, negative and zero eigenvalues.

Now we deduce the spatial compliance matrix of a cantilever beam and a leaf spring by applying above transformation.

As shown in figure 11a, if the coordinate frame is at the centroid of a beam, considering at the same time the actions of a wrench at the centroid C of the beam, the compliance matrix of the beam can be given in axis coordinates [52] as

$$\boldsymbol{C}_C = diag\begin{bmatrix} \frac{l^3}{12EI_y} & \frac{l^3}{12EI_x} & \frac{l}{EA} & \frac{l}{EI_x} & \frac{l}{EI_y} & \frac{l}{GI} \end{bmatrix} \tag{21}$$

where the first three elements denote linear compliance elements and the last there elements are torsional compliance elements. Furthermore, $A = tb$ the cross-sectional area, $I_x = tb^3/12$, $I_y = bt^3/12$ the moments of inertia at the cross section with respect to axes x and y, G the shear modulus and $I = I_x + I_y = tb(t^2 + b^2)/12$ the polar moment of inertia of the cross section.

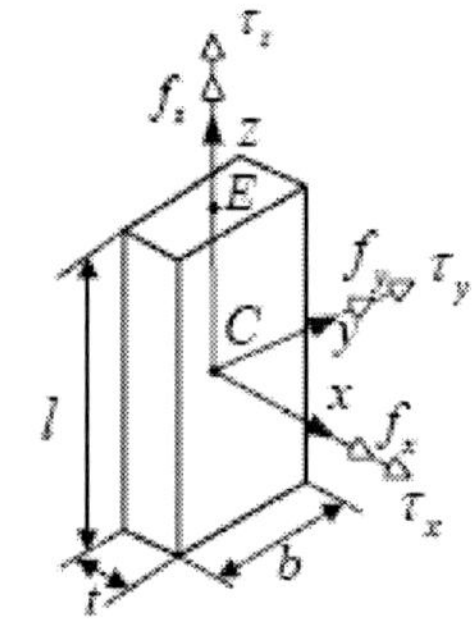

Figure 11. A prismatic beam or a leaf spring.

In most cases, the wrench is exerted on the end of the beam. Thus, the coordinate transformation of compliance matrix is necessary. In order to shift the coordinate frame to the end of the beam E, an adjoint transformation expressed in Eq. (9) is employed as

$$\text{Ad}_g = \begin{bmatrix} \boldsymbol{I} & \boldsymbol{0} \\ \hat{\boldsymbol{t}} & \boldsymbol{I} \end{bmatrix} \tag{22}$$

where

$$\hat{\boldsymbol{t}} = \begin{bmatrix} 0 & -l/2 & 0 \\ l/2 & 0 & 0 \\ 0 & 0 & 0 \end{bmatrix} \tag{23}$$

Thus, from Eq. (10), the compliance matrix represented in a new coordinate frame is

$$\boldsymbol{C}_E = \text{Ad}_g^{\text{T}} \boldsymbol{C}_C \text{Ad}_g = \begin{bmatrix} c_{11} & 0 & 0 & 0 & c_{15} & 0 \\ 0 & c_{22} & 0 & c_{24} & 0 & 0 \\ 0 & 0 & c_{33} & 0 & 0 & 0 \\ 0 & c_{42} & 0 & c_{44} & 0 & 0 \\ c_{51} & 0 & 0 & 0 & c_{55} & 0 \\ 0 & 0 & 0 & 0 & 0 & c_{66} \end{bmatrix} \tag{24}$$

In which $c_{11} = l^3/3EI_y$, $c_{22} = l^3/3EI_x$, $c_{33} = l/EA$, $c_{44} = l/EI_x$, $c_{55} = l/EI_y$, $c_{66} = l/GI$, $c_{15} = c_{51} = l^2/2EI_y$, $c_{24} = c_{42} = -l^2/2EI_x$.

If the beam is designed as a leaf spring, i.e. $b/t > 10$, it follows that $I_y/I_x = t^2/b^2 < 0.01$. This results in the negligible translational displacement along y-axis and negligible twisting motion about x-axis. Similarly, the extension along the axial direction is negligible. Hence, the simplified leaf-spring compliance matrix can be expressed as

$$\boldsymbol{C}_E = \mathrm{Ad}_g^{\mathrm{T}} \boldsymbol{C}_C \mathrm{Ad}_g = \begin{bmatrix} c_{11} & 0 & 0 & 0 & c_{15} & 0 \\ 0 & 0 & 0 & 0 & 0 & 0 \\ 0 & 0 & 0 & 0 & 0 & 0 \\ 0 & 0 & 0 & 0 & 0 & 0 \\ c_{51} & 0 & 0 & 0 & c_{55} & 0 \\ 0 & 0 & 0 & 0 & 0 & c_{66} \end{bmatrix} \tag{24}$$

B. Accuracy of Motion

Ideally, no motion in a CJ is desirable in the DOC direction; however, due to imperfect nature of flexures such as the distributed deformation, there are always undesired motions or parasitic error motions associated with the rotation or translation in a compliant joint.

For a flexural pivot, the parasitic error motions refers as *precision of rotation*, which is characterized as a change of rotational radius and a translation of rotational center called center-shift or axis-shift. Several criterions are thus proposed to evaluate the rotational precision of a flexural pivot in this paper. There, d indicates the distance of the center-shift. If the direction of center-shift needs to be considered, the center-shift can be written in the vector form as $\boldsymbol{d} = \overrightarrow{\boldsymbol{OO}'}$.

(1) When the flexural pivot deflects, as shown in figure 12a, the point O on the designed center (always its geometric center) of the pivot moves to the O'. The center shift can be denoted as the offset d of the center point [23].
(2) For a cantilever, the pivot point is defined as the intersection of the tangent to the free end of the cantilever with the unloaded beam axis. It is clear that the pivot point O, as shown in figure 12b, is located from the free end of the undistorted beam with a distance s [53]. When the deflection or the applied loads is changed, the change of s can be used to evaluate the center shift.
(3) Replacing the flexural pivot with a pin joint at the center of the pivot [46], as shown in figure 12c, the distance d between the pseudo-rigid-body and the actual deflected flexure body can be used to evaluate the rotational precision.
(4) Similar to criterion (46), choose point O on the designed center of the unloaded pivot. Instead of adhering to the flexure part of the pivot, the point O is fixed on rigid body at the free end, as shown in figure 12d. Define the center shift as the distance d between point O' (after deflection) and point O [23].
(5) Adopt the concept of instantaneous rotational center O, which is the absolute center for the free body rotation. If the body makes a pure rotation, the instantaneous center

remains fixed; otherwise, the instantaneous center will move along a curve. Thus the center shift can be defined as the distance between the point on the centrode and the center of the unloaded pivot.

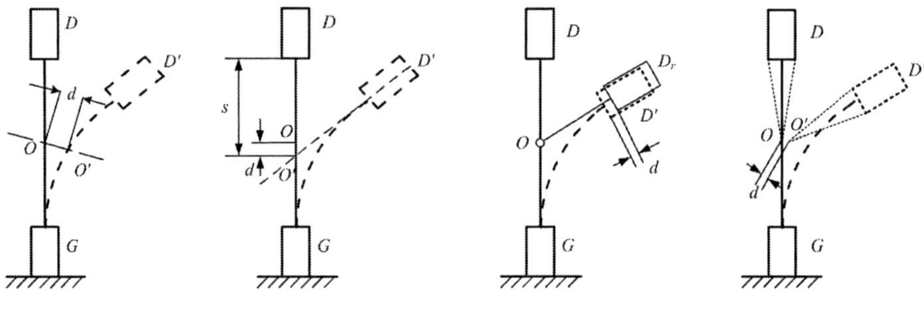

Figure 12. Four kinds of criterions to evaluate the rotational precision.

All of the five criterions defined above can be used to evaluate the rotational precision of a flexural pivot, but their applications have a little difference. Criteria (1), (2), (5) ignore the change of radius, and (4) fixes up the radius. All of them concern only the position of rotational center. On the other hand, criterion (3) fixes the position of center and concerns only the change of radius. Usage of criteria (1) and (2) will lead to a coarser result than others and is available in evaluating notch hinges or single leaf flexures. Of all five criterions, criterion (5) is the most precise since it may reflect the essential of rigid-body rotation. But the center shift calculated by this method is hard to measure through experiments.

For a CTJ, the parasitic error motions usually indicate its displacement δ_y along the traverse axis compared to its primary motion δ_x, as illustrated in figure 13. In fact, due to the curvilinear motion of the output link in 4-bar parallelogram, this kind cross-axis coupling error motion is unavoidable, which leads to decrease in high precision of motion. As a result, in some ultra-precision applications, it is prefer to calling for a CTJ achieving a precise linear motion while eliminating such an undesirable motion. Maybe by means of combination of several 4-bar parallelogram CTJs based on geometry-symmetry principle, as like figure 7e.

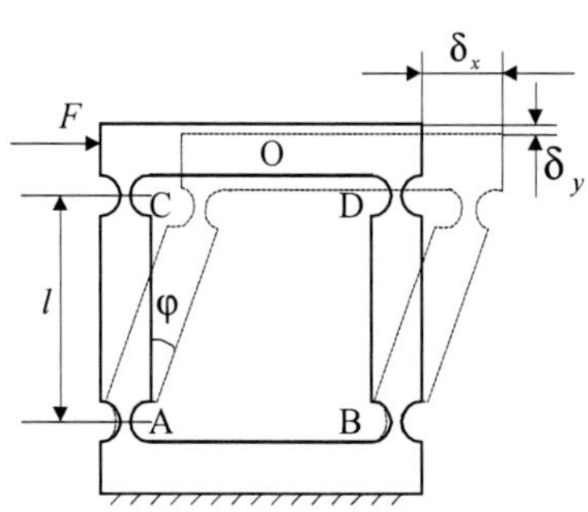

Figure 13. Cross-axis coupling error motion of a parallelogram-type CTJ.

3.3. Design of Novel Compliant Joints

Generally speaking, the range of motion and motion precision (usually rotation precision) are always antagonistic for most of current compliant joints. As well known, a small-length notch flexural pivot has a high precision of rotation but a very limited angular displacement. On the contrary, a leaf-type flexure has a large-deflection stroke but a serious axis drift. By artful design of the geometry, however, a desired performance tradeoff can be obtained. For example, a cross-axis pivot allows a large rotational angle and can provide a relatively stable position of the pivot axis; a cartwheel hinge can offer a relatively smaller deflection than the cross-axis pivot (about 50%), but represents a factor of 5 improvement in the stability of the center of rotation [46]; however, a butterfly flexural pivot invented by CSEM [48], which is a nearly perfect combination of a large deflection angle (±15°) and a high precision (axis shift is smaller than 1*um* for ±10° rotation, much better than the cartwheel hinge).

In addition, although the advantages of some complex CJs still exit including a large stroke, or a high precision of rotation, or a high ratio of off-axis stiffness to axial stiffness to make up for part of drawbacks existed in the flexure primitives. The main disadvantage of this CJ classes lies in their complexity and prone to voluminous in structure. Therefore, it is very significant to design more and more high-performance compliant joints.

At ARMS lab of BUAA, two different kinds of novel compliant joints are developed recently. One of them is an initial-curvature-profile flexure family derived from the conventional notch-type flexures and based on large-deflection axial buckling theory (figure 14), which is capable of high rotational precision and a compact structure. The common feature of these new flexures is that the cross section of each pivot is different from those of conventional counterparts, the front pivot has a hollow section with two curved leaves but the rear has a solid section with two curved edges. The more detailed description and performance analysis can be found in reference [54].

Figure 14. Configuration of a novel flexural pivot.

A leaf-type isosceles-trapezoidal flexural (LITF) pivot regarded as a kind of flexure primitive can provide an idea to design some novel flexural pivots, also including the cross-axis, cartwheel, and butterfly flexural pivot above mentioned. The LITF pivot consists of two compliant beams and two rigid-bodies as shown in figure 15(a). Due to complexity of the structures, both of the cartwheel hinge and the butterfly pivot lack accurate modeling for investigating their behavior, which is also necessary for design purpose. For this purpose, The PRB model of the LITF pivot has been established in reference [55] and shown in figure 15(c). This model is used not only to analyze the rotational precision, but also to evaluate the force-deflection relationship. The results showed such a model is accurate and simple enough to be used in the case of larger-deflection. Another advantage of the LITF pivot is that it can achieve a remote center of motion, because the vacancy in the pivot point enables LITF pivots to be easily superimposed to construct new complex flexure joints. For instance, a cartwheel hinge can be considered as the combination of two symmetrical LITF pivots, and a butterfly flexural pivot by four. In the authors'view, however, these two combinations are not enough at all.

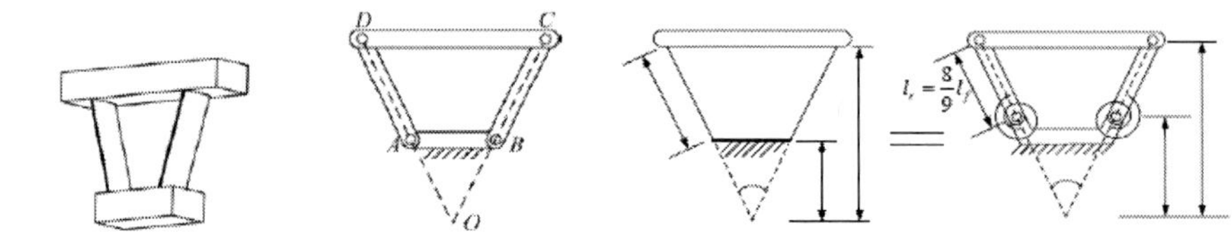

Figure 15. An isosceles trapezoidal CJ and its PRBM.

A family of novel complex compliant joints are developed by regarding above-mentioned LITF pivots as flexure building blocks and combining them in series. Different combination of two single LITF pivots may lead to a variety of double LITF pivots types. Figure 16 illuminates a general procedure of designing novel complex compliant joints by taking a concrete instance.

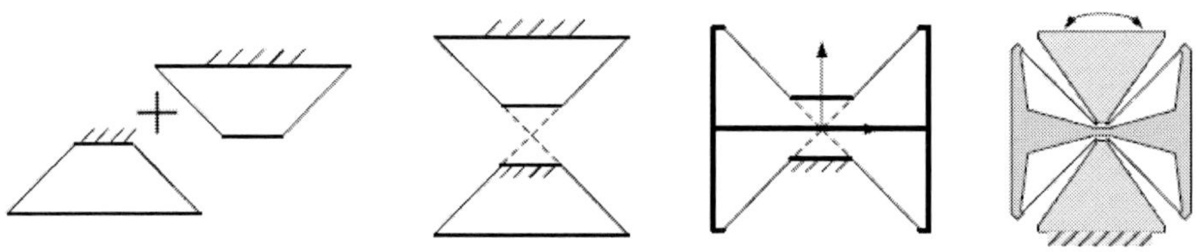

Figure 16. Combination of two LITF pivots in series. IE indicates intermediate element; S indicates stand; ME indicates moveable element.

4. Three FCPKM Prototypes

The final aim of our work is to benefit from the advantages of both compliant joints and parallel kinematics to obtain high-performance multi-DOF FCPKM. Some representatives of FCPKM prototypes have been developed at the ARMS lab of Robotics Institute, Beihang University.

4.1. Planar 3-RRR Fcpkm

As shown in figure 17, a planar 3-RRR FCPKM, whose rigid-body counterpart is the well-known planar 3-DOF $[X,\ Y,\ \theta Z]$ 3-RRR parallel mechanism, has such advantages as compact structure, simple kinematics and dynamics. In this in-plane motion stage, all kinematic pairs are designed as circular notch-type flexure hinges. Moreover, it is indeed a monolithic structure as a consequence of fully wire EDM machining from one single aluminum-bronze alloy, thus a very high resolution (up to 10nm) is obtained. As one component of this compliant PKM, three piezoelectric stack (Tokin AD0505E16D) devices are used as actuators. By applying a voltage up to 150V to each PZT, the PZT can extend in itself 15μm at a maximum level and further generate an output force to each link of the PKM. But this extension is not large enough to finish some micromanipulation tasks such as cell injection, thus a flexure leverage is utilized to amplify the PZT's displacement several times (as shown in figure 18).

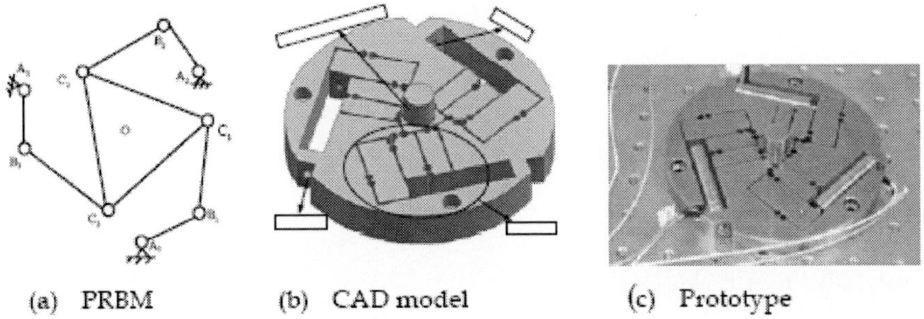

Figure 17. A planar 3-RRR FCPKM.

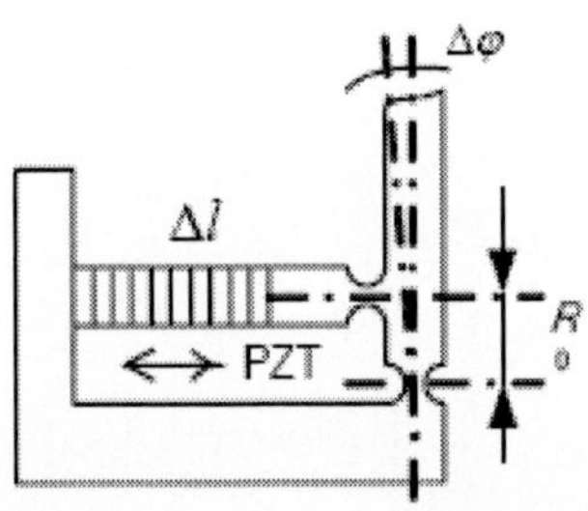

Figure 18. Schematic of a motion amplifier.

4.2. Translational Delta-Type FCPKM

The topology of the spatial compliant PKM illustrated in figure 19a is derived from a parallel-architecture manipulator presented by Tsai [56]. This mechanism is substantially a modified DELTA mechanism. Three identical limbs connect the moving platform to the base platform. Each limb consists of an upper arm and a lower arm. Each upper arm is a planar four-bar parallelogram made up of four revolute joints, and its function is equivalent to a prismatic pair. For each limb, the upper and lower arms, and the two platforms are connected with three parallel revolute joints. The axes of these revolute joints are perpendicular to those of the four-bar parallelogram for each limb. This mechanism has the following advantages: (i). All pairs in this mechanism are made up of revolute joints to avoid more costly compound joints; (ii). It can implement pure translation theoretically; (iii). It has close-form direct and inverse kinematics solutions.

A prototype of the spatial compliant manipulator was fabricated creating a compact, three-actuator system that provides translational motion for some micromanipulation tasks. This prototype is shown in figure 19b. The actuators implemented on this prototype are piezoelectric drivers, consisting of a PZT embedded in a compliant transmission realized through a simple level-arm amplification structure (figure 19c). As shown in figure 19d, a preload spring was applied in the actuation system. The functions of the preload spring include that: (i). The static stiffnesses of the mechanism under preload can be increased in some sense; and (ii). Preload also significantly reduces the hysteresis of the piezoelectric actuator.

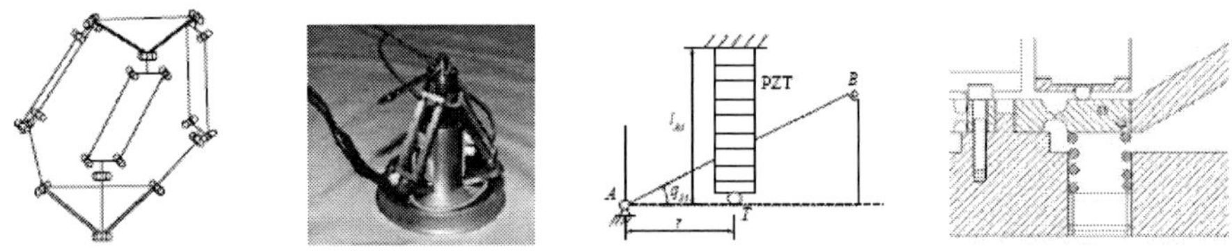

(a) Schematic diagram (b) Prototype (c) Motion amplifier (d) Spring preloading

Figure 19. A 3-RRPR FCPKM.

4.3. 3-RPS FCPKM

As illustrated in figure 20a, the upper plate is connected to the links by means of ball joints which are equally spaced at 120° and at radius from the center of the own plate. The other ends of the links are connected to the base plate through equally spaced pin joints at a radius from the center of the own plate. By varying the link lengths, the upper plate can be manipulated with respect to the base plate. Based on its rigid topology, a 3-RPS FCPKM has been designed, its CAD model and prototype are illustrated in figure 20b and figure 20c, respectively.

The displacement generated by a piezoelectric actuator is usually very small. Therefore, when a large motion is needed, we should use amplifying mechanisms to increase the displacements of actuators. Figure 20d shows the model of the top stage and its amplifying structure, i.e., a four-bar mechanism.

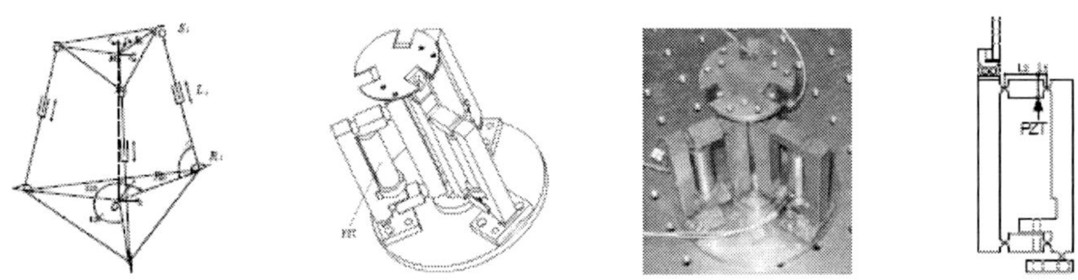

(a) Schematic diagram (b) 3D CAD model (c) Prototype (d) Motion amplifier

Figure 20. Model of 3-RPS FCPKM and its amplifying mechanism.

From fabrication point of view, spherical joint is very difficult to manufacture and the machining cost would also be high. Furthermore, it is hard to analysize the machining errors of this kind of joint in order to modify control model. To avoid machining spherical pivots (see figure 21a), we developed a new type of hybrid hinge for a replacement. It comprises three revolute pairs whose axes intersect at one point (see figure 21b). Theoretically, it can realize three revolute motions just as conventional spherical joint does. So, this kind of spherical joints was chosen while fabricating the prototype (see figure 21c).

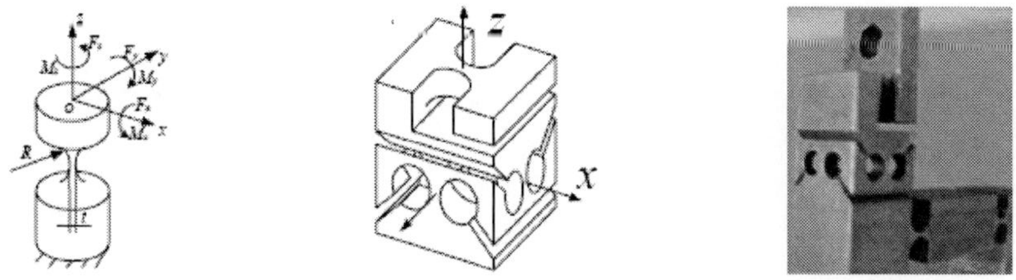

(a) Conventional model (b) Equivalent model (c) Prototype

Figure 21. Two types of compliant spherical joints.

The above three FCPKMs will acted as the examples for analysis and design at the following sections.

5. Kinetostatic Model

Research on kinetostatic model of FCPKMs mainly includes two parts: kinematics, statics, and both. In the kinematics analysis for a planar compliant mechanism, for the purpose of simplification, the PRBM method has been commonly used in that both a flexural pivot and a CTJ have its equivalent PRBM description. However, it is very difficult to establish a concise but precise kinematic mapping between a spatial FCPKM with its PRBM due to compound CJ existing in the mechanism. In this case, stiffness matrix method becomes an alternative to realize the kinetostatic analysis of the spatial one, it is actually a general approach for any a FCPKM. In this section, we will take the FCPKM shown in figure 19b as an example.

5.1. Kinematics Analysis of a FCPKM Based on its PRBM

Mathematically, the task of kinematics is to find equations that relate the motion of the end-effector with that of the actuators. Note that the forward kinematics evaluates the end-effector motion from the known motion of the actuators, and the inverse kinematics calculates the motion of the actuators from the known end-effector motion, both of which can be expressed by the Jacobian matrix. The following procedures should lead to the Jacobian of this mechanism.

Due to only 1-DOF flexural pivot existed in 3-RRPR compliant mechanism, thus we can use the PRBM methodology to realize its kinematics analysis. Figure 22 shows the geometrical relations in the PRBM of the compliant micromanipulator.

For any a kinematic limb in the PRBM, the vector equation can be established as follows

$$\boldsymbol{P} = \boldsymbol{l}_0 + \boldsymbol{l}_1 + \boldsymbol{l}_2 + \boldsymbol{l}_3 \tag{25}$$

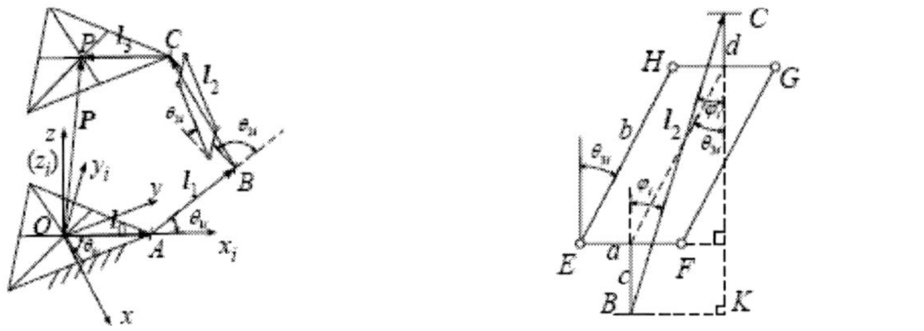

(a) Geometrical relation in the mechanism (b) Geometrical relation in the parallelogram

Figure 22. Geometrical relation in the PRBM.

Considering that Tsai has discussed the inverse and forward displacements of this mechanism, we will simplify the following derivation.

Substitution of the related coordinates and expressions into Eq. (25) yields

$$\begin{bmatrix} p_x \cos\theta_{0i} + p_y \sin\theta_{0i} \\ -p_x \sin\theta_{0i} + p_y \cos\theta_{0i} \\ p_z \end{bmatrix} = \begin{bmatrix} l_0 + l_1 \cos\theta_{1i} + (e + b\cos\theta_{3i})\cos(\theta_{1i} + \theta_{2i}) - l_3 \\ b\sin\theta_{3i} \\ l_1 \sin\theta_{1i} + (e + b\cos\theta_{3i})\sin(\theta_{1i} + \theta_{2i}) \end{bmatrix} \tag{26}$$

where e=c+d, here both c and d are the small offsets.

By differentiating both sides of Eq. (26), the following equation can be found

$$\mathrm{d}\theta_{1i} = h_{i1}\mathrm{d}p_x + h_{i2}\mathrm{d}p_y + h_{i3}\mathrm{d}p_z \tag{27}$$

where

$$h_{i1} = (\cos\theta_{0i}\cos(\theta_{1i} + \theta_{2i})\cos\theta_{3i} - \sin\theta_{0i}\sin\theta_{3i})/(l_1 \sin\theta_{2i}\cos\theta_{3i})$$

$$h_{i2} = (\sin\theta_{0i}\cos(\theta_{1i} + \theta_{2i})\cos\theta_{3i} + \cos\theta_{0i}\sin\theta_{3i})/(l_1 \sin\theta_{2i}\cos\theta_{3i}) \quad h_{i3} = \sin(\theta_{1i} + \theta_{2i})/(l_1 \sin\theta_{2i})$$

$$\theta_{3i} = \arcsin\frac{-p_x \sin\theta_{0i} + p_y \cos\theta_{0i}}{b}$$

$$\theta_{2i} = \arccos(\frac{p_x \cos\theta_{0i} + p_y \sin\theta_{0i} - l_0 + l_3 - l_1 \cos\theta_{1i}}{e + b\cos\theta_{3i}}) - \theta_{1i}$$

For all above equations, i=1, 2, 3.

Eq. (27) can also be written in a matrix form, which leads to

$$\mathrm{d}\boldsymbol{Q} = \begin{bmatrix} h_{11} & h_{12} & h_{13} \\ h_{21} & h_{22} & h_{23} \\ h_{31} & h_{32} & h_{33} \end{bmatrix} \mathrm{d}\boldsymbol{P} \tag{28}$$

where $\mathrm{d}\boldsymbol{Q} = [\mathrm{d}\theta_{11} \ \ \mathrm{d}\theta_{12} \ \ \mathrm{d}\theta_{13}]^{\mathrm{T}}$, $\mathrm{d}\boldsymbol{P} = [\mathrm{d}p_x \ \ \mathrm{d}p_y \ \ \mathrm{d}p_z]^{\mathrm{T}}$.

In robotics, the relationship between the input of the mechanism and the output of the end-effector is described by a general relationship. In the case of this PRBM, we have

$$\mathrm{d}\boldsymbol{P} = \boldsymbol{J} \cdot \mathrm{d}\boldsymbol{Q} \tag{29}$$

where $\boldsymbol{J}$ is Jacobian matrix.

Therefore, if $\boldsymbol{J}$ is not a singular matrix, from Eqs. (28) and (29), we have

$$\boldsymbol{J} = \begin{bmatrix} h_{11} & h_{12} & h_{13} \\ h_{21} & h_{22} & h_{23} \\ h_{31} & h_{32} & h_{33} \end{bmatrix}^{-1} \tag{30}$$

Because the displacement of the end-effector is small enough compared with its geometric size, $\boldsymbol{J}$ will remain essentially constant throughout the workspace of the micromanipulator. In this case, $\boldsymbol{J}$ can be calculated at the initial configuration of the mechanism.

As mentioned above, the inputs of the system are Δl_{Ai} (i=1,2,3) of the actuators rather than $\Delta\theta_{1i}$ (i=1,2,3). However, their relationship described in matrix form can be obtained as

$$\mathrm{d}\boldsymbol{Q} = -\frac{1}{r}\mathrm{d}\boldsymbol{L}_A \tag{31}$$

where $\mathrm{d}\boldsymbol{L}_\mathrm{A}=[\mathrm{d}l_{\mathrm{A1}}, \mathrm{d}l_{\mathrm{A2}}, \mathrm{d}l_{\mathrm{A3}}]^\mathrm{T}$.

Substitute Eq. (31) into Eq. (29), it can be derived as

$$\mathrm{d}\boldsymbol{P} = \boldsymbol{J}_\mathrm{A} \cdot \mathrm{d}\boldsymbol{L}_\mathrm{A} \tag{32}$$

where $\boldsymbol{J}_\mathrm{A}$ is defined as the Jacobian matrix that relates the motion of the end-effector with the motion of the actuators, and $\boldsymbol{J}_\mathrm{A}$ can be represented by

$$\boldsymbol{J}_\mathrm{A} = -\frac{1}{r}\boldsymbol{J} \tag{33}$$

Since $\boldsymbol{J}$ is a constant matrix, $\boldsymbol{J}_\mathrm{A}$ is also a matrix covering constant values.

Because the displacement of the end-effector is sufficiently small, Eq. (32) can be expressed by

$$\Delta\boldsymbol{P} = \boldsymbol{J}_\mathrm{A} \cdot \Delta\boldsymbol{L}_\mathrm{A} \tag{34}$$

As given in table 3, the geometric parameters of this mechanism are obtained by further optimization. Here the objective function for the optimization is chosen so as to achieve the maximum workspace of the end-effector. In addition, its initial position is set to be $[0, 0, 80]^\mathrm{T}$ (measurements: mm).

Table 3. Structural parameters of the PRBM (mm)

l_0	l_1	L_3	a	b	c	d
8	40	9.3	16	53.3	8	8

Therefore,

$$\boldsymbol{J}_{\mathrm{A}} = \begin{bmatrix} 12.9130 & -6.4565 & -6.4565 \\ 0.0000 & 11.1830 & -11.1830 \\ -3.5895 & -3.5895 & -3.5895 \end{bmatrix} \tag{35}$$

5.2. Statics Analysis of a FCPKM Based on Stiffness Matrix Method

In our FCPM shown in figure 19b, the compliant elements include two kinds: leaf spring and circular notched flexure-hinge. Compliance matrix of a leaf spring has been given in section 3. As for the right circular notched flexure hinge, almost compliance parameters are formulated by Paros and Welsbord using linear beam theory. The compliance parameters were computed at the end of a fictitious beam associated with the hinge, not at the hinge center. In addition, the compliance matrix of the notched flexure hinge has the identical parameters with that of the prismatic beam, as shown in Eq. (24), but the corresponding parameter expressions are quietly different. Hereafter, a few parameters in the compliance matrix are corrected by some researchers such as Koseki *et al.* [57]. Therefore, here all parameters are given directly. It is noteworthy that a part of expressions is different from Koseki.

In Eq. (24) available for circular notched flexure hinge, the compliance parameters are enumerated as follows under the condition of $t/2R << 1$.

$$c_{11} = \frac{9\pi r^{5/2}}{2Ebt^{5/2}} + \frac{1}{Gb}(-\frac{2+\pi}{2} + \pi\sqrt{\frac{r}{t}}),\ c_{22} = \frac{12\pi r^2}{Eb^3}(-\frac{1}{4} + \sqrt{\frac{r}{t}}) + \frac{1}{Gb}(-\frac{2+\pi}{2} + \pi\sqrt{\frac{r}{t}}),\ c_{33} = \frac{1}{Eb}(-\frac{2+\pi}{2} + \pi\sqrt{\frac{r}{t}}),$$

$$c_{44} = \frac{12}{Eb^3}[\pi(\frac{r}{t})^{1/2} - \frac{2+\pi}{2}],\ c_{55} = \frac{9\pi r^{1/2}}{2Ebt^{5/2}},\ c_{66} = \frac{3\pi r^{1/2}}{8\lambda Gbt^{5/2}},\ c_{15} = c_{51} = \frac{9\pi r^{3/2}}{2Ebt^{5/2}},$$

$$c_{24} = c_{42} = -\frac{12r}{Eb^3}(\pi\sqrt{\frac{r}{t}} - \frac{2+\pi}{2}) \tag{36}$$

Various approaches can be used to determine the compliance matrix of a general parallel mechanism. A commonly used method is based on the Jacobian matrix. However, on account of structural complexity in the leg of the proposed FCPM, here the adjoint transformation introduced in Section 3 is applied to compute the compliance of this mechanism. In essence, this method is the same as that the matrix method addressed by Koseki.

Analysis for the compliance of the FCPM can begin with any one of its three limbs (each limb is identical). The limb and notations at its nodes are shown in figure 4. In the mechanism, both notched flexure hinges and long links (*OABC, CDEF, GHIJ* and *KL*) can be treated as compliant units, while short link and the moving platform (like *EDFGH, IJK* and $RL_1L_2L_3$ in figure 23) as rigid body. Thus each limb consists of three compliant units, as shown in figure 24. The corresponding compliance matrices $\boldsymbol{C}_{unit1}, \boldsymbol{C}_{unit2}$ and $\boldsymbol{C}_{unit3}$ cover the compliance matrices composed of all compliant elements.

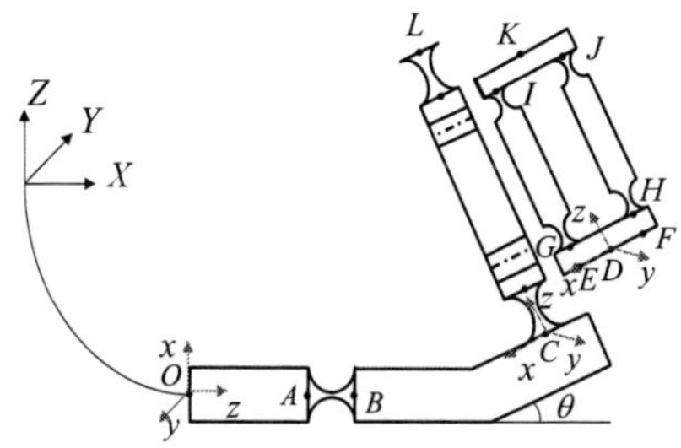

Figure 23. Limb of the compliant mechanism.

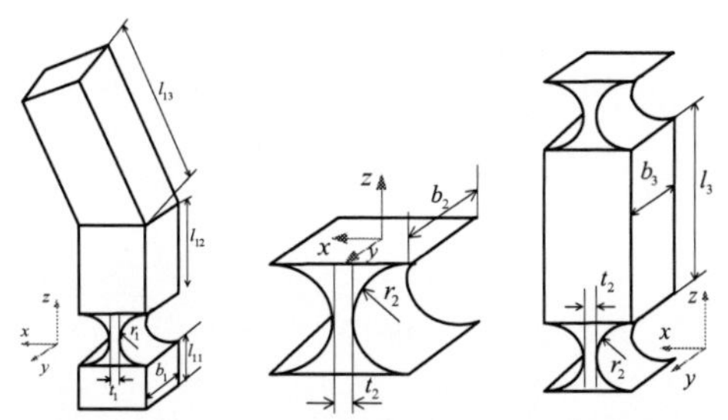

(a) unit1 (b) unit2 (c) unit3

Figure 24. Three compliant units of each limb.

In order to make the analysis more convenient, a global coordinate frame and a series of local coordinate frames at its nodes are defined, as shown in figure 23.

In addition, some symbols are also introduced. $\boldsymbol{C}_{beam}(t, b, l)$ and $\boldsymbol{C}_{flexure}(r, t, b)$ denote the compliance matrices of prismatic beam and right circular notched flexure, respectively. $\boldsymbol{C}_{IJ}$ denotes the compliance matrix of IJ and $\boldsymbol{C}_{unit1}$ the compliance matrix of the first compliant unit. $\mathrm{Ad}_{t(t_x, t_y, t_z)}$, $\mathrm{Ad}_{R(\theta_x, \theta_y, \theta_z)}$ and $\mathrm{Ad}_{g(t_x, t_y, t_z; \theta_x, \theta_y, \theta_z)}$ denote adjoint transformations associated with a pure translation, a pure rotation and a screw motion, respectively, and $\mathrm{Ad}_{t_{AB}}$ an adjoint transformation associated with pure translation from A to B. They are

$$\begin{aligned}\boldsymbol{C}_{unit1} &= \boldsymbol{C}_{beam}(2r_1 + t_1, b_1, l_{11}) + \mathrm{Ad}^{\mathrm{T}}_{t(0,0,l_{11})}\boldsymbol{C}_{flexure}(r_1, t_1, b_1)\mathrm{Ad}_{t(0,0,l_{11})} \\ &+ \mathrm{Ad}^{\mathrm{T}}_{t(0,0,l_{11}+2r_1)}\boldsymbol{C}_{beam}(2r_1 + t_1, b_1, l_{12})\mathrm{Ad}_{t(0,0,l_{11}+2r_1)} + \mathrm{Ad}^{\mathrm{T}}_{g(0,0,l_{11}+2r_1+l_{12};0,\theta,0)}\boldsymbol{C}_{beam}(2r_1 + t_1, b_1, l_{13})\mathrm{Ad}_{g(0,0,l_{11}+2r_1+l_{12};0,\theta,0)}\end{aligned} \tag{37}$$

$$\boldsymbol{C}_{unit2} = \boldsymbol{C}_{flexure}(r_2, t_2, b_2) \tag{38}$$

$$\boldsymbol{C}_{unit3} = \boldsymbol{C}_{flexure}(r_3, t_3, b_3) + \mathrm{Ad}^{\mathrm{T}}_{t(0,0,2r_3)}\boldsymbol{C}_{beam}(2r_3 + t_3, b_3, l_3)\mathrm{Ad}_{t(0,0,2r_3)} + \mathrm{Ad}^{\mathrm{T}}_{t(0,0,l_3+2r_3)}\boldsymbol{C}_{flexure}(r_3, t_3, b_3)\mathrm{Ad}_{t(0,0,l_3+2r_3)} \tag{39}$$

$$\boldsymbol{C}_C = \mathrm{Ad}^{\mathrm{T}}_{R(\pi,3\pi/2,0)}\boldsymbol{C}_{unit1}\mathrm{Ad}_{R(\pi,3\pi/2,0)} \tag{40}$$

$$\boldsymbol{C}_D = \mathrm{Ad}^{\mathrm{T}}_{t(0,0,2r_2)}\boldsymbol{C}_C\mathrm{Ad}_{t(0,0,2r_2)} + \mathrm{Ad}^{\mathrm{T}}_{R(0,\theta,0)}\boldsymbol{C}_{unit2}\mathrm{Ad}_{R(0,\theta,0)} \tag{41}$$

Since *EDFGH* is a rigid body, relative displacements don't occur among *E*, *D*, *F*, *G* and *H*. Thus equilibriums of twists and wrenches between *D* and *GH* can be established and the compliance matrix of GH is derived accordingly.

The formulation about equilibriums of twists can be written as

$$\begin{bmatrix} \mathrm{Ad}_{t_{DG}} \\ \mathrm{Ad}_{t_{DH}} \end{bmatrix} \varsigma_D = \begin{bmatrix} \varsigma_G \\ \varsigma_H \end{bmatrix} = \boldsymbol{C}_{GH} \begin{bmatrix} \boldsymbol{F}_G \\ \boldsymbol{F}_H \end{bmatrix} \tag{42}$$

$$\varsigma_D = \boldsymbol{C}_D \boldsymbol{F}_D \tag{43}$$

The equation about equilibriums of wrenches is

$$\begin{bmatrix} \mathrm{Ad}^{\mathrm{T}}_{t_{DG}} & \mathrm{Ad}^{\mathrm{T}}_{t_{DH}} \end{bmatrix} \begin{bmatrix} \boldsymbol{F}_G \\ \boldsymbol{F}_H \end{bmatrix} = \boldsymbol{F}_D \tag{44}$$

Combining (42), (43), and (44) yields

$$\boldsymbol{C}_{GH} = \begin{bmatrix} \mathrm{Ad}_{t_{DG}} \\ \mathrm{Ad}_{t_{DH}} \end{bmatrix} \boldsymbol{C}_D \begin{bmatrix} \mathrm{Ad}^{\mathrm{T}}_{t_{DG}} & \mathrm{Ad}^{\mathrm{T}}_{t_{DH}} \end{bmatrix} \tag{45}$$

Thus $\boldsymbol{c}_{IJ}$ can be obtained as

$$\boldsymbol{C}_{IJ} = \begin{bmatrix} \mathrm{Ad}^{\mathrm{T}}_{t_{GI}} & \mathbf{0} \\ \mathbf{0} & \mathrm{Ad}^{\mathrm{T}}_{t_{HJ}} \end{bmatrix} \boldsymbol{C}_{GH} \begin{bmatrix} \mathrm{Ad}_{t_{GI}} & \mathbf{0} \\ \mathbf{0} & \mathrm{Ad}_{t_{HJ}} \end{bmatrix} + \begin{bmatrix} \boldsymbol{C}_{unit3} & \mathbf{0} \\ \mathbf{0} & \boldsymbol{C}_{unit3} \end{bmatrix} \tag{46}$$

Since *IJK* is also a rigid body, $\boldsymbol{C}_K$ can be obtained in the similar manner to the computation of $\boldsymbol{C}_{GH}$.

$$\boldsymbol{C}_K = \begin{bmatrix} \mathrm{Ad}^{\mathrm{T}}_{t_{KI}} & \mathrm{Ad}^{\mathrm{T}}_{t_{KJ}} \end{bmatrix} \boldsymbol{C}_{IJ} \begin{bmatrix} \mathrm{Ad}_{t_{DG}} \\ \mathrm{Ad}_{t_{DH}} \end{bmatrix} \tag{47}$$

Therefore, the compliance matrix of *L* in term of *O* can be obtained as

$$\boldsymbol{C}_L = \mathrm{Ad}^{\mathrm{T}}_{t_{KL}}\boldsymbol{C}_K\mathrm{Ad}_{t_{KL}} + \mathrm{Ad}^{\mathrm{T}}_{R(0,\alpha,0)}\boldsymbol{C}_{unit2}\mathrm{Ad}_{R(0,\alpha,0)} \tag{48}$$

Considering the symmetry of the parallel structure, the compliance matrix of all three limbs can be obtained as

$$C_{L_1L_2L_3} = diag\begin{bmatrix} C_{L_1} & C_{L_2} & C_{L_3} \end{bmatrix} \tag{49}$$

where $C_{L_i} = \mathrm{Ad}^{\mathrm{T}}_{R(0,0,2\pi i/3)} C_L \mathrm{Ad}_{R(0,0,2\pi i/3)}$, $(i = 0, 1, 2)$.

Note that $RL_1L_2L_3$ is a rigid body, the compliance matrix of the whole mechanism $\boldsymbol{C}$ can be obtained as follows.

$$C = \begin{bmatrix} \mathrm{Ad}^{\mathrm{T}}_{t_{RL_1}} & \mathrm{Ad}^{\mathrm{T}}_{t_{RL_2}} & \mathrm{Ad}^{\mathrm{T}}_{t_{RL_3}} \end{bmatrix} C_{L_1L_2L_3} \begin{bmatrix} \mathrm{Ad}_{t_{RL_1}} \\ \mathrm{Ad}_{t_{RL_2}} \\ \mathrm{Ad}_{t_{RL_3}} \end{bmatrix} \tag{50}$$

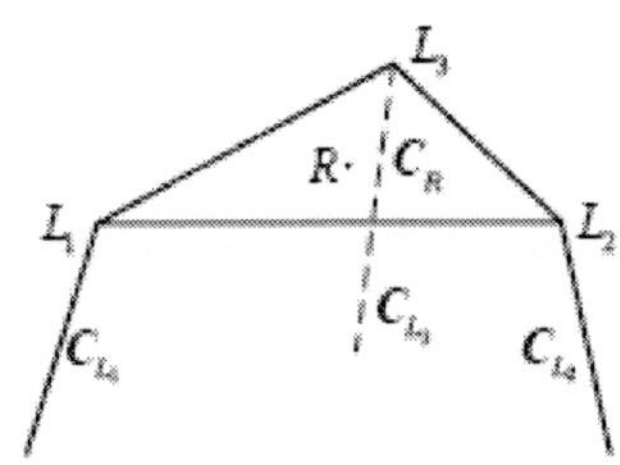

Figure 25. Three limbs and the moving platform.

Table 4. Design parameters of the FCPKM (Unit: mm)

θ (°)	α (°)	l_{11}	l_{12}	l_{13}	l_3	t_1=t_2=t_3	r_1=r_2=t_3	b_1=b_2=b_3
30	90	6	8	30	49	0.5	4	5

Table 4 shows all parameters of the FCPKM. The compliance matrix of the FCPKM can be then written as

$$C = \begin{bmatrix} 56.3724 & -2358.12 & 86.8846 & 185.245 & 8.39896 & 106.944 \\ -2358.12 & 661187 & -2762.64 & -52107 & -311.677 & -30084.9 \\ 86.8846 & -2762.64 & 135.498 & 216.765 & 13.023 & 125.137 \\ 185.245 & -52107 & 216.765 & 4106.5 & 24.4725 & 2370.92 \\ 8.39896 & -311.677 & 13.0233 & 24.4725 & 1.25559 & 14.1283 \\ 106.944 & -30084.9 & 125.137 & 2370.92 & 14.1283 & 1368.92 \end{bmatrix} \tag{51}$$

5.3. Mobility and Constraint Analysis of a FCPKM Based on Screw Theory

There have been some mature approaches to calculate the degree of freedom (DOF) of the rigid-body manipulator, such as the Grüber's mobility criterion used to compute the mobility of a rigid-body manipulator quantitatively. A FCPKM, however, is not a mechanism in a strict sense. It is indeed a kind of compliant mechanism characterized by a complicate topological structure and multiple degrees of freedom (DOFs). Compared with its rigid-counterpart, a FCPKM is more complicate in mobility and constraint analysis. Therefore, the method to evaluate the DOF of a rigid-body manipulator is no longer valid for the FCPKM.

Due that twist and wrench in screw theory can just reflect such a reciprocal relationship between DOF and constraint both in a rigid mechanism and in a compliant one, the screw theory is an effective tool available for the mobility analysis of a general FCPKM. In addition, according to the assumption for a general FCPKM, it can be regarded as that a rigid body is fully supported by a spatial spring system. The motion capability i.e. the mobility of the rigid body may be determined by examining the elastic suspensions. In this way, the rigid body has 6 DOFs in a strict sense. However, most of current FCPKMs step from the topological structures of their rigid-body counterparts. Therefore, if a degenerate-DOF rigid-body parallel manipulator is designed as a compliant mechanism, the desired motion of the moving platform is surely a partially-constrained motion accordingly. As for these partially-constraint compliant manipulators, only some of six DOFs in the manipulator will play a primary role.

In this case, there is a need for a new concept to evaluate the motion capability of this class of compliant manipulator. For this purpose, the concept of primary mobility of a FCPKM is proposed and a method based on the eigencompliance defined in screw theory for determining the primary mobility of a partially-constraint FCPKM is addressed.

The primary mobility of a FCPKM is defined as the mobility that plays a principal role in the independent motions of the mechanism. In addition, the definition of such a new concept involves in an important significance, that is, due to the limitation of factors such as manufacturing capability, insufficient flexibility around rotation axis in each flexure hinge or insufficient stiffness around other axes causes an undesired motion in the moving platform, which disagrees with initial intent of the designer. In this case, the real motion of this manipulator can be evaluated and calculated in the phase of conceptual design by analyzing the primary mobility of the FCPKM.

A. The Eigenscrews and Eigencompliance

The analytical representation of the mapping of compliance can also be a general collineation of screws that assumes the form

$$\boldsymbol{\$} = \boldsymbol{cF} \tag{52}$$

where $\boldsymbol{\$}$ is ray coordinates associate with the twist.

Using the operator $\tilde{\Delta}$, the twist can be converted from axis coordinates ς to ray coordinates $\boldsymbol{\$}$.

$$\boldsymbol{\$} = \tilde{\Delta}\varsigma \tag{53}$$

where

$$\tilde{\Delta} = \begin{bmatrix} \mathbf{0} & \boldsymbol{I}_{3\times3} \\ \boldsymbol{I}_{3\times3} & \mathbf{0} \end{bmatrix} \tag{54}$$

The $\tilde{\Delta}$ operator interchanges the first and last three components of a screw, and thus this changes a screw in axis coordinates to one in ray coordinates and vice versa. The operator has several special properties including orthogonality, $\tilde{\Delta}^{-1} = \tilde{\Delta}^{\mathrm{T}} = \tilde{\Delta}$, and three positive and three negative eigenvalues. Thus,

$$c = \tilde{\Delta}C \tag{55}$$

The formulation of the compliance matrix eigenvalue problem requires that both quantities should be stated in the same coordinate frame, which aims at preserving the integrity of the units and ensures that the results are independent of the coordinate frame.

Combining (52) and (55) yields

$$\boldsymbol{\$} = \tilde{\Delta}\boldsymbol{CF} \tag{56}$$

which is an equation in ray coordinates.

Consider that a twist on a screw is chosen to be transformed under the mapping of compliance, with a result that the wrench obtained acts on the same screw. This screw is accordingly defined as an eigenscrew of compliance, and the ratio of intensities of the twist and that of the wrench is referred to as the corresponding eigencompliance, denoted as λ_i. All six eigenscrews and their corresponding eigencompliances can be determined from the eigenvectors and eigenvalues of $\boldsymbol{c}$ or $\tilde{\Delta}\boldsymbol{C}$.

The corresponding eigenvalue problem is developed by assuming that the wrench and twist are scalar multiples of the same screw, for which

$$\lambda \boldsymbol{e} = \tilde{\Delta}\boldsymbol{Ce} \tag{57}$$

In a general case, this formulation yields six eigenvalues λ_i and six eigenvectors $\boldsymbol{e}_i$. The necessary and sufficient condition for the existence of the eigenvectors $\boldsymbol{e}$ is

$$\left|\lambda - \tilde{\Delta}\boldsymbol{C}\right| = 0 \tag{58}$$

Invariance is always the key issue in geometry. With this mind, consider that a coordinate frame has been changed. Thus, in an analogy with the derivation of Eqs. (6)~(9), a new representation of the collinear form of the mapping of compliance can be obtained by substituting $\boldsymbol{B\$'}$ into $\boldsymbol{\$'}$ and $\boldsymbol{BF'}$ into $\boldsymbol{F'}$ in Eq. (56), thus

$$\boldsymbol{\$} = \begin{bmatrix} \boldsymbol{R} & \boldsymbol{0} \\ \hat{\boldsymbol{t}}\boldsymbol{R} & \boldsymbol{R} \end{bmatrix} \boldsymbol{\$'} = \boldsymbol{B\$'} \tag{59}$$

$$\boldsymbol{F} = \begin{bmatrix} \boldsymbol{R} & \boldsymbol{0} \\ \hat{\boldsymbol{t}}\boldsymbol{R} & \boldsymbol{R} \end{bmatrix} \boldsymbol{F'} = \boldsymbol{BF'} \tag{60}$$

$$\boldsymbol{c} = \boldsymbol{B}^{-1}\boldsymbol{c'}\boldsymbol{B} = \mathrm{Ad}_g^{-1}\boldsymbol{c'}\mathrm{Ad}_g \tag{61}$$

In contrast with Eq. (8) which relates to congruent matrices, Eq. (61) relates to similar matrices. This reflects the fact that both these eigenvalues and eigenvectors are invariant with respect to the transformation from ray coordinates to axis coordinates and vice versa. At the physical meaning, the eigenvalues is the signed magnitude ratio of angular deformation to force, and the eigenvectors are referred as eigenscrews and are in ray coordinates.

Based on the duality in spatial compliance and stiffness, the above eigencompliances and eigenscrews can be converted into a physical description of screw spring with an appropriate spring constant. This can be realized in a parallel mechanism having six screw springs along the six eigenscrews $(\boldsymbol{s}_0 ; \boldsymbol{s})$. The direction of the screw spring axis $\boldsymbol{s}$ is given by the dual part of the eigenscrews, and the distance $\boldsymbol{r}$ to the line of action is given by $\boldsymbol{s} \times \boldsymbol{s}_0$. The eigenstiffness can be obtained from the eigencompliances as

$$\lambda_{ki} = \frac{2h_i}{\lambda_i} \tag{62}$$

Each screw spring is defined by its spring constant $k_i = 1/\lambda_i = \lambda_{ki}/2h_i$.

It should be noted that the eigenscrews of stiffness matrix are identical to eigenscrews of compliance matrix, which is also the principal screws of the potential defined by Ball [50].

B. An Approach to Calculate Primary Mobility of a FCPKM

According to reference [58], eigenscrews discussed above can represent the fundamental modes of the moving platform of a FCPKM. Thus all possible motions of the moving platform may be denoted as the linear combinations of these fundamental eigenvectors.

A special case occurs when one or more eigencompliances are equal to zero. In this case, non-zero eigenscrews with respect to zero eigencompliances can be used to gain some insight into the motion capability of a partially constrained FCPKM. Considering that the space $\boldsymbol{\Gamma}_v$ spanned by all non-zero eigenscrews with respect to zero eigencompliances is equivalent to the space $\boldsymbol{\Gamma}_r$ spanned by all constraint wrenches reciprocal to motions of the platform. Therefore, one can find the space $\boldsymbol{\Gamma}_r$ spanned by a manipulator's wrenches by finding the space $\boldsymbol{\Gamma}_v$ spanned by eigenscrews of the $\tilde{\Delta} \boldsymbol{C}$ with respect to zero eigencompliances. While the space $\boldsymbol{\Gamma}_m$ spanned by all twists is correspondingly derived from $\boldsymbol{\Gamma}_r$ according to the reciprocal relationship between the twists and the wrenches, i.e. $\boldsymbol{F}^{\mathrm{T}} \varsigma = 0$. All twists in the space $\boldsymbol{\Gamma}_m$ represent the independent motions of this manipulator.

For a degenerate-DOF FCPKM, its motion capability is surely constrained partially. As a result, some DOFs in the manipulator may be lost. This yields that some eigencompliances are equal to zero or closer to zero. Thus the primary mobility or principal motion of a FCPKM can be determined accordingly.

Thus, the procedures for determining the primary mobility or principal motion of a FCPKM can be achieved through the following steps:

Step 1: determine the manipulator's compliance matrix $\boldsymbol{C}$.

Step 2: calculate $\tilde{\Delta} \boldsymbol{C}$ and the corresponding eigenvalues and eigenvectors.

Step 3: determine the space $\boldsymbol{\Gamma}_r$ spaned by all wrenches $\boldsymbol{e}$ reciprocal to motions of the moving platform by finding the space $\boldsymbol{\Gamma}_v$ spanned by eigenscrews of $\tilde{\Delta}C$ with eigencompliances equal to zero or closer to zero.

Step 4: compute the primary mobility or principal motion of the manipulator by making a solution to the space $\boldsymbol{\Gamma}_m$ spanned by all twists reciprocal to all wrenches $\boldsymbol{e}$.

C. Case Study

According to the method proposed above, the primary mobility of the 3-legged FCPKM shown in figure 19b can be determined.

Combining (51) and (54) yields

$$\tilde{\Delta}C=\begin{bmatrix} 185.245 & -52107. & 216.765 & 4106.5 & 24.4725 & 2370.92 \\ 8.39896 & -311.677 & 13.0233 & 24.4725 & 1.25559 & 14.128 \\ 106.944 & -30084.9 & 125.137 & 2370.92 & 14.1283 & 1368.92 \\ 56.3724 & -2358.12 & 86.8846 & 185.245 & 8.39896 & 106.944 \\ -2358.12 & 661187. & -2762.64 & -52107. & -311.677 & -30084.9 \\ 86.8846 & -2762.64 & 135.498 & 216.765 & 13.023 & 125.137 \end{bmatrix} \tag{63}$$

Therefore, from Eq. (58) the eigencompliance $\boldsymbol{\lambda}$ can be calculated as

$$\lambda=\begin{pmatrix}-15.939 & 15.662 & -2.332 & 0.052 & -0.041 & 0.008\end{pmatrix} \tag{64}$$

As expected, there are three positive and three negative eigencompliances. The matrix $\boldsymbol{E}$ corresponding to eigenscrews is

$$\boldsymbol{E}=\begin{bmatrix}e_1 & e_2 & e_3 & e_4 & e_5 & e_6\end{bmatrix}=\begin{bmatrix} -0.078474 & 0.078491 & 0.044611 & 0.011553 & -0.019184 & 0.010909 \\ -0.000582 & 0.000379 & -0.069066 & -0.044210 & 0.013051 & 0.077858 \\ -0.045309 & 0.045316 & 0.026672 & -0.087103 & 0.106096 & -0.029424 \\ -0.004021 & 0.003231 & -0.455088 & -0.248363 & 0.148569 & 0.540971 \\ 0.995864 & -0.99587 & -0.501855 & 0.796377 & -0.982418 & 0.321385 \\ -0.005300 & 0.003254 & -0.730456 & -0.542608 & 0.031434 & 0.772668 \end{bmatrix} \tag{65}$$

From $\boldsymbol{\lambda}$ and $\boldsymbol{E}$ shown in above equations, it is very difficult to determine the motion capability of the compliant manipulator. However, if some data that are almost equal to zero in $\boldsymbol{\lambda}$ and $\boldsymbol{E}$ are regarded as zero, some valuable conclusions may be drawn.

$$\bar{\lambda}=\begin{pmatrix}-15.939 & 15.662 & -2.332 & 0 & 0 & 0\end{pmatrix} \tag{66}$$

$$\bar{\boldsymbol{E}}=\begin{bmatrix}\bar{e}_1 & \bar{e}_2 & \bar{e}_3 & \bar{e}_4 & \bar{e}_5 & \bar{e}_6\end{bmatrix}=\begin{bmatrix} 0 & 0 & 0 & 0 & 0 & 0 \\ 0 & 0 & 0 & 0 & 0 & 0 \\ 0 & 0 & 0 & 0 & 0 & 0 \\ 0 & 0 & -0.46 & -0.25 & 0.15 & 0.54 \\ 1 & -1 & -0.50 & 0.80 & -0.98 & 0.32 \\ 0 & 0 & -0.73 & -0.54 & 0 & 0.77 \end{bmatrix} \tag{67}$$

Therefore, the space $\boldsymbol{\Gamma}_r$ is spanned by the three wrenches as

$$\boldsymbol{\Gamma}_r = \begin{bmatrix} \bar{e}_4 & \bar{e}_5 & \bar{e}_6 \end{bmatrix} = \begin{bmatrix} 0 & 0 & 0 \\ 0 & 0 & 0 \\ 0 & 0 & 0 \\ 0.25 & 0.15 & 0.54 \\ 0.80 & -0.98 & 0.32 \\ -0.54 & 0 & 0.77 \end{bmatrix} \tag{68}$$

Thus the space $\boldsymbol{\Gamma}_m$ spanned by all twists reciprocal to all wrenches $\$_r$ is represented in a canonical form,

$$\boldsymbol{\Gamma}_m = \begin{bmatrix} 0 & 0 & 0 \\ 0 & 0 & 0 \\ 0 & 0 & 0 \\ 1 & 0 & 0 \\ 0 & 1 & 0 \\ 0 & 0 & 1 \end{bmatrix} \tag{69}$$

Eqs. (68) and (69) indicate that the constrained wrenches applying in the FCPKM are three couples. Then the primary mobility of this mechanism is still a three translational motion, which is identical to the mobility of its rigid counterpart. In other words, mobility characteristics of this FCPKM with current design parameters can satisfy the designer's original desire from the theoretical view.

Above analytical example reveals that design issue on a FCPKM is more complex than that of its rigid-body counterpart. In the design of a FCPKM, an unreasonable alternative to its structural parameters may lead to the depravation of the mobility capability of this manipulator, even lead to the failure of the whole design.

6. Optimal Design

Like other class of mechanisms, structural optimization is one of the most important topics in design of a FCPKM. For example, the workspace of a compliant stage is decided by the travel range on each axis. In order to maximize the workspace, actuators with large force or displacement output may be utilized. Another univeral method is to use lever-arm transmissions or some specified motion-amplification compliant mechanisms to magnify the displacement output of the actuators. The most important of all, however, structural optimization for the design parameters of the stage is an effective alternative to enlarge the workspace. In addition, other performances including accuracy or stiffness, etc. can be improved greatly by means of an optimal design.

The goal of optimization is to design the topology, size, and shape of the FCPKM that best performs the given design requirements. To adequately design a FCPKM for an application requires that the following criteria be addressed: (i). Required kinematic motion (such as magnitude, direction, mobility, and workspace); (ii). Required stiffness to an external load; (iii). Materials properties; (iv). Stress limitations; (v). Size or weight limitations; (vi). Dynamic considerations; (vii). Actuator characteristics.

In fact, designing the topology, size and shape of a FCPKM while addressing all of the above design criteria is formidable task that has not been accomplished. For instance, the stiffness of the mechanism and its workspace is usually a paradox, usually the larger the stiffness of the mechanism is, and the smaller its workspace is. Consequently, during the structural optimization for the FCPKM, only parts of the above criteria is chosen as the objective optimized. Additionally, in most of case, the optimization for the kinetostatic performances dominates in current optimal design.

There are a large majority of studies on FCPKM with reference to micro-motion device development, but its optimal design method is rarely referred in the literatures. Ryu [59] presents a method for designing and optimizing a planar FCPKM, which is based on its pseudo-rigid-body model (PRBM) and taken its kinetostatic features as the design objectives.

This section will introduce two new methods to optimize the design parameters of the FCPKM by taking a planar case i.e. 3-RRR mechanism for instance.

6.1. PMSS-Based Optimal Design

With respect to a 3-RRR FCPKM, the motion performance is heavily influenced by structural parameters of every limb. This section will introduce a new approach to design the parameters of this moving platform, whose performances covers a larger workspace based on its PRBM. Of all the optimal design methods, the physical model of the solution space (PMSS) is by far the most effective in that it can transfer the problems from multi-dimensional design space to two or three-dimension one. By this means, we can solve such an aporia problem of decreasing the design parameters of parallel manipulators in the field of optimal design of mechanisms. Due to its very limited motion range, it is necessary to optimize the size parameters of 3-RRR FCPKM to maximize its workspace. This method is summarized below.

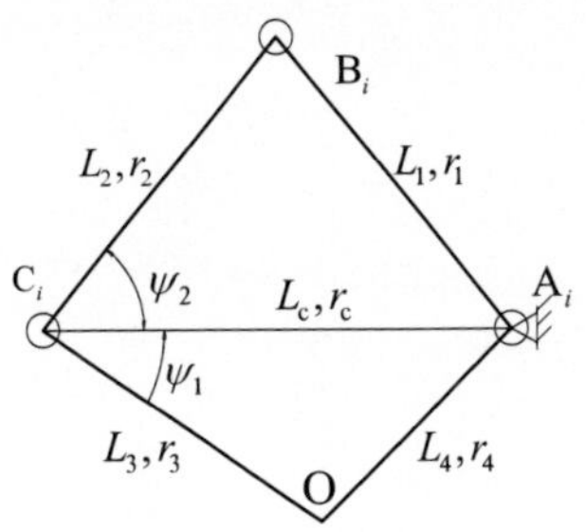

Figure 26. Structural parameters of one Limb.

1. Set Up the Physical Model of the Solution Space of the 3-RRR FCPKM

The theory of the physical model of the solution space is an efficient method that uses the dimensionless parameters instead of the actual sizes of the mechanism, thus a 4D or multi-dimensional problem can be simplified to a 3D issue. As far as 3-RRR platform is concerned, its three limbs are identical. Taking one limb for instance from figure 17, list its parameters L_1, L_2, L_3, L_4 shown in figure 26. Its PMSS is established as follows.

Step 1: Normalization of all kinematic parameters.

Let

$$L = (L_1 + L_2 + L_3 + L_4)/4 \tag{70}$$

and

$$r_i = L_i / L \quad (i=1,2,3,4) \tag{71}$$

where L_i is the real length of the link i, thus r_i is unitary.

Step 2: Determination of constraint conditions.

Obviously,

$$r_1 + r_2 + r_3 + r_4 = 4 \tag{72}$$

$$0 \le r_1,\ r_2 \le 4,\ 0 \le r_3,\ r_4 \le 2 \tag{73}$$

From Eq. (70), it can be obtained as

$$r_1 + r_2 + r_3 = 4 - r_4 \tag{74}$$

Now, a 4D optimal problem reduces to a 3D one.

Step 3: PMSS of 3-RRR platform.

By taking r_1, r_2, r_3 as the axes orthogonal to each other, and r_4 as a variable, we can obtain the 3D PMSS of the bottom stage (see figure 27).

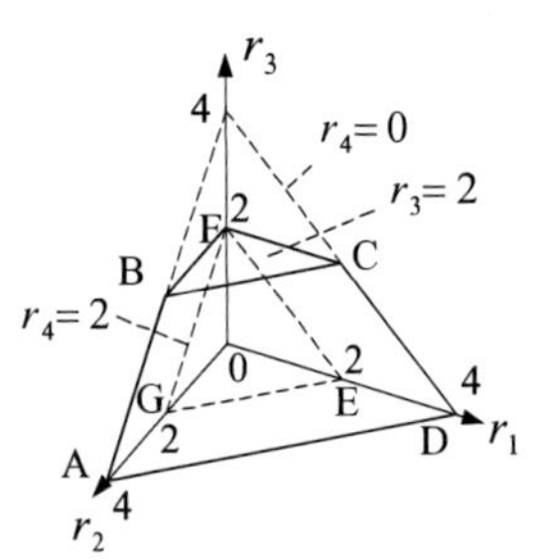

Figure 27. PMSS for the botttom stage.

2. Determine the Optimization Objective

The maximum value of the working space and the maximum output angle value of the ending of the manipulator are selected as the optimization objective. It can be also implemented by combination of the optimization for velocity index and Jacobian of the mechanism. The former is used to minimize the searching range of the dimensions and the latter is used to determine the optimized structural parameters [60].

From reference [10], we can obtain the Jacobian of the stage. Notice that Jacobian of micro-motion platform can be considered to be approximately constant.

$$\begin{pmatrix} \Delta x \\ \Delta y \\ \Delta \gamma \end{pmatrix} = \boldsymbol{J}^0 \begin{pmatrix} \Delta\theta_1 \\ \Delta\theta_2 \\ \Delta\theta_3 \end{pmatrix} \tag{75}$$

where

$$\boldsymbol{J}^0 = \begin{bmatrix} -\frac{\sqrt{3}\sin(\psi_1+\psi_2)+\cos(\psi_1+\psi_2)}{3} & \frac{2\cos(\psi_1+\psi_2)}{3} & \frac{\sqrt{3}\sin(\psi_1+\psi_2)-\cos(\psi_1+\psi_2)}{3} \\ \frac{\sqrt{3}\cos(\psi_1+\psi_2)-\sin(\psi_1+\psi_2)}{3} & \frac{2\sin(\psi_1+\psi_2)}{3} & -\frac{\sqrt{3}\cos(\psi_1+\psi_2)+\sin(\psi_1+\psi_2)}{3} \\ \frac{1}{3R\sin(\psi_1+\psi_2)} & \frac{1}{3R\sin(\psi_1+\psi_2)} & \frac{1}{3R\sin(\psi_1+\psi_2)} \end{bmatrix} \tag{76}$$

3. Determine the Optimized Parameters

The micro-motion stage has one more parameters than its corresponding macro-motion stage. For the 3-RRR micro-motion stage discussed in this paper, five parameters, i.e. r_1, r_2, r_3, r_4, and r_c are used as generalized coordinate to represent the independent structural parameter of the micro-motion stage (shown as figure 26), where r_c is used for determine the initial pose of the mechanism.

$$\sin\psi_1 = \frac{\sqrt{((r_3+r_4)^2-r_c^2)(r_c^2-(r_3-r_4)^2)}}{2r_3r_c} \tag{77}$$

$$\sin\psi_2 = \frac{\sqrt{((r_1+r_2)^2-r_c^2)(r_c^2-(r_1-r_2)^2)}}{2r_2r_c} \tag{78}$$

4. Restrict in the Searching Range of the Size Parameter

The 4D searching for the parameter is decreased to 2D searching according to the results of the physical model of the solution space and the restriction of the structure and stain status, then the searching for r_c is added.

$$|r_1-r_2| < r_c < |r_1+r_2| \tag{79}$$

$$|r_3 - r_4| < r_c < |r_3 + r_4| \tag{80}$$

$$\sqrt{r_4^2 - r_3^2} < r_c \tag{81}$$

5. Calculatet the Objective Function

The determinant value $\det(\boldsymbol{J}^0)$ of the Jacobian matrix and the output angle γ of the mechanism with respect to the dimensionless parameters are calculated.

$$\det(\boldsymbol{J}^0)=\frac{\sqrt{3((r_1+r_2)^2-r_c^2)(r_c^2-(r_1-r_2)^2)^3}}{36r_2^3r_3\sin\left(\sin^{-1}\sqrt{\frac{((r_3+r_4)^2-r_c^2)(r_c^2-(r_3-r_4)^2)}{r_3r_c}}+\sin^{-1}\sqrt{\frac{((r_1+r_2)^2-r_c^2)(r_c^2-(r_1-r_2)^2)}{r_2r_c}}\right)} \tag{82}$$

$$\gamma=\frac{\sqrt{((r_1+r_2)^2-r_c^2)(r_c^2-(r_1-r_2)^2)}}{24r_2r_3\sin\left(\sin^{-1}\sqrt{\frac{((r_3+r_4)^2-r_c^2)(r_c^2-(r_3-r_4)^2)}{r_3r_c}}+\sin^{-1}\sqrt{\frac{((r_1+r_2)^2-r_c^2)(r_c^2-(r_1-r_2)^2)}{r_2r_c}}\right)} \tag{83}$$

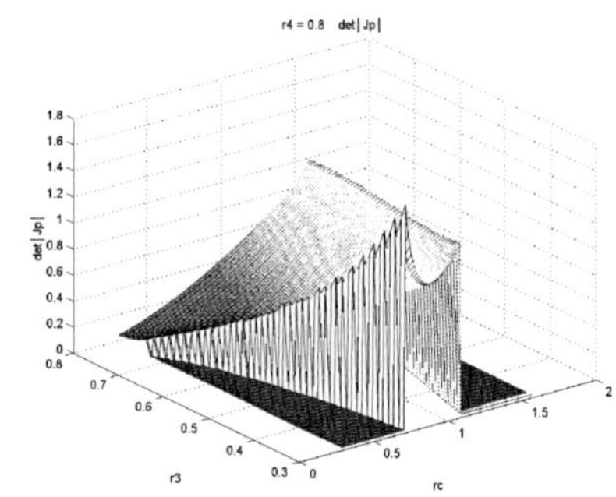

Figure 28. $\det(\boldsymbol{J}^0)$ varies with r_3 and r_c

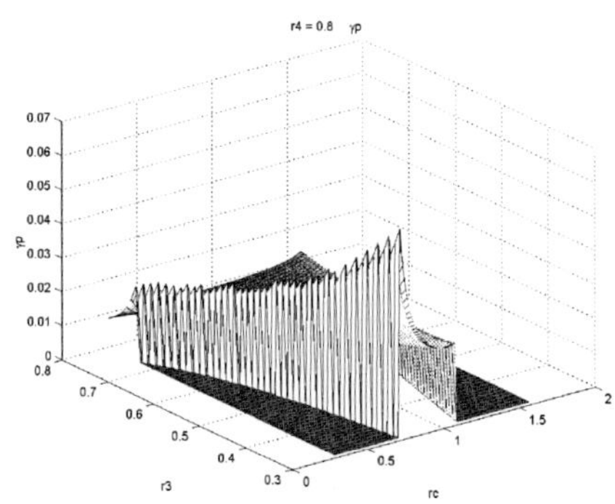

Figure 29. γ varies with r_3 and r_c

6. Select the Kinematic Parameter

The dimensionless parameters are selected according to the results calculated by the program (figures 28, 29), both of the determinant value of the Jacobian matrix and the output angle of the mechanical and also the restriction of the structure and stain status are considered. Thus, the selected dimensionless parameters of the 3-RRR micro-motion stage are: r_1=1.35, r_2=1.35, r_3=0.5, r_4=0.8, r_c=0.8, the actual size parameter can be determined accordingly.

7. Calculate the Working Space of the Manipulator

According to Eq. (82), the working space of the micro-motion stage is calculated in the case of the given maximum driving angle ($\theta_{i\max} = 0.38^{\circ}$), the result is shown in table 5. Theoretical speaking, the selected size parameters can meet the desired performance requirement.

Table 5. Theoretical workspace of the Stage

x_{max}	y_{max}	γ_{max}
121.9μm	134.5μm	1.001°

Based on the design process addressed above, we designed the bottom stage. Its unitary parameters and real values of every parameter are listed in table 6.

Table 6. Parameters of the stage

Parameters	Size(no unit)	Parameters	Size(unit: mm)
r_1	1.35	L_1	27
r_2	1.35	L_2	27
r_3	0.5	L_3	10
r_4	0.8	L_4	16
r_c	0.8	L_c	16

The flowchart of the optimal method shown in figure 30 describes the details.

6.2. ADAMS-Based Optimal Design

It is easy to derive the position of O' and orientation of the platform given the three input angles θ_i (figure 31). The coordinates of point B_i can be obtained from the geometry relationship.

$$\begin{Bmatrix} x_{B_i} \\ y_{B_i} \end{Bmatrix} = g(A_i B_i, D_i E_i, \theta_i) \tag{84}$$

With the three calculated coordinates of the B_i, the position of O' can be calculated.

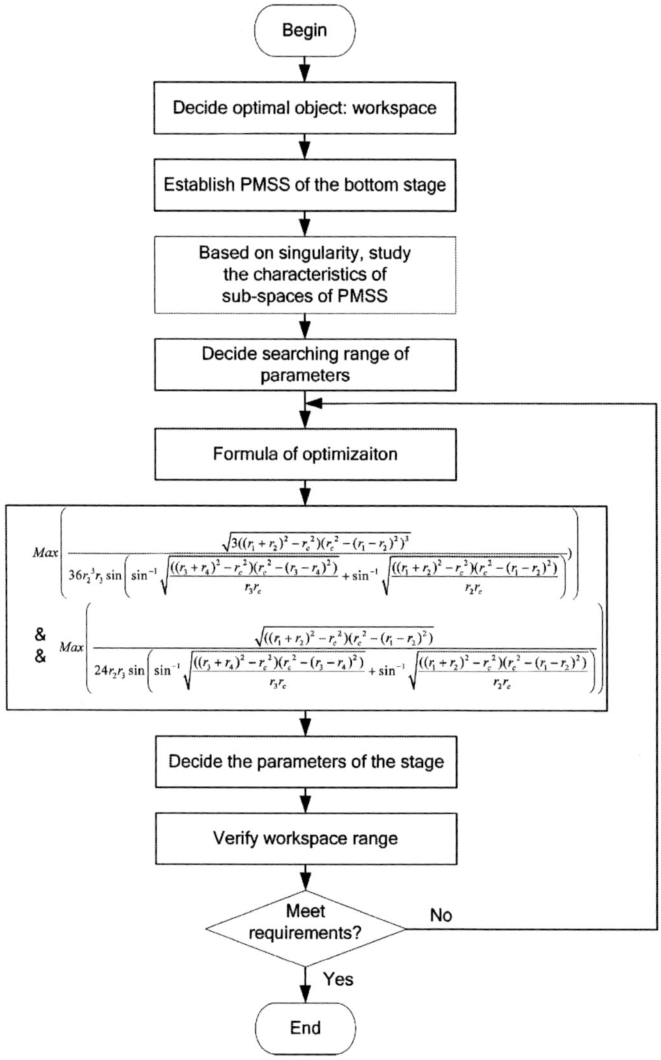

Figure 30. Flowchart of optimal design method.

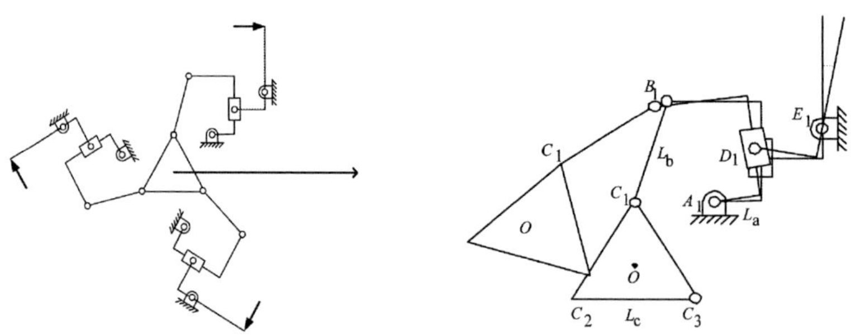

(a) (b)

Figure 31. (a) PRBM of 3-RRR FCPKM with motion-amplifying mechanism, (b) Kinematic diagram of one leg.

$$\begin{Bmatrix} x_{O'} \\ y_{O'} \end{Bmatrix} = f(l_3, a, x_{B_i}, y_{B_i}) \qquad i = 1,2,3 \tag{85}$$

which the new position of the center of the platform and the C_i, the orientation can also be computed.

The combination of all the link length and locations of the revolute joint decides the position and orientation of the tool point. These could be the design variables of the optimization of the workspace. Depending on the application of this PM, the objective function could be to maximize the displacement of O or the rotation angle of the platform.

With three identical applied motions at joints E_i being inputs and the angle limits of the joints being constraints, the optimizations were performed with ADAMS as shown in figure 32.

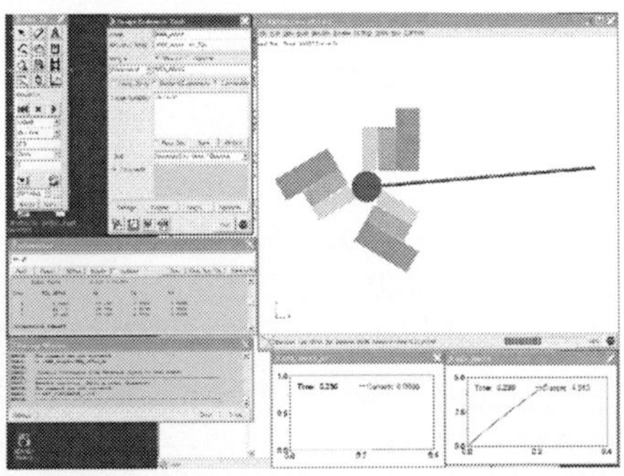

Figure 32. ADAMS optimization snapshot.

If the objective function is to maximize the rotation of the platform with three identical inputs at joints E_i, the results are shown in table 7. L_a, L_b and L_c are the location of the joints in their local frames.

Table 7. Optimization results of maximum of rotation angle

Design Variables	L_a(mm)	L_b(mm)	L_c(mm)
Initial Value	7.5	28.25	2.5
Final Value	6.5 (-13.3%)	29.25(+3.54%)	1.515(-39.4%)

If the objective function is to maximize the displacement of the tool point in X direction with an input at joints E_1, the results are show in table 8.

Table 8. Optimization results of maximum of displacement

Design Variables	L_a(mm)	L_b(mm)	L_c(mm)
Initial Value	7.5	28.25	2.5
Final Value	7.5(+0%)	28.25(+0%)	2.515(+0.6%)

7. DESIGN METHODOLOGY

This section starts with a summarization of a general rules for designing a FCPKM from the viewpoints of material, geometry, actuator, motion, and manufacture, etc. and each design rule is articulated and explained. Three design methods available for a general FCPKM is then discussed in a descriptive way.

7.1. Generalized Design Rules

As far as most FCPKMs are concerned, a high resolution and a good repeatability is a primary requirement. In addition, it should also have a maximized structural stiffness and a high natural frequency to ensure a good dynamic properties and a capability of resisting external disturbance. To satisfy above requirements, some broad design rules are outlined, and are organized into five aspects in the following context.

A. Material Selection

(1). *Choose a material with a high ratio of strength over the modulus of elasticity* (*E*). The elastic limit of the material restricts the mobility of the compliant joints, and an amount of the driving forces to be applied depends upon the stiffness of the compliant joints. It is therefore reasonable to use materials of high strength and moderate elasticity. References [61] and [62] have enumerated some materials with high value of σ_s/E, where TiAl alloys and CuBe2 etc. are all the desired metal materials, while polypropylene is used for non-metal compliant material in that it has a low elastic modulus. Additionally, QSn6.4-1, stainless steel, nickel-iron alloy, Brass, QA17, and others can be also used as the preferred compliant materials.

(2). *The fatigue of a given material should be taken more concerns.* A compliant joint uses the elastic properties of material, but its motion range is limited by the admissible stresses inside the material. It is well known that these stresses depend on the fatigue of the material. Unfortunately, this fatigue is unavoidable because of cyclical loading applied in the compliant member. Thus fatigue criterion is sure to be a factor under a designer's consideration. In general case, the fatigue resistance life of non-metal materials is much shorter than that of metal materials.

B. Geometry

(1). *Mechanical structure should be simple and compact, at the meantime, it is necessary to shorten the kinematic loops*. In order to decrease the errors and the calibration times, the mechanical structure of the FCPKM should be compact and simple, and each parts of the mechanism should be connected rigidly. Thus it can simplify kinematics and dynamics of the mechanism, which makes control and calibration easy. It is also necessary to shorten the kinematic loops to reduce vibration, to minimize heat deformation, and to maximize structural stiffness to realize a high-precision machine. On the other hand, to be free from backlash and Coulomb friction, the monolithic structure is the best design.

(2). To improve the characteristics of the FCPKM, both a rational arrangement for each of its chain structure and an optimization for the corresponding structural parameters are necessary.

C. Motion and Loading

(1). *Requirement for workspace of the FCPKM.* One of the current concerns of FCPKMs is the reconciliation of submicronic repeatability with relatively large workspace. One is often obtained to the detriment to the other. Therefore, it is necessary to trade off between large workspace and high resolution under the condition of large enough workspace. On the other hand, considering the arithmetic inaccuracy, the uncompleted deflection in compliant joints, the undesired performance of the actuators, and other error accumulations, the actual workspace of the FCPKM is smaller than the expected one affirmatively. Therefore, in designing the FCPKM, it is very significant to optimize its parameters by taking the maximum workspace as an objective.

(2). *The positions and orientations of input forces applied by actuators and those of output force attached to end-effecor heavily influence the motion of a FCPKM.* It is not the case in its rigid-body counterpart. In rigid-body mechanisms, the motion is only dependent on the geometry of the links. Whereas the motion of a FCPKM is not only determined by the geometry of the rigid links, but by the placement and orientation of the input and output forces as well. For instance, if a force into a flexure hinge directly, it may lead to an undesired motion and stress existing in the compliant joints.

(3). *Anti-vibration capability of the FCPKM is a remarkable issue.* Generally, in order to minimize vibration of the end-effector of a FCPKM, a high structural frequency is required. Nevertheless, even though the body has a high structural frequency, vibration of the FCPKM is hardly avoided due to some external factors including environmental noise etc. In order to increase its anti-vibration capability, the FCPKM should be placed on the anti-vibrating table.

D. Machining Method

(1). *Try to make the FCPKM fully compliant and free from assembly.* It will be better if a FCPKM is built up in a single setup. In order to realize the monolithic machining, some methods including wire electrical discharge machining (EDM), rapid prototype machining, CNC milling, laser cutting, abrasive water jet cutting, stereo lithography, and SPM processing technology etc. are adopted. Actually, because of the difficulties in machining, some spatial FCPKMs are hardly machined monolithically. In this case, all components should be assembled and tightly fixed by means of solid pins in order to eliminate any possible relative motion.

(2). *Try to eliminate the residual stress produced in the course of manufacturing. In particular, the residual stress attached to compliant joints should be eliminated.* The residual stress attached to compliant joints, particularly the residual axial stress is an important factor of preventing the compliant joints from deflecting. In order to keep the motion of the compliant joint out of the range of the residual stress of material, therefore, the residual stress attached to compliant joints should be eliminated as possible. The reason has been described in proceeding section.

E. Actuation System

(1). *It is necessary to select a proper actuator.* To match the properties of the actuator with the characteristics of the mechanical structure in a FCPKM, the following criteria should be addressed: (i). A small shape and size, (ii). A suitable working stroke, (iii). A nanometer range motion resolution, (iv). A very fast response frequency. In addition, other performances such as speed, DOF, stiffness, motional linearity and cost, etc. are all the possible criteria to be considered as well. Thus Piezoelectric actuators, shape memory alloy, voice coil motor, inchworm motor, supersonic motor, and step motor can be employed in performing precise motions.

(2). *Because of nonlinearity of the high-resolution actuators, such as hysteresis and drift of piezoelectric actuators, closed-loop control is a necessity.*

(3). *In order to tradeoff high resolution and large stroke in the actuators for a FCPKM, an amplification device is usually required.*

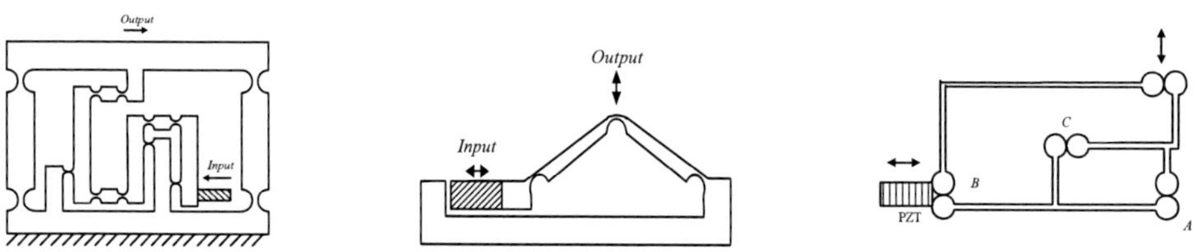

(a) Lever type (b) Frame type (c) mechanism type

Figure 33. Three types of displacement amplifiers.

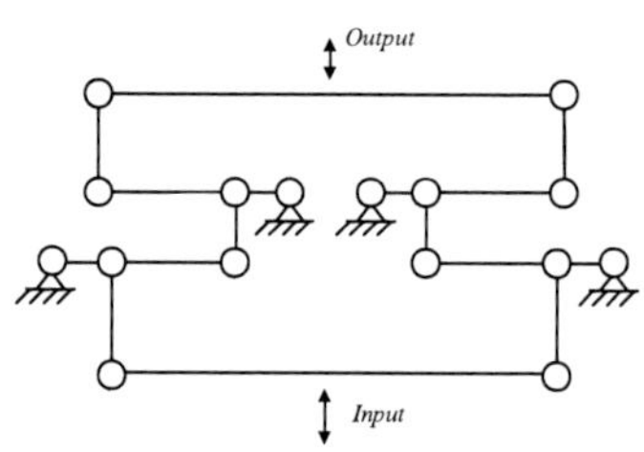

Figure 34. Schematic of amplifier

In order to overcome the shortcomings of a limited stroke commonly existed in a high-resolution actuator, a mechanical amplifier is required to be located between the manipulator structure and the actuator, especially for the applications where larger movements are demanded. Nevertheless, the desired displacement amplifier should also have a high resolution, and a high amplification factor or the mechanical advantage. Amplifying linkage topologies can be divided into three general classes: levers (figure 33a), frames (figure 33b) and mechanisms (figure 33c). A simple lever topology is selected because of its high efficiency, a frame topology can yields more compact design, and the utilization of some particular mechanism such as Scott-Russell mechanism [63] as the displacement amplifier is to realize a relatively large amplification times. As a transmission system, the displacement amplifier always damages a mount of energy to some extent. Material, structure, and flexure hinge are all the factors that may lead to the loss of energy. However, the loss should be minimized to maximize the efficiency of the displacement amplifier. Two methods can increase the efficiency of the displacement amplifier, one method is the use of a symmetrical structural design (figure 34), the other is to preload the displacement amplifier (figure 35). The former can eliminate the Y-directional parasitic error effectively, while the latter can eliminate the large stress existing in the rigid linkages.

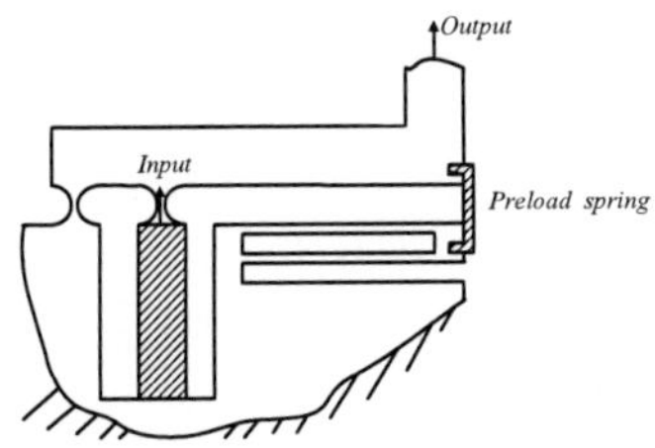

Figure 35. Preloading.

7.2. Design Methodology

During the phase of concept design of compliant mechanisms, also synthesis of the mechanism, a effective design method is definitely important for generating a creative design. Currently, these design methods can boil down to three categories: creative thinking stepped from designer's intuition and experience, systematic design method based on topological synthesis or PRBM, and a "nature" design method in terms of kinematic building block (figure 36).

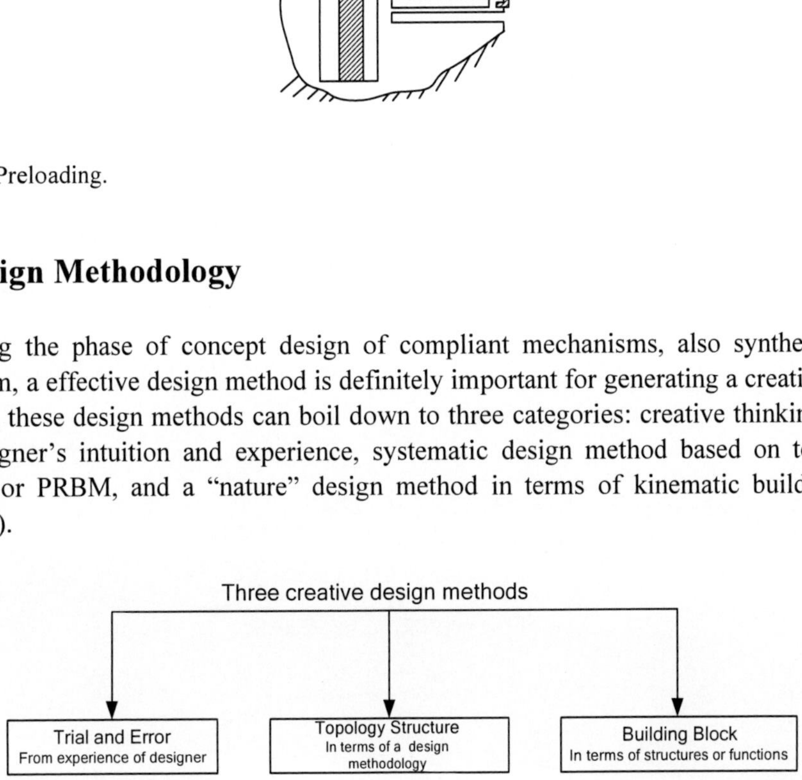

Figure 36. Three creative design methods.

There is a traditional approach that is based on creative thinking and intuition. Although this approach has resulted in several new topologies, it has its clear limitation. Due to the lack of an analytical basis and a systematic approach, only a percentage of the designs that are thus conceived have acceptable performance.

Typically, there are two approaches known in the literature for the systematic synthesis of compliant mechanisms: *Topology optimization* (continuum-based approach) or *Pseudo-Rigid-Body Model* (kinematics-based approach).

The continuum-based approach provides an optimal topology, shape and size of single piece compliant mechanisms for given deformation requirements. Optimal topology design methods such as homogenization methods, level-set methods and frame ground structure methods refer to the appropriate material distribution in a fixed region. This optimal material distribution is usually obtained with combined finite element analysis (FEA) for objective evaluation and optimization procedures for improvement. Although topology optimization is used at the conceptual stage of design, there is little actual input from the designer to guide the process. As a result, research in topology optimization has yielded varied technologically unrealistic structures.

On the other hand, the kinematics-based approach considers a compliant mechanism as an assembly of rigid links and compliant joints. This approach makes use of mechanism theories to design rigid links and flexible joints assembly (pseudo rigid-body model). The kinematics-based approach is natural for engineers in that it takes place within a more global design process and allows a more realistic design taking into account technological constraints. With the PRBM, however, systematic conceptual design has been limited to a graph theory akin to the corresponding methodology in traditional mechanism design. While graph theory effectively enumerates mechanisms, it fails to account for mechanism function.

Therefore, these two basic approaches can be seen as complementary.

A general PRBM kinematics-based approach has been listed in figure 37.

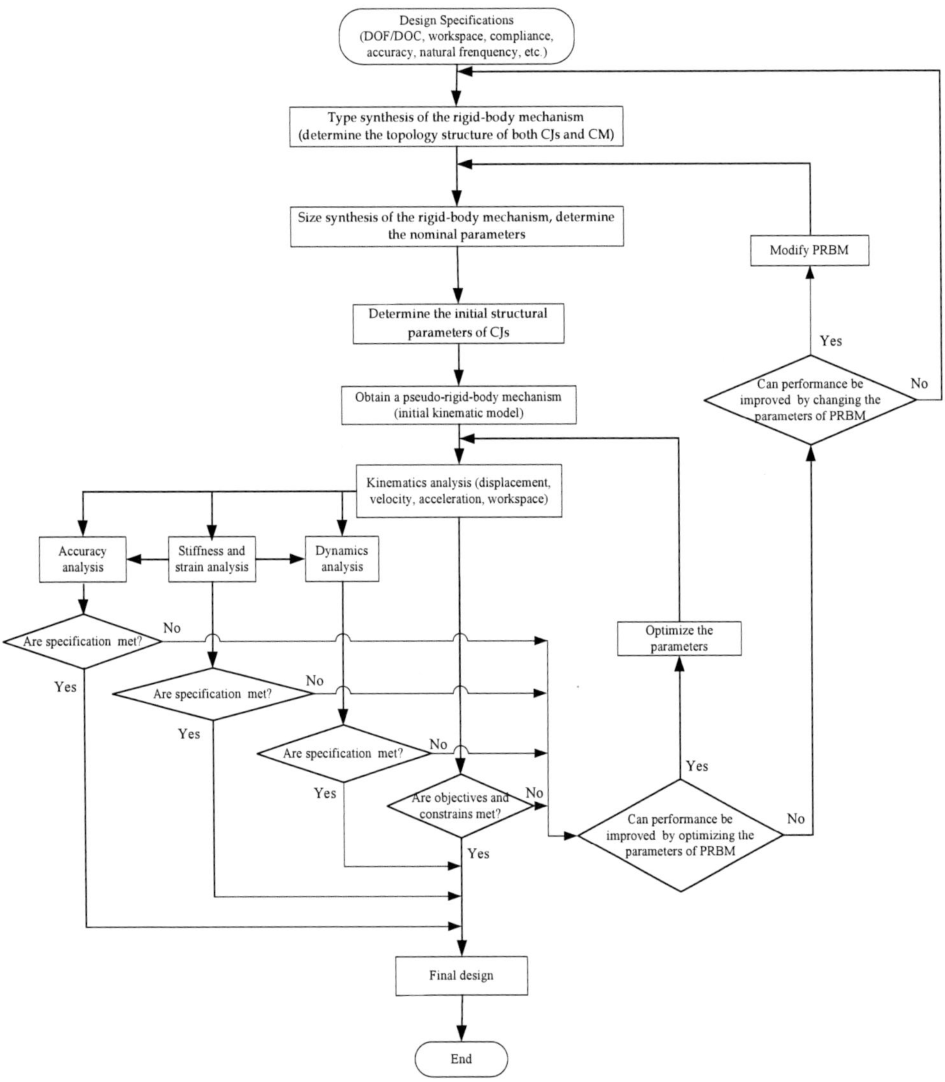

Figure 37. A flow chart for a compliant mechanism design process based on PRBM.

Kinematic building block method combines the advantages of both two kinds of approaches. This method considers a compliant mechanism as an assembly of compliant building blocks. The usage of blocks as elementary functional elements is a natural approach for engineers who can incorporate in their definition their own experience. While the generation of elementary functional elements may come from some analytical tools and optimization techniques, as systematic synthesis method does. In fact, an effective combination of creative thinking, aided with mathematic tools is probably an available recipe for a good FCPKM design.

8. Applications: An Automatic Packaging System for Waveguides

8.1. Background

In recent years, due to the increased demands for micro-optical-mechanical-electrical-systems (MOEMS) in telecommunications, automatically assembling equipments for such devices are receiving special attentions from both research and industrial communities. In the process of packaging, challenges are alignment and attachment of such components for they require submicron accuracy. Waveguide-to-waveguide alignment is one of the most important techniques to achieve high quality packaging for the fiber-optical components (See figure 38). Conventional alignment methods need people to participate, which may encounter some problems such as low yields, poor quality, etc. It is of great importance to develop a system that can package waveguides fully automatically.

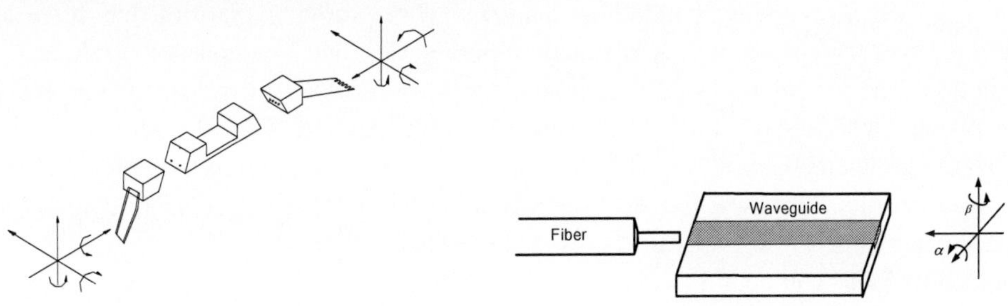

(a) Scheme diagram (b) DOF requirement

Figure 38. Scheme of waveguide-to-waveguide alignment.

For this purpose, an automated packaging system for waveguides (APSW) has been developed at the ARMS lab, which can package waveguides fully automatically. To do so, a high resolution vision system is set up to form a global vision servo system, which leads the motion control system to align waveguides coarsely, to get initial coupling light. In succession, micro light power servo system is employed to realize fine alignment between waveguides. Bonding process is implemented by bonding system. Whether in the process of coarse alignment or fine one, a macro-micro positioning stage is always a necessity.

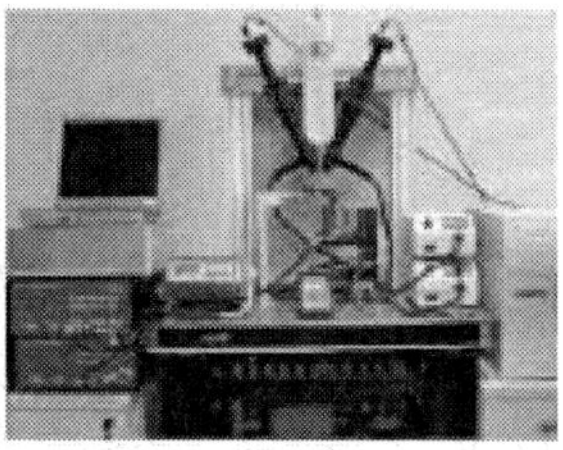

Figure 39. Prototype of APSW.

8.2. System Makeup

The APSW consists of 4 sub-systems: a positioning system, a high resolution stereo vision system, a micro light power servo system and a bonding system.

A. Macro-Micro Motion System

The coarse positioning system consists of 3-DOF motorized translational stage based on sliding-screw transmission and actuated by three stepper-motors (RM series of M24241S/D, RD-0234M as driver), the role of which is to cover a relatively large workspace area and in the meanwhile provide enough position accuracy in order to bring it into the field of the vision system and fine positioning system. The accuracy of this stage is 10μm along X, Y, Z with a resolution of 0.1μm. Due to the limited repeatability of the coarse manipulator, a fine motion stage is needed to fulfill the final alignment between waveguides.

Thus a 6-DOF serial-parallel FCPKM is designed as a fine positioning system. The top stage is a 3-RPS compliant parallel mechanism, whose moving platform has three spatial DOF, i.e. two rotations (around X axis and around Y axis) and one translation (along Z axis). The bottom one is a planar 3-RRR compliant parallel mechanism, whose movable plat has three planar DOF (move along X, Y axis and rotate around Z axis). Both of them have equilateral symmetry geometries, which make them free of temperature gradient. These two compliant parallel stages are connected in series through a tapered pin existing at the bottom stage and a tapered hole machined at the base plate of the top stage. Consequently, the combined FCPKM can obtain fully mobility (6 DOFs)

To achieve high precision, piezoelectric ceramic elements (PZT, PSt150/7/40 VS12) equipped with strain gauges as feedback sensors are selected as actuators of the fine positioning stage. HPV series high voltage amplifiers are chosen as drivers for low price and good performance. In addition, sensing plays an important role in the application of high-precision. As far as this system is concerned, sensing is done at the joint level through strain gauges attached to PZTs. However, these do not measure the real pose at the end-effector due to link and joint flexibility, and kinematic model errors. In this case, calibration becomes an inevitable process. By the calibration method addressed in reference [], calibration results show that the fine motion stage possess an accuracy of 50nms resolution along X, Y, Z, and 18 arcsec resolution around X, Y, Z, respectively, within the workspace of 129.4μm×141.6μm×130.5μm×18.3mrad×18.3mrad×18.3mrad.

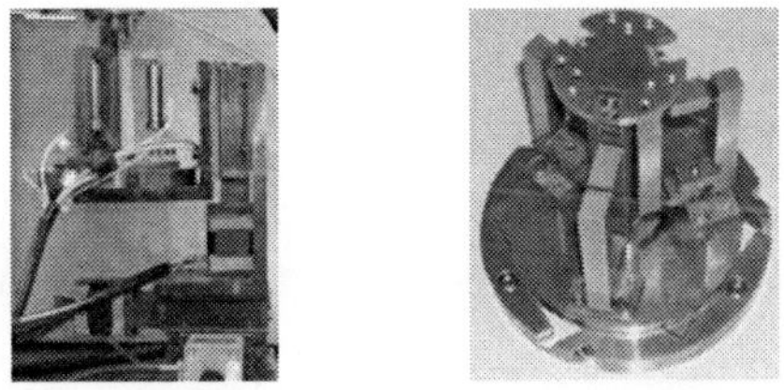

(a) Macro-micro system (b) 6-DOF FCPKM

Figure 40. Prototype.

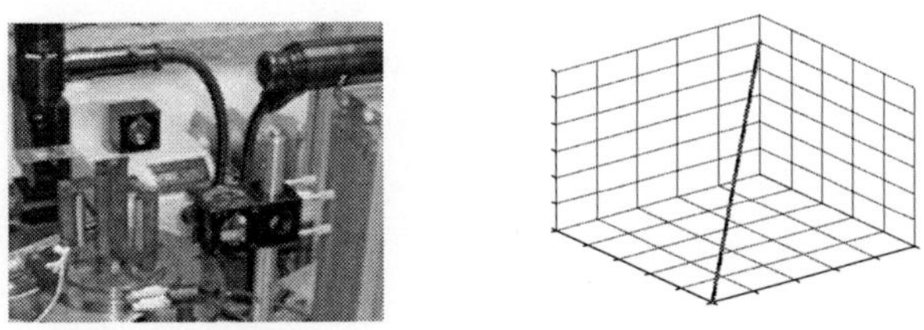

(a) Calibration (b) Line tracking

Figure 41. Verification experiment by means of a laser interferometer.

B. High Resolution Stereo Vision System

The detailed configuration of vision system is as follows: Optics: Z125C system and its corresponding parts (Thales-Optem, USA); High-intensity illuminator： VSI 220 (Thales-Optem company, USA); Camera: MVC1000 (MicroView company, China); vision-process PC: Pentium IV 2.40GHz CPU， ASUS mainboard, 512MB RAM, 80G hard disk.

A completely whole stereo vision system includes image capture, camera calibration, characteristic abstraction, stereo match, depth recovery and surface interpolation. Of all these issues, camera calibration and stereo match are of most challenge in the use of stereo vision system.

The objective of camera calibration is to determine the relationship among world coordinate, camera coordinate and image coordinate, thus to identify the corresponding parameters. Accuracy of calibration results will affect the succeeding image-process sharply. Stereo match is a key and difficulty technology as respect to a stereo vision system, especially for micro stereo vision system. It aims at establishing the mapping relationship between match units in a series of images. a stereo vision system in terms of integration of all above six parts can achieve the goal of detecting the spatial 3D information of the waveguides. The resolution of this vision system is about 2μm.

B. Micro Light Power Servo System

Micro light power servo system, which consists of a laser source, a light power meter, a search algorithm and so on, is used to adjust the relative pose differences between waveguides. As mentioned above, vision system is performed to instruct the motion system to align the waveguides coarsely first until the initial light coupling is achieved. Then, the control computer using a sophisticated motion control system executes the search algorithm to obtain the optimum coupling effect. The most important part of this system is the search algorithm. As a gradient based algorithm, hill-climbing method has the advantages of quickly ascending the optimum point and easy implementation. The motion control system depicted in the former part, which possesses repeatability and absolute position accuracy of 50nm, makes it possible for the climber to come extremely close to the theoretical peak thus resulting in an optimum optical power coupling efficiency, where is the optimum position of waveguides.

C. Bonding System

Bonding is absolutely a very vital issue in the process of packaging for it will determine the final quality of waveguides. In this system, the bonding part of Newport company's AutoAlignTM product is employed. After optimum position of waveguides is attained, a precise amount of adhesive will be injected. Then, a re-optimization process is performed to compensate for index changes. Finally, the UV illuminators are lit up for a certain time and the curing process ends.

8.2. Auto-Alignment Experiment

In order to test the feasibility of APSW, we carry out an experiment that packages a fiber to another fiber (fiber is the most simple waveguide) and packaging a waveguide to another waveguide, as illustrated in figure 42a. figure 42b and figure 42c are two photos. Image process computer calculate the relative position and orientation relations between them, which are listed in table 9 (based on the established world coordinate).

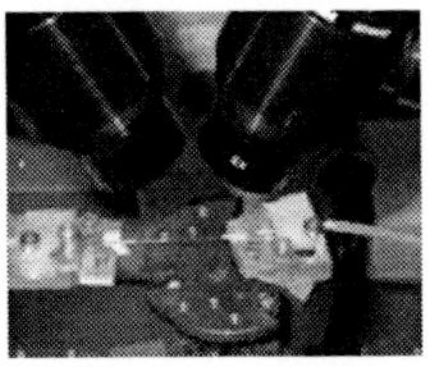

(a) packaging two fibers

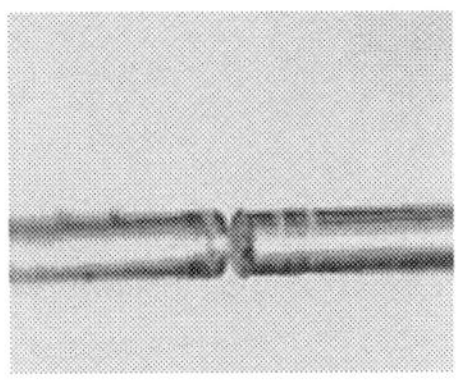

(b) Alignment of two fibres

(c) Alignment of two waveguides

Figure 42. Experiment of packaging two fibers or two waveguides.

Table 9. Pose difference between fibers

Position differences	Values	Orientation differences	Values
X	50 (μm)	α	0 (mrad)
Y	10 (μm)	β	7 (mrad)
Z	8 (μm)	γ	3 (mrad)

Using APSW, we carry out an experiment that packages a fiber to another fiber (fiber is the most simple waveguide), and packaging a waveguide to another waveguide. Experimental results show that packaging operation can be accomplished successfully (table 10). More important, all these operations are processed automatically.

Table 10. Values of light power (Unit: μW)

Initial value	Input value	After vision servo	After light power servo
0	570	21.4	564.7

9. Summary

This paper demonstrates a comprehensive design process for a general FCPKM, which covers topology and size synthesis of the FCPKMs based some modern geometrical tools such as the displacement Lie group, design of compliant joints especially large-stroke novel flexural pivots employed in the FCPKMs, kinetostatic analysis for the FCPKMs in terms of matrix method and screw theory, and design methodology for the general FCPKMs. At last, based on the proposed theoretical analysis and design method, three macro-scale compliant in-parallel stages are designed, and part of them has been used in a specified application as the high-precision micromanipulator for the automatic alignment of two fibers or two waveguides.

Acknowledgements

This work was supported by the National Natural Science Foundation of China under Grant No. 50405007, No. 50475002, No. 50675007, No. 50775007, and Beijing Nova Program (2006A13).

References

[1] I. Her. Methodology for compliant mechanism design, *Ph.D. Dissertation*, Purdue university, 1986

[2] I. Her, J.C. Chang. A Linear Scheme for the Displacement Analysis of Micropositioning Stages with Flexure Hinges, *Transactions of the ASME, Journal of Mechanical Design*, 1994, 116(3): 770-776

[3] M. Frecker, G. Ananthasuresh, S. Nishiwaki, *et al*. Topological synthesis of compliant mechanisms using multi-criteria optimization, *J. of mechanical design*, 119: 238-245,1997

[4] J.C. Hudgens, D. Tesar. A fully-parallel 6-DOF micromanipulator: kinematic analysis and dynamic model, *ASME Mechanical Conference*, 1988, DE-15(3), 29-37

[5] S.S. Bi and G.H. Zong. Accuracy analysis of the serial-parallel micromotion manipulator, *IEEE International Conference on System, Man and Cybernetics*, Orlando, USA, 1997, 2258-2263

[6] J.E. McInroy, J.C. Hamann. Design and control of flexure jointed hexapods, *IEEE Transactions on Robotics and Automation*, 2000,Vol.16, No.4, 372-381

[7] J. Hesselbach, N. Plitea, R. Thoben. Advanced technologies for micro assembly, *Proceedings of SPIE International Symposium on Intelligent Systems and Advanced Manufacturing*, Pittsburgh, USA, 1998, 178-190

[8] M.L. Culpepper, G. Anderson, P. Petri. HexFlex: a planar mechanism for six-axis manipulation and alignment. *Proceedings of the 17th Annual ASPE Meeting*, November 2002

[9] S. Chang, C. Tseng and H. Chien. An ultra-precision *XYθZ* piezo-micropositioner — Part I: Design and analysis, *IEEE Transactions on Ultrasonics, Ferrorelectrics and Frequency Control,* 1999, 46(4): 897–905

[10] W.J. Zhang, J. Zou, L.G. Watson, W. Zhao *et al.* The constant-Jacobian method for kinematics of a three-DOF planar micro-motion stage, *Journal of Robotic Systems*, 2002, Vol.19, No. 2, 63-72

[11] J.W. Ryu, S.Q. Lee, D.G. Gweon, K.S. Moon. Inverse kinematic modeling of a coupled flexure hinge mechanism, *Journal of Mechatronics*, 1999, No.9, 657-674

[12] K.M. Lee, S. Arjunan. A three-degrees-of-freedom micromotion in-parallel actuated manipulator, *IEEE Transactions on Robotics and Automation*, 1991,Vol.7, No.5, 634-641

[13] S.L. Canfield and J. Beard, Development of a spatial compliant manipulator, *International Journal of Robotics and Automation* 2002, **17**(1): 63–71

[14] Pernette, S. Henein, I. Magnani, and R. Clavel. Design of parallel robots in microrobots, *Robotica*, 1997, Vol.15, 417-420

[15] T. Tanikawa, T. Arai. Development of a micro-manipulation system having a two-fingered micro-hand, *IEEE Transactions on Robots and Automation*, 1999, Vol.15, No.1, 152-162

[16] J.P. Bacher, S. Bottinelli, J.M. Breguet, R. Clavel. Delta3: design and control of a flexure hinges mechanism, Microrobotics and Microassembly, *Proceedings of SPIE Vol. 4568*, 2001, 135-142

[17] J.J. Yu, Y.D. Hu, S.S. Bi, G.H. Zong, *et al.* Kinematics feature analysis of a 3 DOF in-parallel compliant mechanism for micro manipulation, *Chinese Journal of Mechanical Engineering*, 2004， Vol. 17, No. 1, 127-131

[18] X.Y. Tang, I-Ming Chen, and Q. Li. Design and nonlinear modeling of a large-displacement XYZ flexure parallel mechanism with decoupled kinematic structure, *Review of Scientific Instruments*, 2006, 77: 115101

[19] S. Awtar, and A.H. Slocum. Constant-based design of parallel kinematic XY flexure mechanisms, *ASME Journal of Mechanical Design*, 2007, 129: 816-830

[20] S.T. Smith. *Flexures: elements of elastic mechanisms*. New York: Gordon and Breach Science, 2000

[21] S. Awtar, and A.H. Slocum. Characteristics of beam-based flexure modules, *ASME Journal of Mechanical Design*, 2007, 129: 625-638

[22] J. Paros, L. Weisbord. How to design flexure hinges, *Machine Design*, 1965, Vol.37, No.27, 151-156

[23] N. Lobontiu. Compliant Mechanisms: Design of Flexure Hinges, CRC, Boca Raton, FL. 2003

[24] L.L. Howell. *Compliant mechanisms*, Wiley-interscience Publication, 2001

[25] J.J. Yu. Methodology for analysis and design of fully compliant robotic mechanisms, *Ph.D. Dissertation*, Beihang university(In Chinese), 2004

[26] S. Zhang, E.D. Fasse. A finite-element-based method to determine the spatial stiffness properties of a notch hinge, *Transactions of the ASME, Journal of Mechanical Design*. 2001, Vol.123, 141-147

[27] X.J. Liu, J.S. Wang, F. Gao, and L.P. Wang. On the design of 6-DOF parallel micro-motion manipulators, *Proceedings of the 2001 IEEE/RSJ International Conferences on Intelligent Robots and Systems*, 21 October–3 November 2001, 343

[28] J.J. Yu, S.S. Bi, and G.H. Zong. The dynamic characteristic-based optimal design of a fully compliant mechanism. *IFToMM2004,* April 1-4, 2004, Tianjin, China, 1416-1421

[29] Kawaja, F. Arai, T. Fukuda. Calibration for contact-type of micro-manipulation, Proceeding of the 1999 IEEE/RSJ International Conference on Intelligent Robots and Systems, 715-720

[30] J.E. McInroy. Dynamic modeling of flexure jointed hexapods for control purposes, *Proceedings of IEEE Conference of Control Applications*, Kona, HI, 1999, 508-513

[31] Kimball, L.W. Tsai. Modeling and batch fabrication of spatial micro-manipulators, *Proceedings of DETC'2000, ASME Design Engineering Technical Conferences*, Baltimore, USA, 2000

[32] K.H. Hunt. Structure kinematics of in-parallel actuated robot arms. *Journal of Mechanism, Transmissions and Automation in Design*, 1983, 105: 705-712

[33] L.W. Tsai. The enumeration of a class of three-DOF parallel manipulators. In *Proc. of the 10th World Congress on the Theory of Machine and Mechanisms*, Oulu, Finland, 1999, 1121-1126

[34] J.M. Hervé. Analyze structurelle des mécanismes par groupe des déplacements. *Mech. Mach. Theory*, 1978, 13: 437–450

[35] J. Meng, G.F. Liu, and Z.X. Li. A geometric theory for synthesis and analysis of sub-6 DoF parallel manipulators. *ICRA2005*, Barcelona, Spain, 2005, 2938-2943

[36] Q. Jin, and T.L. Yang. Theory for topology synthesis of parallel manipulators and its application to three-dimension-translation parallel manipulators, *ASME Journal of Mechanical Design*, 2004, 126(4): 625-639

[37] Gao, *et al*. New kinematic structures for 2-, 3-, 4-, and 5-DOF parallel manipulator designs. *Mechanism and Machine Theory*, 2002, 37: 1395-1411

[38] J.S. Dai, Z. Huang, and H. Lipkin. Screw system analysis of parallel mechanisms and applications to constraint and mobility study. *28th ASME Biennial Mechanisms and Robotics Conference*, September, Salt Lake City, DETC2004-57604, 2004

[39] Z. Huang, and Q.C. Li. General methodology for the type synthesis of lower-mobility symmetrical parallel manipulators and several novel manipulators, *International Journal of Robotics Research*, 2002, 21(2): 131-145

[40] Y.F. Fang, and L.W. Tsai. Structure synthesis of a class of 4-DoF and 5-DoF parallel manipulators with identical limb structures. *International Journal of Robotics Research*, 2002, 21(9): 799-810

[41] X. Kong, and C.M. Gosselin. A class of 3-DOF translational parallel manipulators with linear input-output equations. *Proc. of the Workshop on Fundamental Issues and Future Research Directions for Parallel Mechanisms and Manipulators*, Quebec, Canada, 2002, 25-32

[42] T.S. Zhao, J.S. Dai, and Z. Huang. Geometric Synthesis of Spatial Parallel Manipulators with Fewer Than Six Degrees of Freedom. *Journal of Mechanical Engineering Science*, 2002, 216(C12): 1175-1186

[43] J.J. Yu, S.S. Bi, G.H. Zong, W. Zhao. Research on type synthesis of micromanipulation mechanisms for bioengineering. *Journal of Beijing University of Aeronautics and Astronautics*(in Chinese), 27(3): 356-360

[44] J.J. Yu, J.S. Dai, X.J. Liu, S.S. Bi, *et al.* Type synthesis of 3-DOF orthogonal translational parallel manipulators based on atlas of DOF characteristic matrix, *ASME International DETC2006, 30th Biennial Mechanisms and Robotics Conference*, 2006

[45] Y.M. Moon B.P. Trease, S. Kota. Design of large-displacement compliant joints, *Prodeeding of DETC'02: 27th Biennial Mechanisms and Robotics Conference*, 2002

[46] B.D. Jensen, L.L. Howell. The modeling of cross-axis flexural pivots. *Mechanism and Machine Theory*, 2002, 37: 461-476

[47] M. Goldfarb and J. Speich. The development of a split-tube flexure, *Proceedings of the ASME Dynamics and Control Div.*, 2000, 861-866

[48] http://www.csem.ch/detailed/e211flexpivot.html

[49] K. Ragulskis, M. Arutunian, A. Kochikian, M. Pogosian. A study of fillet type flexure hinges and their optimal design, *Vibration Engineering*, 1989, 112, 447-452

[50] R.S. Ball. *A Treatise on the theory of screws*, Cambridge University Press, London, 1900

[51] T. Patterson, H. Lipkin. Structure of robot compliance, *ASME Journal of Mechanical Design*, 1993, Vol.115, No.3, 576-580

[52] M. Von. Motorrechnung: ein neues hilfsmittel in der mechanic, *zeitschrift fur angewandte mathematic und mechanic*, 1924, 4(2): 155-181.

[53] S. Zelenika, F.D. Bona. Analytical and experimental characteristics of high-precision flexural pivots subjected to lateral loads. *Journal of the International Societies for Precision Engineering and Nanotechnology*, 2002, 26: 381-388

[54] J.J. Yu, G.H. Zong, Z.W. Yu, and S.S. Bi. A new family of large-displacement flexural pivots, *ASME International DETC2007*, Las Vegas, USA.

[55] X. Pei, J.J. Yu, G.H. Zong, and S.S. Bi. The modeling of leaf-type isosceles-trapezoidal flexural pivots, *ASME International DETC2007*, Las Vegas, USA.

[56] L.W. Tsai, G.C. Walsh, and R.E. Stamper. Kinematics of a Novel three DOF Translational Platform, Proc. of the 1996 *IEEE Int. Conf. on Robotics and Automation (ICRA'1996)*, 1996, 3446-3451

[57] Y. Koseki, T. Tanikawa, T. Arai, *et al.* Kinematic analysis of translational 3-DOF micro parallel mechanism using matrix method, *Proc. IEEE/RSJ IROS*, 2000, 786-792

[58] J. Yu, S. Bi, and G. Zong, *et al.* A method to evaluate and calculate the mobility of a general compliant parallel manipulator, *Proc. of DETC'2004*, 743-748

[59] J.W. Ryu, D.G. Gweon, K.S. Moon. Optimal design of a flexure hinge based $XY\theta_z$ wafer stage, *Precision Engieering*, 1997, Vol. 21, 18-28

[60] R. Stoughton, T. Arai. Kinematic optimization of a chopsticks-type micromanipulator, *Proc. of the Japan U.S.A. Symposium on Flexible Automation*, San Francisco, 1992, 151-157

[61] Pernette, S. Henein, I. Magnani, *et al.* Design of parallel robots in microrobots, *Robotica*, Vol.15, pp.417-420, 1997

[62] M.D. Berglund, S.P. Magleby, L.L. Howell. Design Rules for Selecting and Designing Compliant Mechanisms for Rigid-Body Replacement Synthesis, *2000 ASME Design Engineering Technical Conferences (DETC'2000)*, 2000

[63] S.H. Chang, B.C. Du. A precision piezodriven micropositioner mechanism with large travel range, *Review of Scientific Instruments*, 1998, Vol. 69, No. 4, 1785-1791

In: Parallel Robotics: Recent Advances in Research... ISBN: 978-1-60456-859-2
Editors: Jinsong Wang and Xin-Jun Liu, pp. 217-251

Chapter 8

DESIGN OF RECONFIGURABLE PARALLEL KINEMATIC MACHINES

Xiaoqiang Tang* and Chunshen Lin
Institute of Manufacturing Engineering,
Department of Precision Instruments and Mechanology,
Tsinghua University, Beijing, 100084, P.R. of China

ABSTRACT

Parallel kinematic machine (PKM) is recognized as an important supplement of traditional serial mechanisms, and will play a key role in reconfigurable manufacturing system (RMS) since they can be quickly and efficiently reconfigured.

A reconfigurable PKM consists of an inventory of interchangeable modules. This chapter presents a kind of structural design of reconfigurable PKMs. The reconfigurable PKM is established in either telescopic drive form or railway linear drive form with the same modules but different assembly modes. In the architecture of the PKM, an angle-variable module is designed to change the setting-angle in X and Z orientations for working more efficiently. Due to the modularity and reconfigurability, a PKM constructed in this mode can have difference structures, degree of freedoms (DOFs), and performance. In the workspace analysis of the PKMs with two drive modes, inverse kinematics solution is given, and the effect of setting-angle of spherical joint on PKM has been investigated and the spherical joint setting-angle selection method has been proposed. A comparison result is supplied to show the difference of workspace between these two kinds of PKMs.

Reconfiguration planning is recognized as an important factor for reducing the cost in manufacturing reconfigurable products, and the associated main task is to generate a set of optimal or approximate optimal reconfiguration sequences by using some effect algorithms. After structure design of PKMs, a method of modularizing and integrating is proposed to design modules. Modules' interface information is described by the combination of interface status number and undirected graph. Based on this description, a Petri Net is applied to the modeling reconfiguration process in reconfiguration task planning. A reconfiguration cost matrix(RCM) is used to figure out dynamic cost in

*E-mail: tang-xq@mail.tsinghua.edu.cn

reconfiguration. Then a one-direction search algorithm and a two-direction search algorithm, which combine graph-subgraph recognition and heuristic algorithms, are presented to search optimal reconfiguration path aiming at minimizing total cost.

In experiment section, reconfiguration and cutting experiments of several reconfigurable PKMs are presented to verify the theory and method provided in this chapter. The results of two searching algorithms are compared with other ones. The same structure searching algorithm is realized by Visual C++.

1. INTRODUCTION

In recent years, changing manufacturing environment caused by intensive competition on a global scale and rapid changes in process technology requires an efficient production system. Therefore, a reconfigurable manufacturing system characterized by modularity, integrability, customization, convertibility and diagnosability, which can be easily reconfigured and response to unpredictable market changes cost-effectively, is needed. The earliest large-scale research of reconfigurable manufacture systems can be found in literature was probably the FMC project launched in 1977 by MITI in Japan. Other reconfiguration projects such as MOSYN, KERNEL and MOTION were focused on modular machine tools. After the foundation of the Engineering Research Center of Reconfigurable Machining Systems(ERC/RMS) at the University of Michigan in 1996, researchers began to develop the technology and theory of reconfigurable manufacturing systems systematically[1,2].

Parallel kinematic machines (PKMs) have many advantages and have been used in machine industry widely. Benefited from its modular structure, PKM is more reconfigurable than the traditional serial machine. With the modular design, several modes of PKMs can be integrated into a module set so that the modules can be reassembled rapidly according to the application.

Today, the research on PKMs has made some substantive progress but there still exist some problems that should be resolved for its further application. The workspace is one of important indices for analyzing and designing the PKM. However, few works have been done for considering the composite effect between workspace and parameters of the PKM, including the setting-angle of spherical joint, which coming from limited allow-half-cone-angle of spherical joint.

Four main research objectives of reconfigurable manufacturing system were proposed by Koren[1] for further systematic research: system level design for RMS, reconfigurable machine design, reconfigurable control design in open-architecture environment, and ramp-up time reduction for new and reconfigured systems. In all of these processes, reconfiguration task planning should be recognized as an important factor for reducing the cost in reconfiguration, and the associated main task is generating a set of optimal or near-optimal reconfiguration sequences by using some effect algorithms. However, few literatures concerned about reconfiguration task planning. The reason is, at least in part, that the systematic research of reconfiguration is just developed recently.

Reconfiguration is the combination processes of assembly and disassembly, and the research of reconfiguration in RMS is mainly based on some results of processes consequently. However, there are some differences between reconfiguration and assembly-disassembly in task planning. They are

(1) The main target of disassembly task planning is primary for material recycle and less for direct reuse; while the objective of disassembly in reconfiguration is reusing parts and sub-assemblies maximally to re-assembly a new configuration.
(2) The objective of assembly task planning is assembling final product with every part, by controlling assembly cost and ensuring product function; the assembly in reconfiguration is assembling according to disassembly result, while assembly unit may be some sub-assemblies.
(3) Since there are interfaces exchange of modules in reconfiguration other than pure assembly or disassembly, a prompt representation method and a task planning strategy are needed in modules reconfiguration.

This chapter is organized as follows: the reconfigurable structure design is presented in section 2 and the corresponding kinematic analysis is introduced in section 3. In section 4, the reconfiguration representation of reconfigurable product is presented. In section 5, two reconfiguration sequence heuristic searching algorithms are proposed. In section 6, reconfigurable PKM example is used to compare several task planning algorithms, and section 7 draws some conclusions of this chapter.

2. Reconfigurable Structure Design

The structure of reconfigurable PKMs is similar to that of some ordinary PKMs. They consist of common parts like: a base platform, a mobile platform on which ending executor is installed, and some parallel-actuated extendible legs or movable legs between the mobile and base platforms. Universal joints are mounted both on the mobile platform and base platform to connect platforms and legs, and provide multi-position and versatile attitude[3].

The workspace of a PKM is determined by the dimension and configuration of these common parts, so are some other performances such as precision, rigidity, and dynamics. When working conditions are changed, reconfigurable PKMs will adjust either dimension or configuration of their parts quickly to meet the new performance requirement. This reconfiguration should be achieved through modular design, so that every module can be easily replaced by another one according to the parameter and configuration needed.

The main contents of modular design of reconfigurable PKMs include strut modules, base platform modules, mobile platform modules, joint modules, and their interfaces.

2.1. Strut Modules

According to driving mode, the PKM can be mainly cataloged into two kinds: telescopic driving PKM, whose struts are extendable with one end mounted on framework or base platform and another on the mobile platform, and linear driving PKM, whose struts have constant length with one end mounted on slider and another on the mobile platform.

Both of the two kinds of PKMs have their advantage and disadvantage. Therefore, they are generally designed to apply in respective cases: the linear driving PKM is inclined to be adopted in machining long-shape parts due to its dominant movement direction, and the telescopic driving PKM is suitable to machine more complicated shape other than long-shape.

The strut module includes 3 sub-modules: strut driving module, joint module, and angle-variable module:

(1) Strut driving module. The strut driving modules provide motion for mechanism. The driving mode has two categories: telescopic driving and linear driving.

(2) Joint module

In PKMs, the 1- to 3-DOF joints are usually used. The following table shows some joints.

Table 1. Joints

DOF	Joint Type
1	R, P, H
2	T
3	S

(3) Angle-variable module

The angle-variable modules are installed at the end of the struts. The joints setup vector can modified through the reconfiguration of these modules.

To meet the requirement of the rapidly changed market, machine tools are requested to possess reconfigurable capacity. A modular strut structure is designed to realize the reconfiguration of the two kinds of struts.

According to Wang [4], shown in figure 1, the strut driving module of a reconfigurable PKM consists of a work-piece module, a constant-length rod, two spherical joint modules, three rod-fix modules, two angle-variable modules, and a linear drive module.

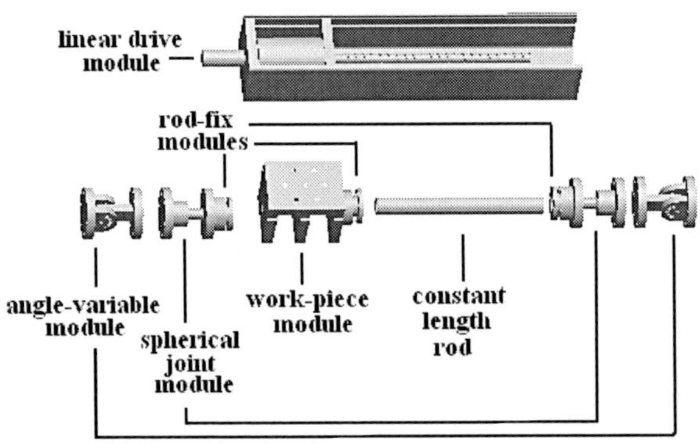

Figure 1. Components of the module strut.

The linear drive module provides conversion of rotary motion to linear translation, which including a railway, a screw-nut, a screw, a coupling and a motor, illustrated in figure 2. At the one end of the railway, a protrusion with the same outside diameter as that of constant-length rod is used to mount the linear module to one rod-fix module.

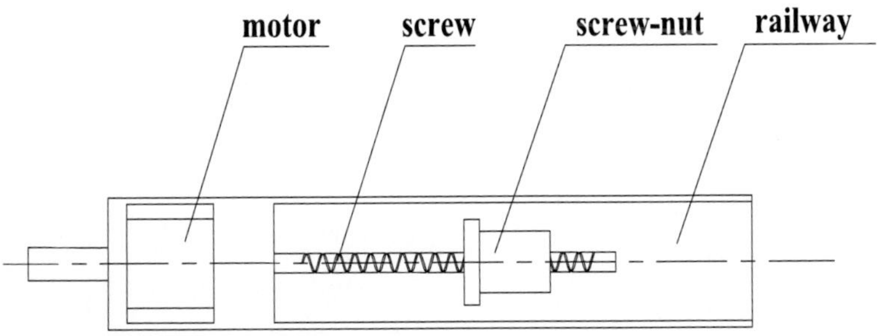

Figure 2. Sketch of linear driving module.

There are three interfaces in the work-piece module which is shown in figure 3: interface 1 is used to connect a linear drive module across its screw-nut, interface 2 to connect an angle-variable module, and interface 3 to connect a rod-fix module.

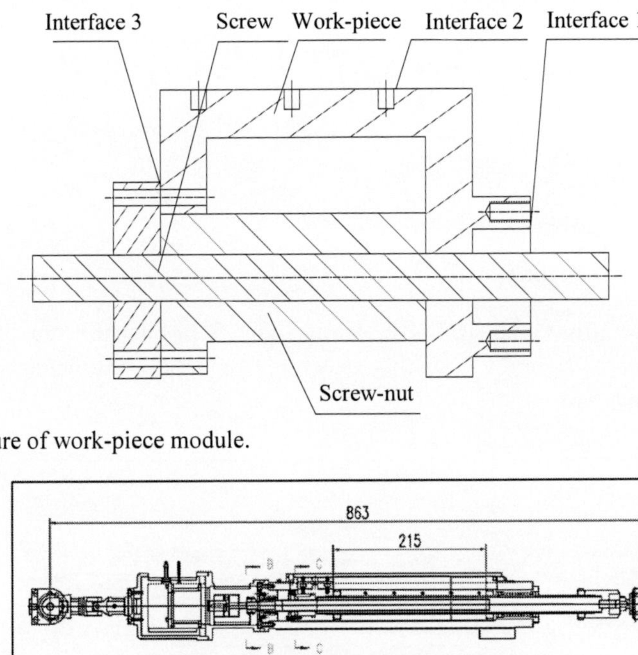

Figure 3. Structure of work-piece module.

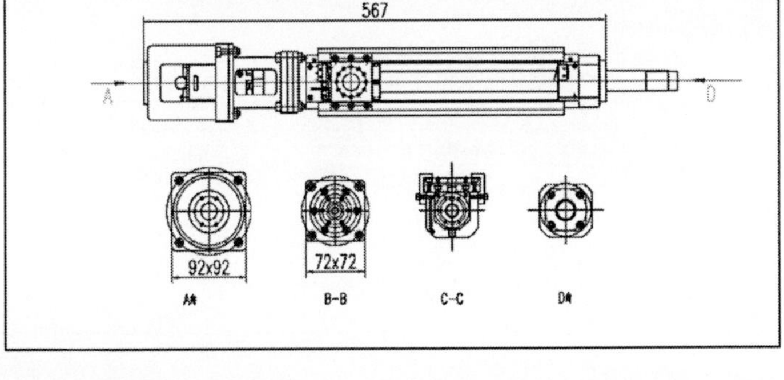

Figure 4. Reconfigurable strut.

Figure 4 is a reconfigurable strut drawn by AutoCAD software, and figure 5 and figure 6 show the real struts of telescopic driving type and linear driving type used in reconfigurable PKMs.

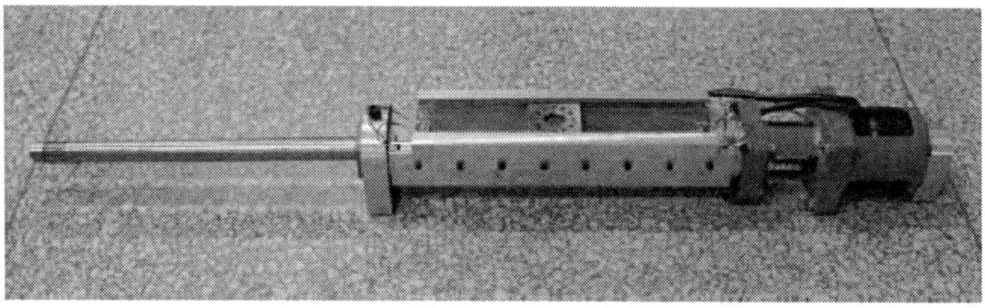

Figure 5. Real strut of telescopic type.

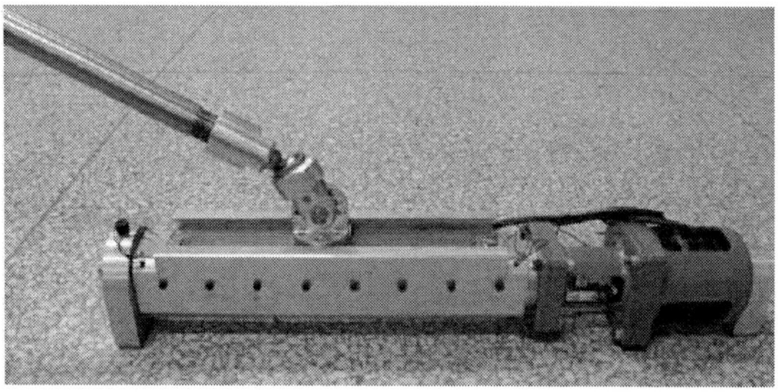

Figure 6. Real strut of linear type.

Another type reconfigurable strut is shown in figure 7 In this strut, the motor axis vector is reverse to the one in figure 4 . By this structure, the ratio of traveling distance to total length of strut will increase.

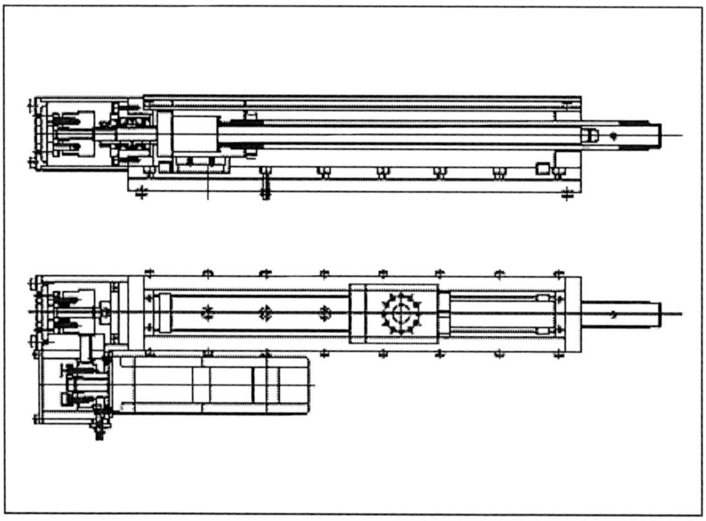

Figure 7. Another type reconfigurable strut.

Figure 8 and Figure 9 show the real struts of telescopic driving type and linear driving type used in reconfigurable PKMs.

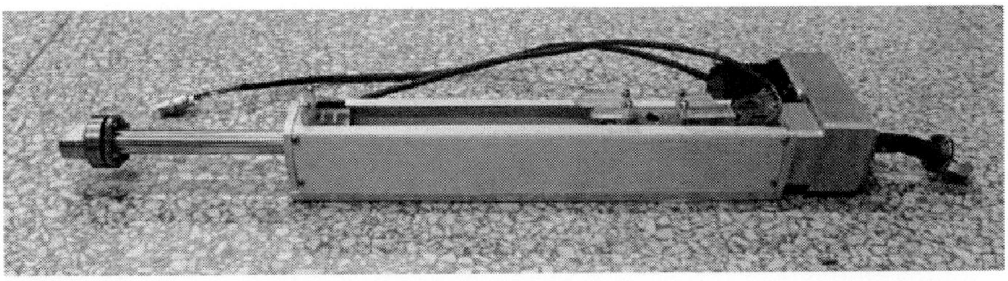

Figure 8. Real strut of telescopic type.

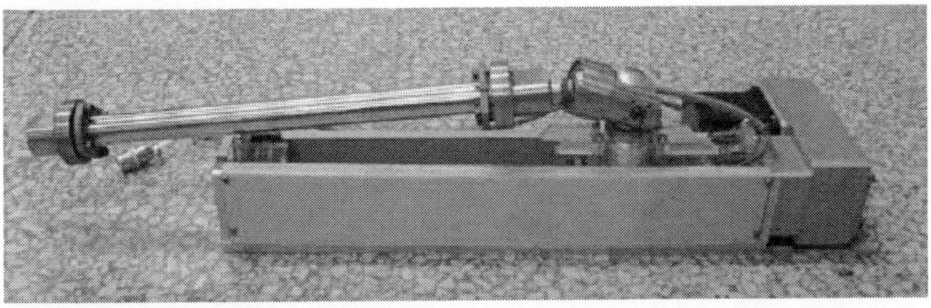

Figure 9. Real strut of linear type.

2.2. Base Platform Module Design

The base platform module is the base of reconfigurable PKM and the struts are installed on it through some interfaces. Figure 10 shows a type base platform.

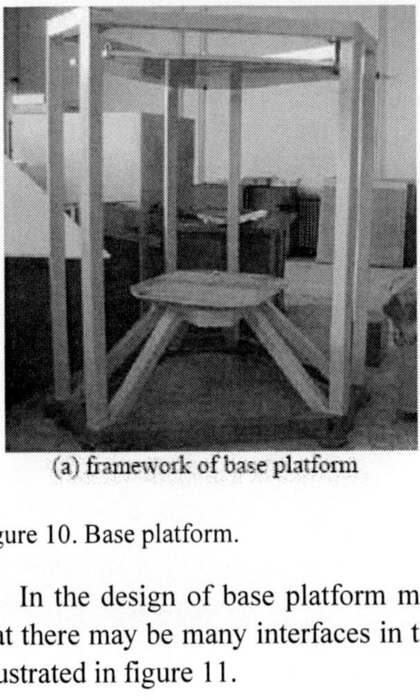

(a) framework of base platform

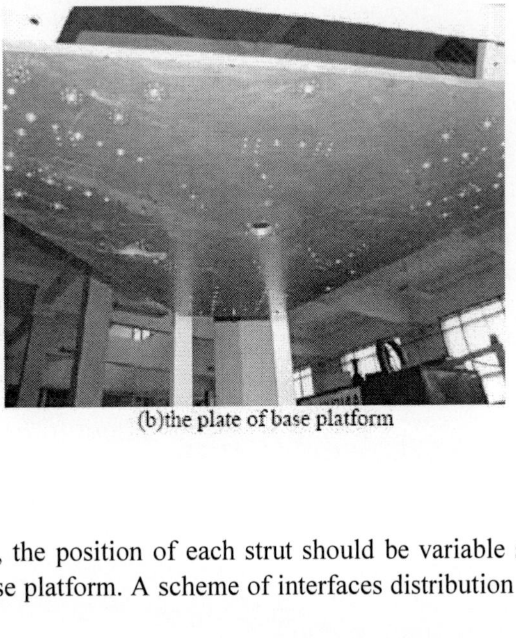

(b)the plate of base platform

Figure 10. Base platform.

In the design of base platform module, the position of each strut should be variable so that there may be many interfaces in the base platform. A scheme of interfaces distribution is illustrated in figure 11.

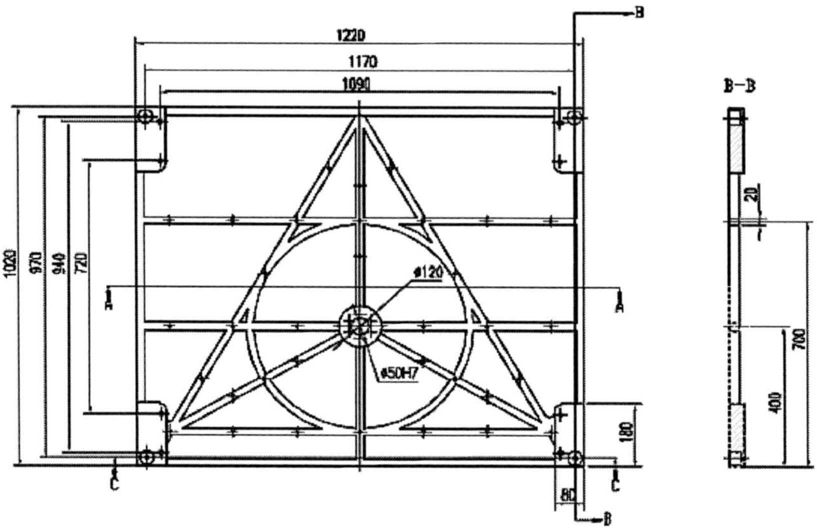

Figure 11. Distribution of interfaces.

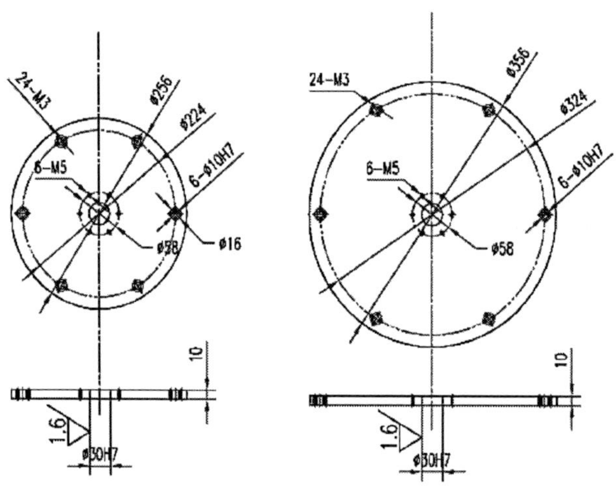

Figure 12. Mobile platforms.

2.3. Mobile Platform Module Design

One of the ends of a strut is fixed in mobile platform. With similar to the base platform, the radius of struts distribution circle is variable by different mobile platform, which shown in figure 12.

2.4. Joint Module Design

In reconfigurable PKM, a universal joint can work with 2 DOFs, and a spherical joint with 3 DOFs. With classic manufacturing process, it is difficult to fabricate a high precision spherical joint. To resolve this problem, a universal joint can work with both 2 DOFs and 3 DOFs is designed, as shown in figure 13.

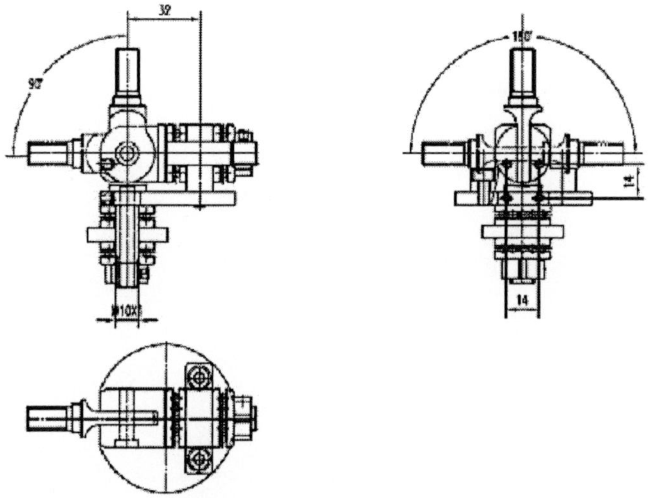

Figure 13. Spherical joint for reconfigurable PKMs.

2.5. Reconfiguration Issues[5-11]

In modular structure of strut, the interface connecting a spherical joint to a rod-fix module has the same connection mode and geometry dimension as the interface connecting the work-piece module to a rod-fix module. This is important for the reconfiguration.

There are two connect modes to realize operation of telescopic driving and linear driving:

In figure 14, the telescopic drive mode is represented: at one end of the linear drive module, a rod-fix module(7), a spherical joint module(5), and an angle-variable module(6) with its position being fixed are in turn mounted; at the other end, the work-piece module(4) is connected to linear drive module across the screw-nut, a rod-fix module(8) is connected to the work-piece module across the interface alone the direction of the screw of linear drive module, then a constant-length rod module(2), a rod-fix module(9), a spherical joint module(1), and an angle-variable module(10) are in turn mounted.

In figure 15, linear drive mode is represented as following: the linear drive module is fixed, the work-piece module(3) is connected to linear drive module(4) across the screw-nut, a angle-variable module(6) is connected to the work-piece module across the interface alone the direction perpendicular to the screw of linear drive module, then a spherical joint module(5), a rod-fix module(8), a const-length rod module(2), a rod-fix module (7), a spherical joint module(1) , and an angle-variable module (10) are in turn mounted. Compared with the former mode, a rod-fix module is saved for this mode.

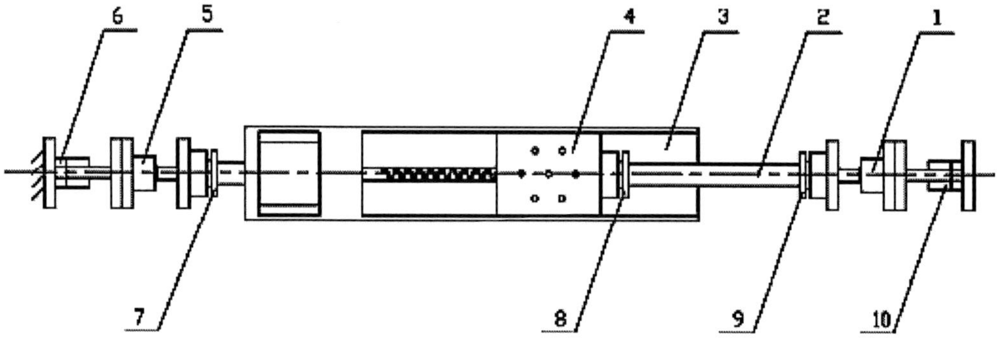

Figure 14. Schematics of telescopic drive mode.

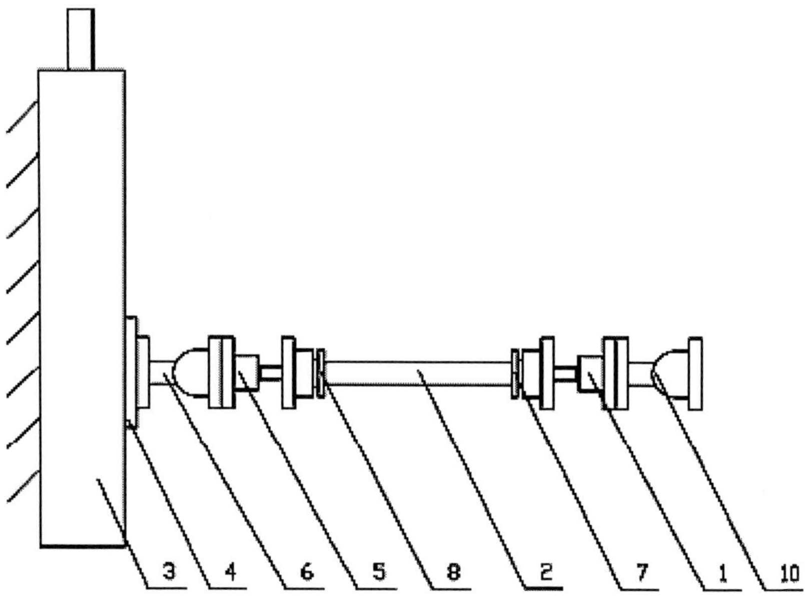

Figure 15. Schematics of linear drive mode.

The angle-variable module provides the function to change the setting-angle at some certain values. It has a movable part and a fixed part. As presented in figure 16, rotation of the fixed part can realize the rotation of setting-angle in the Z orientation, and switching the connect holes between the movable part and the fixed part of the angle-variable module can realize the rotation of setting-angle in the X orientation.

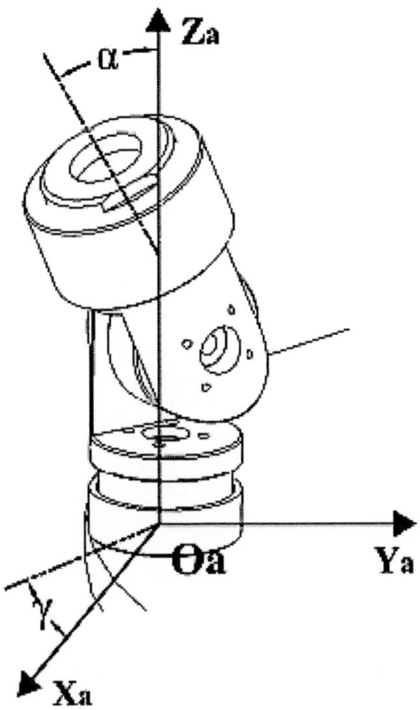

Figure 16. Angle-variable module.

Reconfigurable PKM is assembled by those modules introduced above. Figure 17 and figure 18 show two Pro/Engineer 3D models of the reconfigurable PKM using the linear drives and telescopic drives respectively.

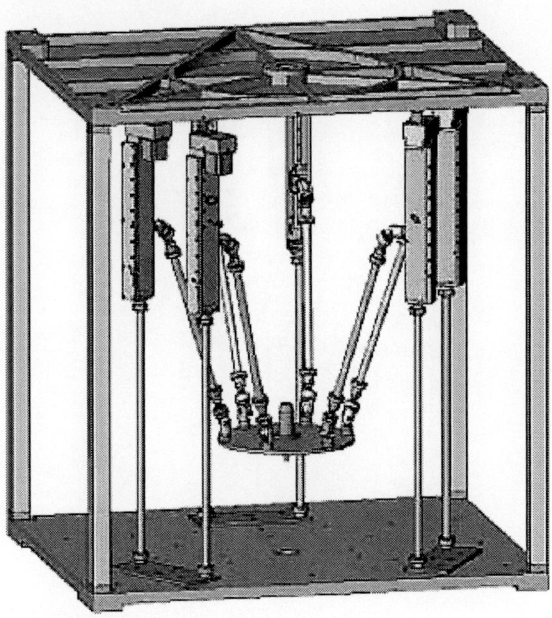

Figure 17. 3D Model of PKM with linear struts.

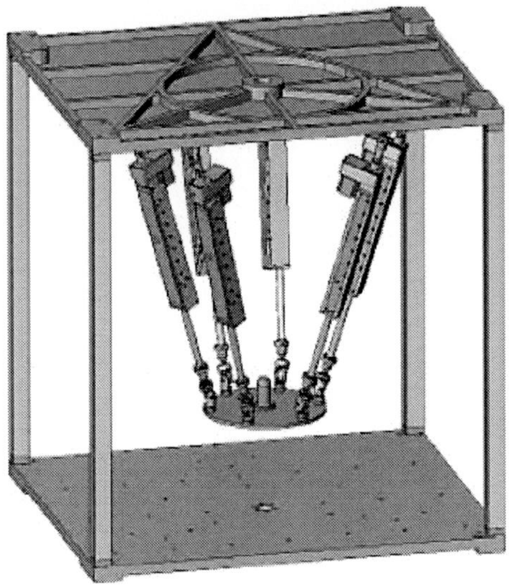

Figure 18. 3D Model of PKM with telescopic struts.

3. Kinematics Analysis

Inverse kinematics is an important factor for workspace analysis. The schematic diagram of these two kinds of PKMs aforementioned in this paper is shown in figure 19, where both PKMs consist of base platform ($\mathbf{B}_1\mathbf{B}_2\mathbf{B}_3\mathbf{B}_4\mathbf{B}_5\mathbf{B}_6$), mobile platform ($\mathbf{A}_1\mathbf{A}_2\mathbf{A}_3\mathbf{A}_4\mathbf{A}_5\mathbf{A}_6$), struts ($\mathbf{C}_i\mathbf{A}_i$ $,i=1,2,...,6$) and struts, as shown in figure 19.

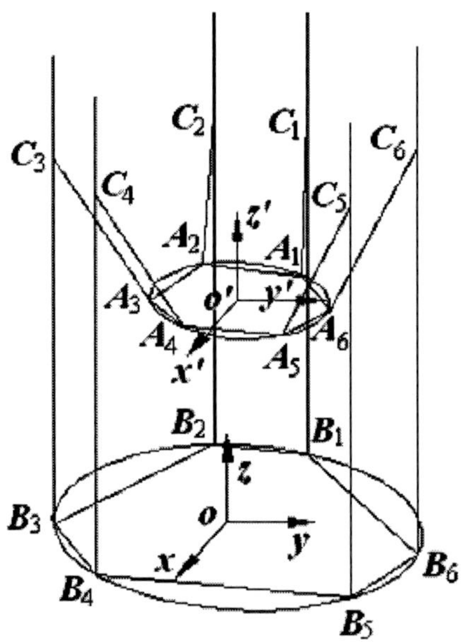

Figure 19. Schematics of PKM.

The coordinate system $\Re : o-xyz$ is fixed to the base platform $\mathbf{B_1B_2B_3B_4B_5B_6}$ where o is the center of the base platform, and another system $\Re^{'}: o'-x'y'z'$ is fixed to the mobile platform $\mathbf{A_1A_2A_3A_4A_5A_6}$ where o' is the center of the mobile platform. $\mathbf{B_1}$, $\mathbf{B_2}$, $\mathbf{B_3}$, $\mathbf{B_4}$, $\mathbf{B_5}$, and $\mathbf{B_6}$ are six vertices of hexagon whose circumscribed circle's radius is presented by R. Points $\mathbf{A_1}$, $\mathbf{A_2}$, $\mathbf{A_3}$, $\mathbf{A_4}$ $\mathbf{A_5}$, and $\mathbf{A_6}$ are six vertices of equilateral hexagon whose circumscribed circle's radius is presented by r. Vectors $[\mathbf{A}_i]$ $(i=1,2,...,6)$ are defined as the position vectors of points $\mathbf{A_i}$ with vectors $[\mathbf{B_i}]$ $(i=1,2,...,6)$ as the position vectors of points $\mathbf{B_i}$. The geometric parameters of the machine tool are $\mathbf{A_iB_i}=\mathbf{L_i}$ $(i=1,2,...,6)$.

The position of point o' in the fixed frame $\Re$ is denoted by the vector

$$[\mathbf{T}]_{\boldsymbol{R}}=[x,y,z]^{\mathrm{T}} \tag{1}$$

The rotation of coordinate system $\Re^{'}$ relative to $\Re$ can be expressed by the matrix

$$\mathbf{R}=\begin{bmatrix} c\phi c\theta & c\phi s\theta s\varphi - s\phi c\varphi & c\phi s\theta c\varphi + s\phi s\varphi \\ s\phi c\theta & s\phi s\theta s\varphi + c\phi c\varphi & s\phi c\theta c\varphi - c\phi s\varphi \\ -s\theta & c\theta s\varphi & c\theta c\varphi \end{bmatrix} \tag{2}$$

The vectors of $[\mathbf{B}_i]$ in the base platform coordinate system $\Re$ can be written as equations

$$\begin{aligned}
[\mathbf{B_1}]_{\Re} &= \left[-R\cdot s\left(\frac{\pi}{2}-\frac{\theta}{2}\right), R\cdot c\left(\frac{\pi}{2}-\frac{\theta}{2}\right), z_1\right]^{\mathrm{T}} \\
[\mathbf{B_2}]_{\Re} &= \left[-R\cdot s\left(\frac{\pi}{2}+\frac{\theta}{2}\right), R\cdot c\left(\frac{\pi}{2}+\frac{\theta}{2}\right), z_2\right]^{\mathrm{T}} \\
[\mathbf{B_3}]_{\Re} &= \left[-R\cdot s\left(\frac{\pi}{2}-\frac{\theta}{2}+\frac{2}{3}\pi\right), R\cdot c\left(\frac{\pi}{2}-\frac{\theta}{2}+\frac{2}{3}\pi\right), z_3\right]^{\mathrm{T}} \\
[\mathbf{B_4}]_{\Re} &= \left[-R\cdot s\left(\frac{\pi}{2}+\frac{\theta}{2}+\frac{2}{3}\pi\right), R\cdot c\left(\frac{\pi}{2}+\frac{\theta}{2}+\frac{2}{3}\pi\right), z_4\right]^{\mathrm{T}} \\
[\mathbf{B_5}]_{\Re} &= \left[-R\cdot s\left(\frac{\pi}{2}-\frac{\theta}{2}+\frac{4}{3}\pi\right), R\cdot c\left(\frac{\pi}{2}-\frac{\theta}{2}+\frac{4}{3}\pi\right), z_5\right]^{\mathrm{T}} \\
[\mathbf{B_6}]_{\Re} &= \left[-R\cdot s\left(\frac{\pi}{2}+\frac{\theta}{2}+\frac{4}{3}\pi\right), R\cdot c\left(\frac{\pi}{2}+\frac{\theta}{2}+\frac{4}{3}\pi\right), z_6\right]^{\mathrm{T}}
\end{aligned} \tag{3}$$

where $z_1 = z_2 = \ldots = z_6 = z_0$ if in the PKM with telescopic drives.

The vectors of $\left[\mathbf{A}_i\right]$ in the mobile platform coordinate system $\Re$ can be written as

$$[\mathbf{A}_\mathrm{i}]_{\Re'} = \left[-r \cdot s\left(\frac{i}{3}\pi\right), r \cdot c\left(\frac{i}{3}\pi\right), 0\right]^\mathrm{T}, \ (i = 1, 2, \ldots, 6) \tag{4}$$

Therefore, the vectors of $\left[\mathbf{A}_i\right]$ has the expression in the coordinate system $\Re$ as follows

$$[\mathbf{A}_\mathrm{i}]_{\Re} = R[\mathbf{A}_\mathrm{i}]_{\Re'} + [\mathbf{T}]_{\Re}, \ (i = 1, 2, \ldots, 6) \tag{5}$$

The relationship between vectors $\left[\boldsymbol{A}_i\right], \left[\boldsymbol{B}_i\right]$ and struts vectors has the following form according to the geometric constraint of the architecture:

$$\left\|[\mathbf{B}_\mathrm{i}]_{\Re} - [\mathbf{A}_\mathrm{i}]_{\Re}\right\| = \left\|\mathbf{L}_\mathrm{i}\right\| = l_i, \ (i = 1, 2, \ldots, 6) \tag{6}$$

where $l_1 = l_2 = \ldots = l_6 = l_0$ as the PKM with linear drives.

The inverse kinematics problem of the manipulator can be solved by Eq. (5), where the unknown items are l_i ($i = 1, 2, \ldots, 6$) for the PKM with telescopic drives and z_i ($i = 1, 2, \ldots, 6$) for the PKM with linear drives.

3.1. Workspace Analysis and Comparison

Workspace of a PKM is a work region which the tool can reach, and is an important indication of the PKM performance. In this section, the parameter used to represent workspace of PKM is its volume.

A numeric method is applied to calculate the PKM reachable points to integrate workspace surface. To simplify analysis, the 3 rotation angles of the mobile platform are all set to zero in this calculation.

The classic workspaces of the two kinds of PKMs are shown in the figures 20 and 21. The geometry of architecture corresponding to Figs. 20 and 21 is listed in table 2.

In this section, the effect of radius of mobile platform and strut length, spherical joint setting-angle will be discussed. As a matter of fact, the two drive modes have same moving distance. In this reason, the following relation exists in the workspace analysis.

$$l_{max} - l_{min} = S_{max} - S_{min} \tag{7}$$

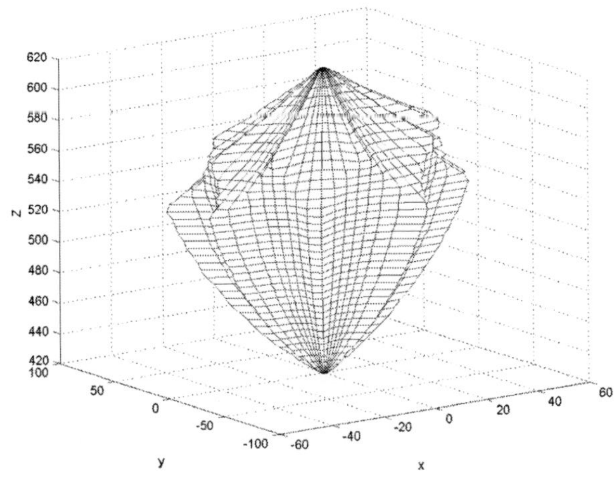

Figure 20. Workspace shape of PKM with linear struts.

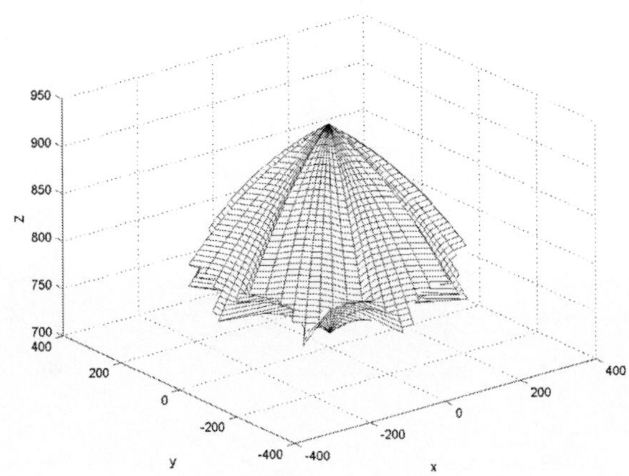

Figure 21. Workspace shape of PKM with telescopic struts.

Table 2. Some Design Variables

Design Variable	Value	Description
R (mm)	500	Radius of base platform
r (mm)	150	Radius of mobile platform
PKM with telescopic drives		
l_{min} (mm)	800	Maximal length of strut
l_{max} (mm)	1000	Minimal length of strut

Table 2. Some Design Variables (Continued)

PKM with linear drives		
S_{min} (mm)	600	Maximal height of slider
S_{max} (mm)	800	Minimal height of slider
l (mm)	600	Length of strut

3.2. Effect of Radius of Mobile Platform

For the purpose of comparing these two configurations, a series of constant-length rods are used. In the linear strut, the strut length is about the length of const-length rod, however, in the telescopic strut, the maximal strut length is the sum of lengths of railway and the rod. Therefore, if a series of constant-length rods have the value (mm):400, 450, 500, 550 and 600, then the corresponding length ranges of struts in the telescopic drives are (mm): 800-1000, 850-1050, 900-1000, 950-1150 and 1000-1200.

Figure 22 shows the relationship between volume of workspace and radius of mobile platform at several strut length ranges in the PKM with telescopic drives.

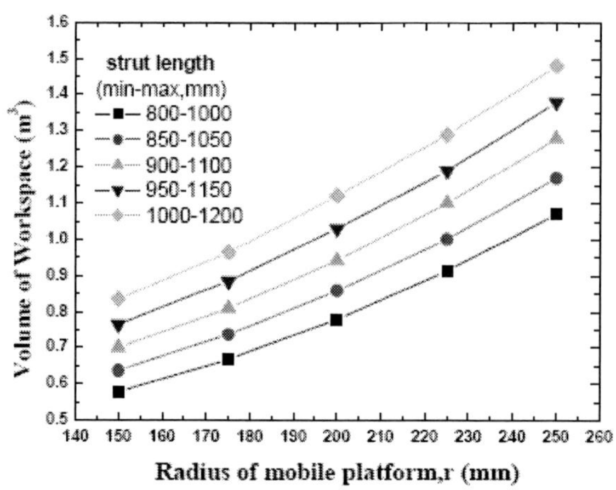

Figure 22. Effect of radius of mobile platform at several strut length ranges in PKM with telescopic drives.

From the figure 22 it is clear that the volume of workspace increases as the radius of mobile platform or the strut length increases.

Figure 23 illustrates the relationship between volume of workspace and radius of mobile platform at several strut lengths.

From the diagram presented in figure 23, the volume of workspace increases as the radius of mobile platform or the strut length increases, but the volume increases more rapidly at large strut length.

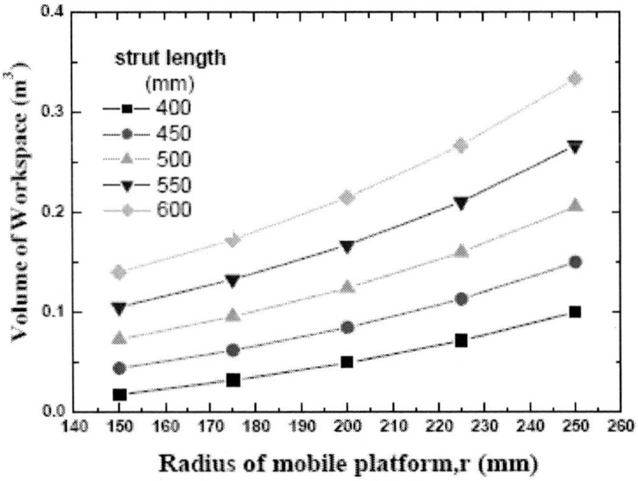

Figure 23. effect of radius of mobile platform at several strut lengths in PKM with linear drives.

Compared to the figure 22, it can be observed that the volume of workspace in PKM with linear drives is smaller than the one with telescopic drives.

3.3. Effect of Spherical Joint Properties

Actually, the spherical joint will constrain the strut's rotation so that it always yields the workspace volume to an extent. Spherical joint setting-angle and allow-half-conic-angle should be taken into account in evaluation of workspace parameters. Figure 24 presents the angle-variable module distribution on the mobile platform.

Figure 24. Angle-variable modules on mobile platform.

In every strut, there are two angle-variable modules to change the spherical joint setting-angle. In theory, the spherical joint's effect on the workspace will be least if the setting angle has same vector as the strut. So by traversing the two angle-variable modules combination which is more close to the strut vector, the practical workspace can be explored.

In this study, allow-half-conic-angle is assumed to 30°. Table 2 shows an example of a PKM with linear drives for the effect of setting-angle on workspace. The geometry data are still those in table 2. In table 3, the values of *Rotation of Z axis* are zeros for this 6-axis architecture for optimal results and the initial status of *Rotation of Z axis* is shown in figure

24 and figure 19. These values would be non-zero in the analysis of other possible assembly models.

From table 3, according to the marked cell, the 30° for the first X-angle and 45° for the second X-angle are chosen for optimal performance in workspace.

Table 3. The relationship between workspace volume and spherical joint setting-angle

Angle-variable module on mobile platform		Angle-variable module on strut		Volume of workspace
Rotaion of X axis(°)	Rotation of Z axis (°)	Rotaion of X axis(°)	Rotation of Z axis (°)	(mm3)
15	0	15	0	4.712E7
30	0	15	0	5.001E7
45	0	15	0	0
15	0	30	0	6.100E7
30	0	30	0	1.400E8
45	0	30	0	5.005E7
15	0	45	0	6.100E7
30	0	45	0	1.401E8
45	0	45	0	1.400E8
If spherical joint is not considered				1.402E8

Table 4. the relationship between workspace volume and spherical joint setting-angle

Angle-variable module on mobile platform		Angle-variable module on strut		Volume of workspace
Rotation of X axis (°)	Rotation of Z axis (°)	Rotation of X axis (°)	Rotation of Z axis (°)	(mm3)
15	0	15	0	0
30	0	15	0	0
45	0	15	0	0
15	0	30	0	9.922E5
30	0	30	0	9.922E5
45	0	30	0	9.922E5
15	0	45	0	3.803E8
30	0	45	0	3.803E8
45	0	45	0	2.632E8
If spherical joint is not considered				5.785E8

Table 4 shows an example of a PKM with telescopic drives for the effect of setting-angle on workspace. The geometry data are as same as table 2 and the values of *Rotation of Z axis* are zero.

From table 4, according to the marked cell, the 15° or 30° for the first X-angle and 45° for the second X-angle are chosen for optimal performance in workspace. Compared to the table 3, it can be observed that the decrease extent in volume of workspace resulting from spherical joint in the PKM with linear drives is more rapidly than in the PKM with telescopic drives.

4. Reconfiguration Representation

The task planning of reconfiguration should be based on the representation. This section summaries previous works about similar spectrum in assembly and disassembly, and develops a reconfiguration state-representation method based on Petri net.

4.1. Reconfiguration State Representation

In the representation of an assembly system proposed by Cao and Sanderson [12], an object in the system can be defined as either a component, or a subassembly, or a final assembly or product. Based on the characteristics of feasible objects, three types of basic high-level operations are defined: assembling, disassembling, and internal state transiting [12]. Suppose that an assembly system consists of n components, $C_1, C_2, ..., C_n$ where C_i represents the ith component. $S(C_i)$ is used to represent the state of component C_i at a given time, assuming a discrete time representation. Each component may occupy a fixed number. So system state represented by integer vectors is:

$$\underline{S}_j = \left(s_j(C_1) s_j(C_2) L s_j(C_n)\right)^T, 0 \le j \le M \tag{8}$$

where $M+1$ is all available states number.

Assembly, disassembly and reconfiguration can be represented as internal state transiting. The process of transiting, or sequences of assembly, disassembly or reconfiguration, is a set of queue state vectors in math. When the system is changed from one sub-state to another, the definition of sub-state is as:

$$\underline{S}_j' = \left(s_j(C_{p1}) s_j(C_{p2}) \cdots s_j(C_{pr})\right)^T \tag{9}$$

where $\left(s_j(C_{p1}) s_j(C_{p2}) \cdots s_j(C_{pr})\right) \subseteq \left(s_j(C_1) s_j(C_2) \cdots s_j(C_n)\right)$.

Three kinds of state-definitions of assembly, disassembly and reconfiguration are as follows:

(1) Assembly: One set of components or subassemblies are combined with or put in geometric contact with one or more other sets of components or subassemblies.

$$A\left(\left\{O_{i_1},O_{i_2},\ldots,O_{i_u}\right\}\right)=\left\{\left\{C_{j_1},C_{j_2},\ldots,C_{j_r}\right\}\right\} \tag{10}$$

where $O_{i_k}=\left\{C_1^{(i_k)},C_2^{(i_k)},\ldots,C_{n_k}^{(i_k)}\right\}, O_{i_k}\subseteq\left\{C_{j_1},C_{j_2},\ldots,C_{j_r}\right\}, 1\le k\le u$, and $O_{ik_1}\cap O_{ik_2}=\varnothing$, $1\le k_1,k_2\le u$

(2) Disassembly: an assembly or a subassembly is separated into a set of components or subassemblies.

$$D\left(\left\{\left\{C_{j_1},C_{j_2},\ldots,C_{j_r}\right\}\right\}\right)=\left(\left\{O_{i_1},O_{i_2},\ldots,O_{i_u}\right\}\right) \tag{11}$$

where $O_{i_k}=\left\{C_1^{(i_k)},C_2^{(i_k)},\ldots,C_{n_k}^{(i_k)}\right\}, O_{i_k}\subseteq\left\{C_{j_1},C_{j_2},\ldots,C_{j_r}\right\}, 1\le k\le u$, and

$O_{ik_1}\cap O_{ik_2}=\varnothing, 1\le k_1,k_2\le u$

(3) Reconfiguration: The internal state of a set of components is modified by changing the relative geometric relations of the components, or by modifying a property of a component subset in this set

$$R\left(\left\{O_p\right\}\right)=\left\{O_q\right\} \tag{12}$$

where $O_p=\left\{s\left(C_{i_1}\right),s\left(C_{i2}\right),\cdots,s\left(C_{i_s}\right),s\left(C_{i_{s+1}}\right),\cdots,s\left(C_{i_{s+d}}\right),\cdots,s\left(C_{i_r}\right)\right\}$,

$O_q=\left\{s\left(C_{i_1}\right),s\left(C_{i2}\right),\cdots,s'\left(C_{i_s}\right),s'\left(C_{i_{s+1}}\right),\cdots,s'\left(C_{i_{s+d}}\right),\cdots,s\left(C_{i_r}\right)\right\}$, and

$O_p-O_q=\left\{s\left(C_{i_s}\right),s\left(C_{i_{s+1}}\right),\cdots,s\left(C_{i_{s+d}}\right)\right\}, O_q-O_p=\left\{s'\left(C_{i_s}\right),s'\left(C_{i_{s+1}}\right),\cdots,s'\left(C_{i_{s+d}}\right)\right\}$.

In the description of three operations proposed by Cao and Sanderson [12], all possible states are defined by a binary arrays, which represent whether a set of parts and sub-assemblies are mated or separated with another set of parts and sub-assemblies. However, in design of reconfigurable systems, modules are connected through interfaces. Some modules have many interfaces, and final product might achieve different function according to which interfaces it used. As illustrated in figure 25, the assembly is composed of three parts – A, B and C, part A has only one interface, Part B has three and C has two. There is only connection type between parts A and B, where two between parts B and C because there are two interfaces between parts B and C.

Following Cao and Sanderson's rule, in State 1(S1), since parts A, B and C are not connected with each other, it is denoted as $A^{(0)}B^{(0)}C^{(0)}$; in State 2(S2) and State 3(S3), since parts A, B and C are connected together, it is denoted as $A^{(1)}B^{(1)}C^{(1)}$, although S2 and S3 have different connection modes. Therefore, the representing rules proposed by Cao and Sanderson[12] will not work in condition of figure 25.

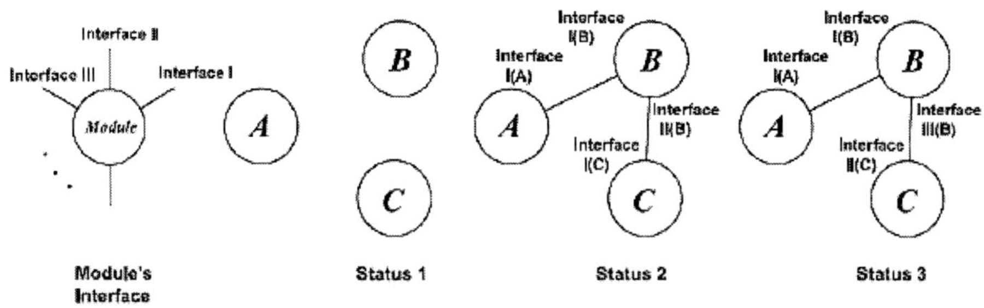

Figure 25. Three parts reconfiguration in Cao's rule.

To solve this problem, a method is proposed as the supplement for Cao's rule, and the steps of the new rule are as follows.

(1) Find all interfaces in every module.
(2) In some system configurations, perhaps more than 1 interface be used in a certain module, so get interface-sets which will be used at one time numbered, like $1,2,\cdots,n$.
(3) Set module state symbol according to its interface-sets used, for example, if module F is not used, its state symbol is $F^{(0)}$; if the No. 1 interface-set is used, its state symbol is $F^{(1)}$, if the No. 2 interface-set is used, its state symbol is $F^{(2)}$, ..., if the No. n interface-set is used, its state symbol is $F^{(n)}$
(4) The process of switching from interface-set n_1 to n_2 as system from State 1 to State 2, can be divided into three sub-processes: disassembly on interface-set n_1, switching from interface-set n_1 to n_2, assembly on interface-set n_2. When economic factor is considered, the cost in interface switching process should be counted. In fact, the sub-process switching between interface-sets is a virtual process only for math representation, and its cost can be ignored, for example, the part unit B switching between two of $B^{(0)}$, $B^{(1)}$, and $B^{(2)}$ in table 5.

Table 5. Assembly state of three parts by new rule

Part/Assembly	State	Description
A	Unused	$A^{(0)}$
	Used	$A^{(1)}$
B	Unused	$B^{(0)}$
	Interface 1 and 2 used	$B^{(1)}$
	Interface 1 and 3 used	$B^{(2)}$
C	Unused	$C^{(0)}$
	Interface 1 used	$C^{(1)}$
	Interface 2 used	$C^{(2)}$
ABC	Not connected (S1)	$A^{(0)}B^{(0)}C^{(0)}$
	Connected (S2)	$A^{(1)}B^{(1)}C^{(1)}$
	Connected (S3)	$A^{(1)}B^{(2)}C^{(2)}$

Every module uses only one interface-set in a system state, so that state of a module is determined not only by whether used state but also by connection mode. The system state in figure 15 can be described in table 5 by this improved rule.

Reconfiguration Petri Net

Petri net is used widely in understanding, representing and reasoning assembly and disassembly process. A Petri net model will be proposed in this section for task planning research.[13]

A Petri net is described as a 5-tuple directed bipartite graph, $RPN = (P, T, I, O, m_0)$, in which

(1) $P = \{p_1, \cdots, p_n\}$ is a finite set of places, where $n > 0$ and n is the place number. In reconfiguration Petri net, places stand for part, sub-assembly and assembly.
(2) $T = \{t_1, \cdots, t_m\}$ is a finite set of transitions, where $m > 0$ and m is for transition number. Transitions stand for processes of assembly and disassembly in system reconfiguration. Every transition has a reconfiguration cost factor $c_i\ (i = 1, 2, \cdots)$, which decided by time, difficulty, tool used, power consumption, environment impact of assembly and disassembly in reconfiguration.
(3) $I : P \times T$ is the input function, a mapping from transition to bags of places.
(4) $O : T \times P$ is the output function, a mapping from places to bags of transitions.
(5) The initial marking is $m_0(p_1) = 1$ and $m_0(p) = 0$, $\forall p \in P - \{p_1\}$; $m_f(p_{END}) = 1$ and the final marking is $m_f(p) = 0, \forall p \in P - \{p_{END}\}$.

When system state is switched from State 2 to State 3, the process can be decomposed into disassembling State 2 to parts and assembling them to State 3. Accordingly, the reconfiguration Petri net is integration of the assembly and disassembly Petri net under the new representing rule. Figure 26(a) is disassembly Petri net from State 2 to parts, and figure 26 (b) is assembly Petri net from parts to State 3.

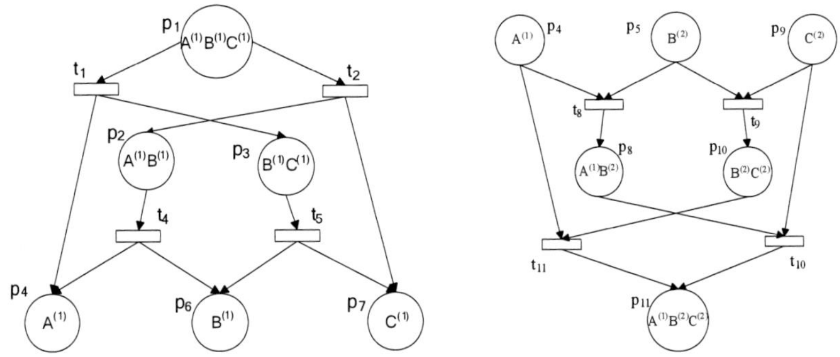

(a) Disassembly Petri net (b) Assembly Petri net

Figure 26. Disassembly Petri net of State 2 and Assembly Petri net of State 3. (* Pn stands for places, tn stands for transitions).

To create the reconfiguration Petri net, firstly, we combine the same units from the bottom of disassembly Petri net and the top of assembly Petri net. In the example in figure 27 - State 2($A^{(1)}$ $B^{(1)}$ $C^{(1)}$) switching to State 3($A^{(1)}B^{(2)}C^{(2)}$), we establish disassembly Petri net of State 2, and assembly one of State 3, and find $A^{(1)}$ is the same unit in two Petri nets, then delete one and redirect companying arc ending point to the same one. Secondly, find the same unit with different state, like $B^{(1)}$ and $B^{(2)}$, and put a transition between them. The final reconfiguration Petri net is shown as figure 27.

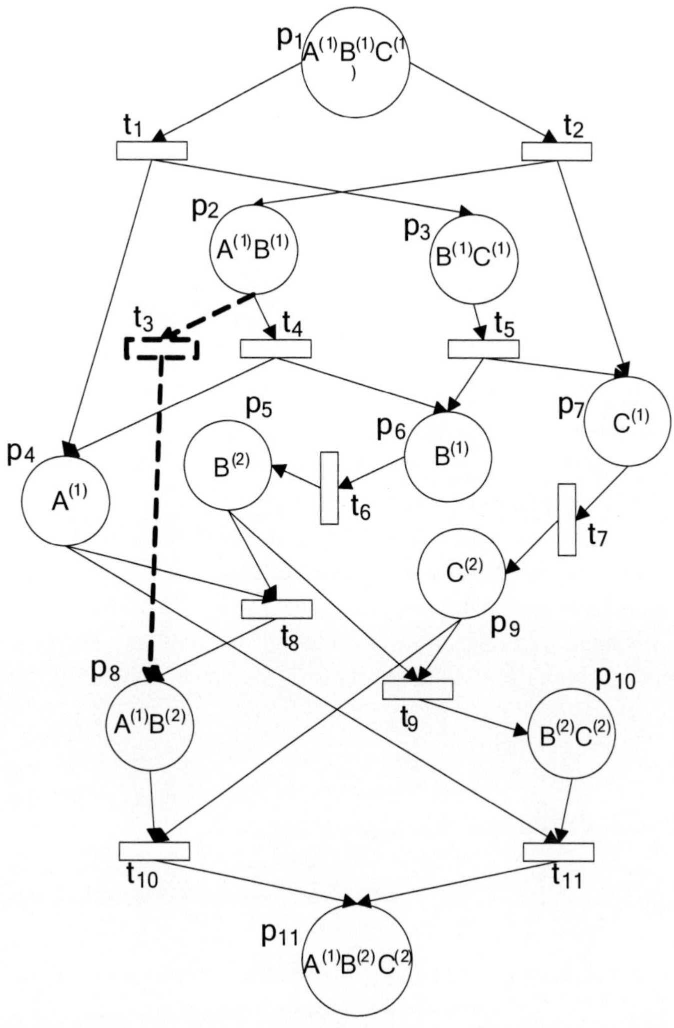

Figure 27. Reconfiguration Petri net from State 2 to State 3.

In figure 27, P_1 to P_{11} are parts, sub-assemblies and assemblies; t_1 to t_{11} are processes in assembly and disassembly, where t_3, t_6 and t_7 are interface-set switching processes. The dashed transition and arcs of reconfiguration Petri net is not a definite one, but it is preferred to be found in search algorithm to improve reconfiguration efficiency.

4.3. Reconfiguration Cost Matrix

Every transition in reconfiguration Petri net has a reconfiguration cost factor $c_i\left(i=1,2,\cdots\right)$. Cost factor is decided by engineering experience, filed condition, etc. In fact, transition firing sequence may cause changing in some cost factors, therefore these factors are arranged in a Reconfiguration Cost Matrix(RCM) to be used in total cost evaluation, whose element RCM_{ij} is the cost of transition j firing after transition i. For example, the costs of transition t_8 and t_9 may be different depending on whether t_4 or t_5 is their previous transition. The RCM should be $RCM=\begin{bmatrix} c_{48} & c_{49} \\ c_{58} & c_{59} \end{bmatrix}$.

5. Reconfiguration Sequence Searching Algorithms

If reconfiguration Petri net is constructed, all possible characteristics of the system can be fully analyzed. An optimal sequence is obtained by generation the reachability graph and finding the optimal path from the initial state to the final state. The path is represented by transition firing sequence of reconfiguration Petri net. Even for a simple Petri net, the reachability graph may be too large to generate in its entirety. A heuristic search algorithm is supposed to be employed instead of blindness search.

Generation of optimal or near-optimal sequence is an important point in this study. In reconfigurable systems, direct reuse of units is an effective means to reduce cost, accelerating product distribution, and decreasing environment impact. Therefore, in the reconfiguration Petri net, there are same parts or sub-assemblies in former disassembly and assembly Petri nets. The search algorithms are designed to find the same units to generate more optimal sequence.

5.1. One-Direction Searching Algorithm

An one-direction search algorithm is listed below, and the flowchart is shown in figure 28.

Step 1. Construct an empty list TLIST, put the initial marking m_0 in it.

Step 2. Construct an empty list SLIST to hold sequence.

Step 3. Define a cost function $c\left(m\right)=p\left(m\right)+h\left(m\right)$, this function will be explained later.

Step 4. If TLIST is empty, then terminate program with failure.

Step 5. Move first marking m from TLIST to SLIST, and check

a) Whether exist same marking in SLIST, if it does, delete all marks between this two same markings, including this mark.

b) Whether exist marking just different in interface-set number, if it does, delete all marks between this two same marks.

Step 6. If m is target marking, the get sequence from first mark and terminate.

Step 7. Find the enabled transitions of the marking m . The dashed transition in figure 3 is possible enabled, but it remains unknown before it is found.

Step 8. Generate successor marking, that is, successor node m^* , set its pointer to parent node m . For every successor m^* of m , calculate $h(m^*)$ and $c(m^*)$, then put m^* in the TLIST.

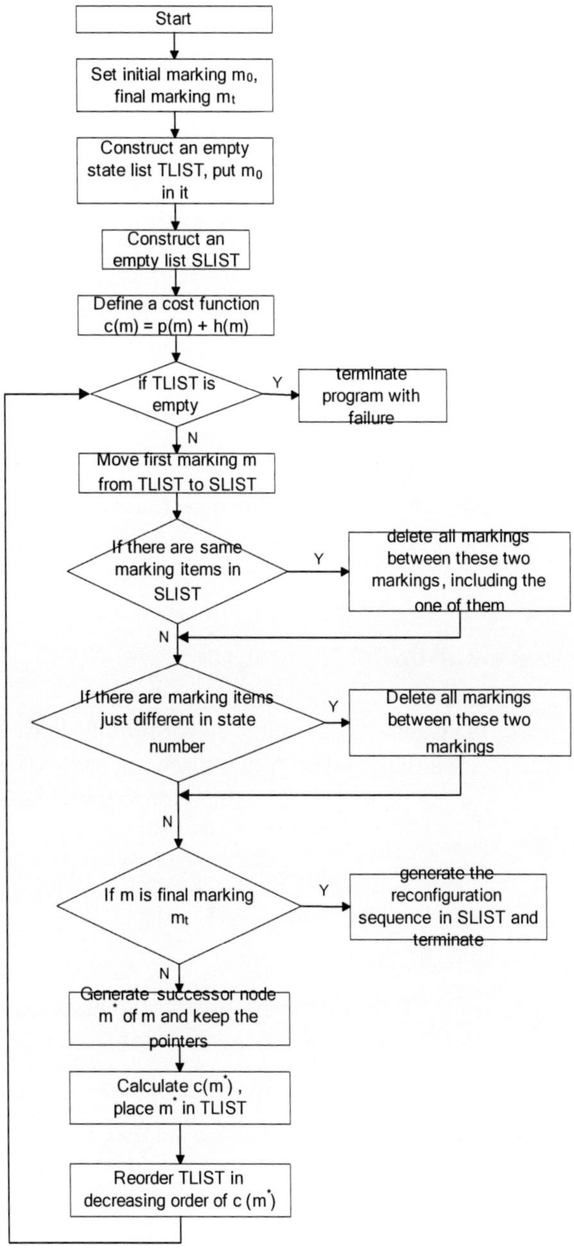

Figure 28. flowchart for one-direction search algorithm.

Step 9. Reorder the TLIST in the decreasing order of the function $c(m)$.

Step 10. Go to Step 4.

The heuristic function

$$c(m) = p(m) + h(m) \tag{13}$$

is used in sequence search,
where $c(m)$ is an estimate of the profit gained during reconfiguration operation from the initial marking to the final marking; $p(m)$ is the current highest profit obtained from initial marking to the current marking m; $h(m)$ is an estimate of the profit from the marking m to final marking along the optimal path that goes through the marking m, it is the key factor to search result.

This algorithm uses

$$h(m) = \alpha \cdot dep(m) \tag{14}$$

as the heuristic item, where $dep(m)$ is the depth of the marking m in the reachability graph and α is the weight factor which should be selected based on some experimental or theoretical explanation or intuition. α is supposed be 1 in this research.

5.2. Two-Direction Searching Algorithm

The same units exist in former disassembly and assembly Petri net, so it is possible to search two directions to find the same ones. A two-direction search algorithm is as following, which uses the same heuristic function as one-direction search one. The flowchart of it is illustrated in figure 29.

Step 1. Construct two empty lists TLIST1 and TLIST2, Put the initial marking m_0 and final marking m_t in them respectively.

Step 2. Construct two empty lists SLIST1 and SLIST2 to hold sequence.

Step 3. Define a cost function $c(m) = p(m) + h(m)$.

Step 4. If TLIST1 and TLIST2 are empty, then terminate program with failure.

Step 5. Move first marking m_1 and m_2 from TLIST1 and TLIST2 to SLIST1 and SLIST2 respectively, and check

a) Whether m_1 and m_2 are same marking, if they are, then delete m_1, reorder the sequence in SLIST2 reversely, and join it into SLIST1, generate the reconfiguration sequence.

b) Whether m_1 and m_2 are just different in interface-set number, if they are, then reorder the sequence in SLIST2, and join it into SLIST1, generate the reconfiguration sequence.

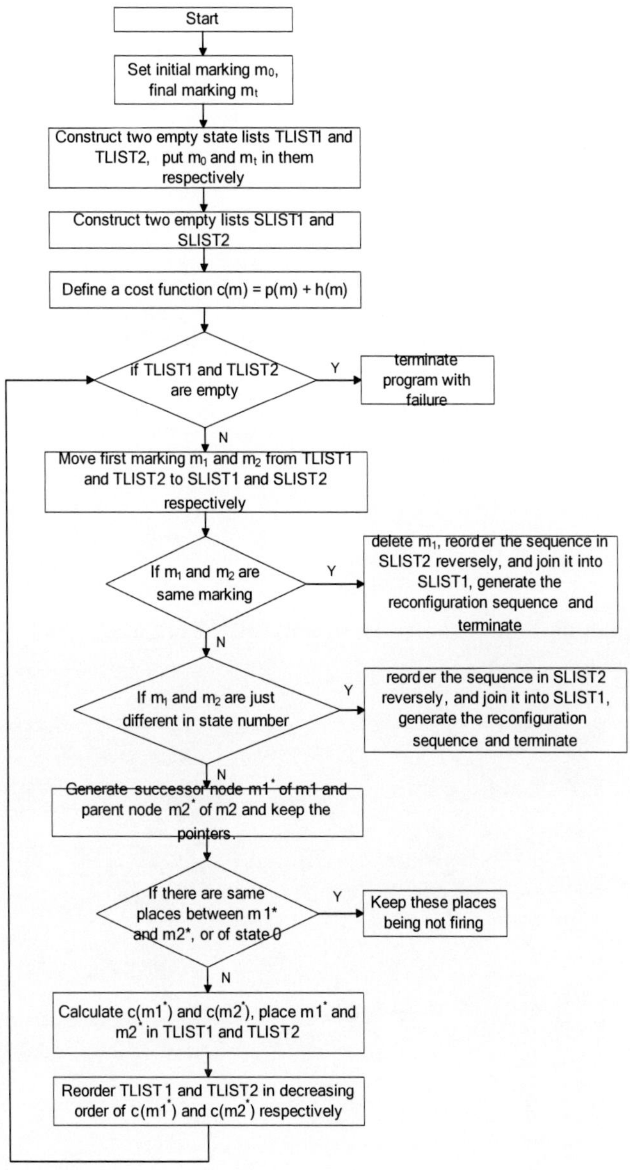

Figure 29. Flowchart for two-direction search algorithm.

Step 6. Find the enabled transitions to the marking m_1 and from the marking m_2.

Step 7. Generate successor marking, that is, successor node m_1^* of m_1 and parent node m_2^* of m_2, and check whether there are same places between m_1^* and m_2^*, or both are of State 0, if there are or both are of State 0, these places are conserved and transition related with them are not allow fired.

Step 8. Calculate $h\left(m_1^*\right)$ and $c\left(m_1^*\right)$, put m_1^* in the TLIST1; calculate $h\left(m_2^*\right)$ and $c\left(m_2^*\right)$, put m_2^* in the TLIST2.

Step 9. Reorder the TLIST1 in the decreasing order of the function $c\left(m_1\right)$; Reorder the TLIST2 in the decreasing order of the function $c\left(m_2\right)$

S*tep 10.* Go to Step 4.

6. EXPERIMENT

A series of experiments are designed to verify the modules for reconfigurable PKM described above. The content includes reconfiguration experiment and cutting experiment. The former experiment is to verify module assembly and interface design, and the latter is to verify workspace defined by Cutting Location Files(CLF).

6.1. Reconfiguration Experiment

The reconfigurable PKMs shown in figures. 30-34 are assembled by struts illustrated in figure 5 and figure 6.

Figure 30. 6-DOF Mechanism I(Stewart Platform).

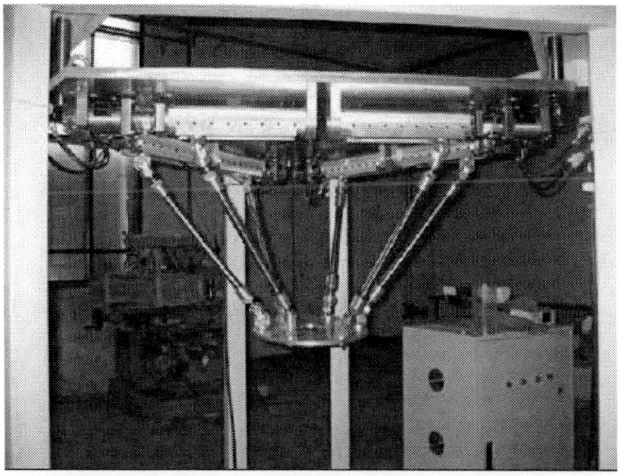

Figure 31. 6-DOF Mechanism II.

Figure 32. 6-DOF Mechanism III.

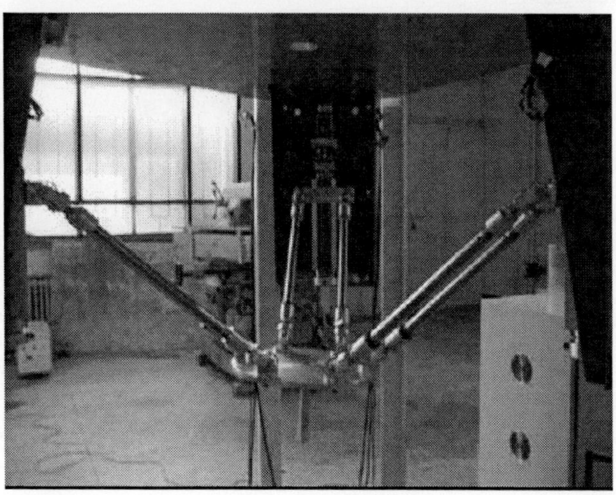

Figure 33. 3-DOF Mechanism I.

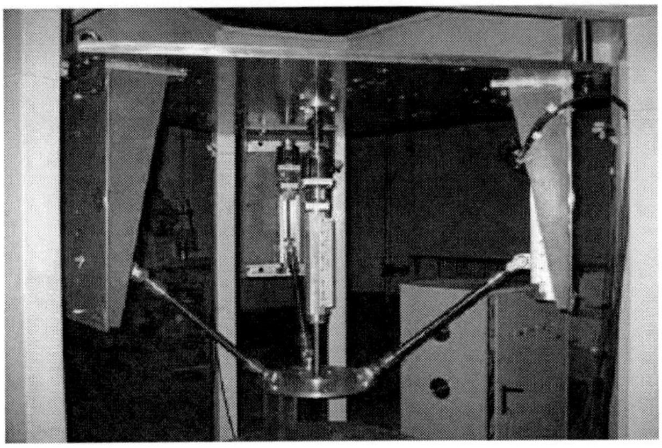

Figure 34. 3-DOF mechanism II(Tricept).

The reconfigurable PKMs shown in figures 35-38 are assembled by struts illustrated in figure 8 and Figure 9

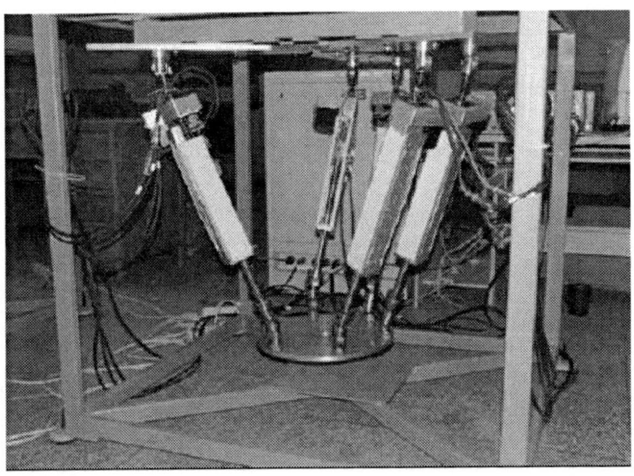

Figure 35. 6-DOF Mechanism I.

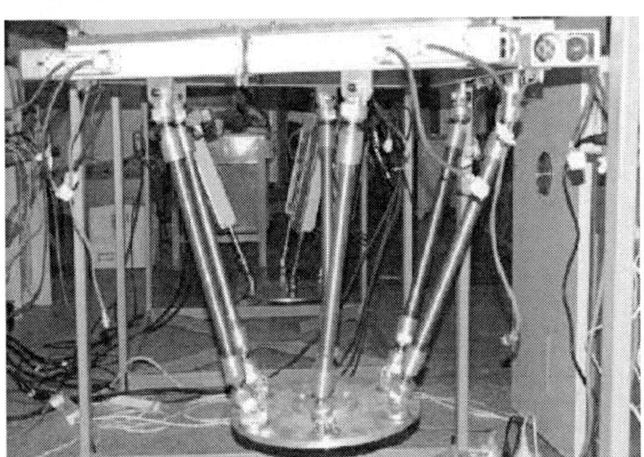

Figure 36. 6-DOF Mechanism.

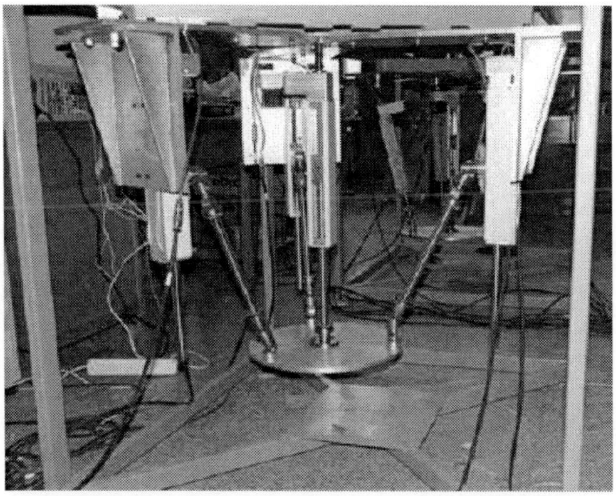

Figure 37. 3-DOF Mechanism.

Figure 38. 3-DOF Mechanism.

When the type is reconfigured from one to another (shown in figure 39), an optimal sequence can be calculated by the algorithm proposed in this chapter. The software interface is illustrated in figure 40.

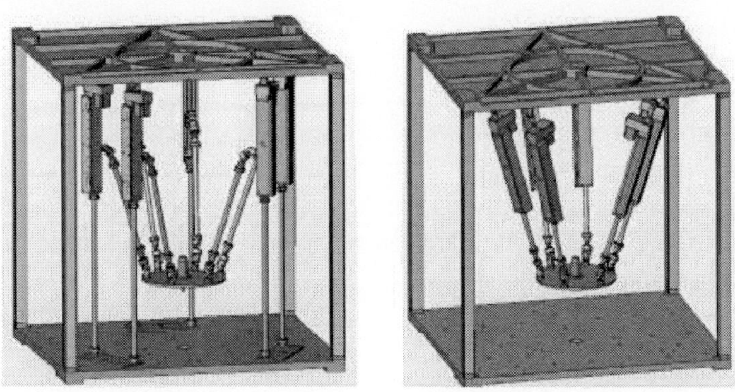

Figure 39. 6PSS and 6SPS PKMs.

The same sub-assembly:

(1) 2-3-4;
(2) 40-54-55;
(3) 6-8-9-10-11-12-13-15-16-17-18-19-20-21-22-23-24-25-26-30-31-32-33-34-52-53-73;
(4) 1-43-48-71;
(5) 28-59-60-61-62-63-64-65-66-67-68;
(6) 41-42.

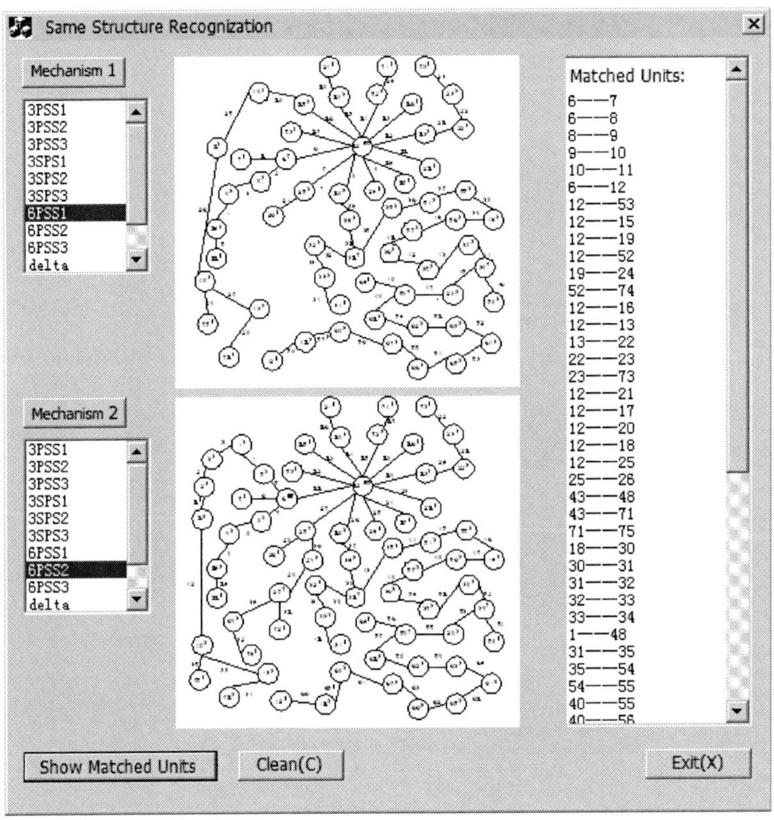

Figure 40. Program Interface.

6.2. Cutting Experiment

The cutting experiment is carried out on the 6-DOF Stewart platform. As shown in figure 41, a spindle is fixed on the mobile platform module. The trail of tool head is controlled by cutting location data filled in CLF.

Two samples cut by Stewart platform which reconfigured from another type PKM is shown in figure 42.

Figure 41. The spindle fixed on mobile platform.

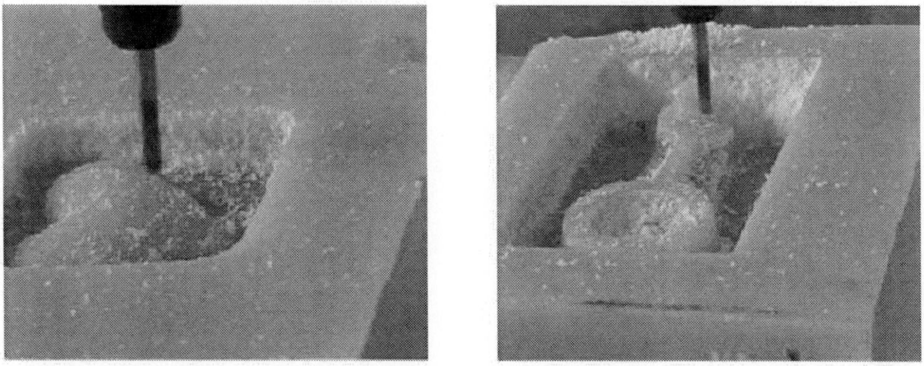

Figure 42. Two machining samples.

7. Analysis and Result

A reconfigurable PKM can be assembled by some interchangeable modules. This chapter presents the structural design of a kind of reconfigurable PKM.

After structure design, inverse kinematics solution is given in the workspace analysis of the PKMs with two driving modes, the effect of setting-angle of spherical joint on PKM has been investigated and the spherical joint setting-angle selection method has been proposed. A comparison result is supplied to show the difference of workspaces of the two kinds of PKMs.

In the reconfiguration planning, modules' interface information is described by the combination of interface status number and undirected graph. Based on this description, a Petri Net is applied to the modeling of reconfiguration process. A reconfiguration cost matrix is used to figure out the dynamic cost in reconfiguration. By combining the graph-subgraph recognition and heuristic algorithm, a one-direction searching algorithm and a two-direction searching algorithm are then presented to search optimal reconfiguration path aiming at minimizing total reconfiguration cost. The results of these two algorithms are compared with others.

Finally, experiments are designed and implemented to verify the structure design and workspace evaluation.

The interfaces in modules will decrease the stiffness of mechanism. The precision and stiffness performance of reconfigurable PKM should be taken into consideration in further research.

ACKNOWLEDGMENTS

This research is sponsored by the State High-Technology Development Program of China (No.2006AA04Z133), the National Natural Science Foundation of China (No.50605035, 10778625), and the National 973 Program of China (No. 2007CB714007).

REFERENCES

[1] Koren, Y, Ulsoy, A G, Reconfigurable Manufacturing Systems, Engineering Research Center for Reconfigurable Machining Systems(ERC/RMS), *Report #1*, The University of Michigan, Ann Arbor, Ml, 1997

[2] Koren, Y, Computer Control of Manufacturing Systems, McGraw-Hill, New York, NY, 1983.

[3] F. Jovane, S. P. Negri, I. Fassi, L. Molinari Tosatti, Design Issues for Reconfigurable PKMs, 2002 *Parallel Kinematics Machines International Conference*, Germany, pp. 69-82. Apr. 2002.

[4] J. Wang, X. Tang, C. Lin, G. Duan, J. Shi, Modular drive strut architecture for reconfigurable parallel kinematics machine . *Chinese Patents*: Apply No.03137758.0, July. 2003.

[5] Stewart, D. 1965. A Platform with Six Degrees of Freedom. *Proceedings of the Institute of Mechanical Engineering.* London. Vol.180. pp.371-386.

[6] Tsai, L.-W., Stamper, R. 1996. A Parallel Manipulator with Only Tranlational Degrees of Freedom. Proceedings of the *ASME Design Engineering Technical Conference and Computers and Information in Engineering Conference.* Irvine. Paper DETC2000/MECH-1152.

[7] Clavel R. DELTA, A Fast robot with parallel geometry. Proc. Int. Symposium on Industrial Robots, 1988:91-100

[8] A. Fattah and G. Kasaei, Kinematics and dynamics of a parallel manipulator with a new architecture, *Robotica,* Vol 18, pp. 535–543, Sept. 2000.

[9] K. H. Hunt, Structural kinematics of in-parallel-actuated robot arm, ASME Transaction, Journal of Transmission, *Journal of Transmission, Automation, Design* 105, Vol 4, pp. 705–712. 1983.

[10] L. W. Tsai, "Kinematics of a three-DOF platform with three extensible limbs," in Recent Advances in Robot Kinematics, J. Lenarcic and V. Parenti-Castelli, Eds. New York: Kluwer Academic, 1996, pp. 401–410.

[11] Shi J S, Tang X Q, Lin C S, et al., Comparison study of two Tricept units for reconfigurable parallel kinematic machines, *High Technology Letters*, 2005, 11(4): 392-396

[12] Cao, T H, Sanderson, A C, Task sequence planning using fuzzy Petri nets[J], *IEEE Transactions on Systems, Man, and Cybernetics,* 1995, 25: 755-768.

[13] Sinha, N, Kumar, S, Rai, R, et al, A Petri Net based approach to determine the disassembly strategy of a product, *International Journal of Production Research,* 2002, 40(5): 1113-1129.

INDEX

#

A

B

C

D

E

N

O

P

Q

R

S

T

U

V

W

X

Y